Macromolecular Reactions

Macromolecular Reactions

Peculiarities, Theory and Experimental Approaches

NICOLAI A. PLATÉ

A. V. Topchiev Institute of Petrochemical Synthesis,
Russian Academy of Sciences
Moscow, Russia

ARKADY D. LITMANOVICH

A. V. Topchiev Institute of Petrochemical Synthesis,
Russian Academy of Sciences
Moscow, Russia

and

OLGA V. NOAH

Chemistry Faculty of the M. V. Lomonosov
Moscow State University
Moscow, Russia

JOHN WILEY & SONS

Chichester · New York · Brisbane · Toronto · Singapore

Copyright © 1995   by John Wiley & Sons Ltd,
Baffins Lane, Chichester,
West Sussex PO19 1UD, England
Telephone   *National* (01243) 779777
International (+ 44) 1243 779 777

Other Wiley Editorial Offices

John Wiley & Sons, Inc., 605 Third Avenue,
New York, NY 10158-0012, USA

Jacaranda Wiley Ltd, GPO Box 859, Brisbane
Queensland 4001, Australia

John Wiley & Sons (Canada) Ltd, 22 Worcester Road,
Rexdale, Ontario M9W 1L1, Canada

John Wiley & Sons (SEA) Pte Ltd, 37 Jalan Pemimpin #05-04,
Block B, Union Industrial Building, Singapore 2057

Library of Congress Cataloging-in-Publication Data

Platė, Nikolaĭ Al 'fredovich.
  Macromolecular reactions: peculiarities, theory, and experimental
approaches / Nicolai A. Platė, Arkady D. Litmanovich, and Olga
V. Noah.
    p.   cm.
  Includes bibliographical references and index.
  ISBN 0-471-94392-4:
    1. Polymers.   2. Chemical reactions.   I. Litmanovich, Arkady D.
II. Noah, Olga V.   III. Title.
  QD381.8.P57   1995   94-30634
  547.7'0459—dc20   CIP

British Library Cataloguing in Publication Data

A catalogue record for this book is available from the British Library

ISBN 0 471 94392 4

Typeset in 10/12 Times by Thomson Press (India) Ltd, New Delhi
Printed and bound in Great Britain by Biddles Ltd, Guildford, Surrey

Contents

Preface

Chemical transformations in polymeric chains is an area of polymer science that began its development even before the scientific foundations of polymerization and polycondensation were laid. It is sufficient to recall the reactions of cotton cellulose modification, e.g. the preparation of celluloid and pyroxylin, or vulcanization of rubber. Strictly speaking, it is this area that should be called 'the chemistry of macromolecular compounds' and not the generally accepted area of polymerization and polycondensation, which is only a specific case of addition and substitution reactions in organic chemistry, although possessing specific features associated with the formation of long-chain molecules.

The study of chemical reactions of polymers in order to obtain information about the behavior of macromolecular particles has so far mainly developed as part of the organic chemistry of macromolecules, accumulating valuable experimental data about new reactions of functional groups and units of the chain, about how to modify synthetic and natural polymers in order to improve and change their important properties and about polymer ageing and reactions that hinder this process and prolong the lifetime of materials. Essential achievements have been made in this direction which not only have broadened our conception of the principal capabilities of the polymer chemistry, but also have led to the design of important technological processes for the preparation of materials ranging from cellulose derivatives to rubber vulcanizates, surface modified polymers, thermoset plastics and coatings, interpenetrated networks and some composites.

In the area of particular chemical reactions of polymers many comprehensive monographs exist. The first one was the famous *Chemical Reactions of Polymers* edited by E. Fettes (1964), followed by *Reactions on Polymers* edited by J. Moore (1973), *Modification of Polymers* edited by Ch. Carraher and J. Moore (1983) and *Chemical Reactions on Polymers* (ACS Series) edited by J. Benham and J. Kinstle (1988). A very interesting review on chemical modification of polymers was published by J.-C. Soutif and J.-C. Brosse in *Reactive Polymers* (1990–1991). Almost all of them (except maybe the last one) as well as thousands of published papers deal with macromolecules as with common reagents, i.e. as with objects containing particular chemical moieties and therefore capable in principle of performing chemical reactions. However, even the simple consideration that an ensemble of uniform units (not to mention chemically different units) may behave not quite like, and sometimes quite unlike, the sum of small molecules of similar structure towards another low-molecular (and moreover high-molecular) reagent makes the picture completely different. Macromolecular chains as reagents are to a much greater extent influenced by 'chemical consequences' caused by their flexibility and capability of making conformational changes, and this forces one to accept that a special approach is required for the development of a theory of macromolecular reactivity. In its turn this requires a physicochemical approach to the estimation and generalization of the existing wealth of experimental and theoretical data collected in a study of macromolecular reactions from the standpoint of the polymeric nature of interacting species.

The present book represents an attempt to fill this gap in the literature and to describe the basic regularities that characterize the chemistry of macromolecules, i.e. those features that distinguish the chemical behavior of long-chain molecules and of units and functional groups incorporated in these chains in various transformations. In this book we shall mainly concern ourselves not with the problem of which reactions can proceed with macromolecules but rather why they behave in such a manner in these reactions and whether it is possible to describe their chemical behavior from the point of view of a physical chemistry, as has been done for small organic molecules.

The authors succeeded in realizing this approach, several years ago publishing a book on this subject (N. A. Platé, A. D. Litmanovich and O. V. Noah, *Macromolecular Reactions* (in Russian), Khimiya Publishers, Moscow, 1977). Interest in this area has only grown during the last decade, but the English reading community has been unable to obtain this review.

Since any reaction of homopolymer transformation (intramolecular or polymer-analogous reactions with a low-molecular reagent) is related to the intermediate formation of chains of mixed (copolymer) structure, considerable attention will be devoted to an analysis of how general the mathematical approaches may be to the quantitative description of the structure of copolymer chains obtained in this way or synthesized from low-molecular monomeric substances. The modern state of the theory and experiment in macromolecular reaction kinetics and of the quantitative description of units distribution and compositional heterogeneity in products of intramolecular and polymer-analogous reactions is an essential part of this book.

The book also includes chapters devoted to methods of experimental estimation of the polymer chain microstructure both from the standpoint of stereochemistry and units sequence distribution. This information, while being basic for the study of the structure of macromolecular reaction products, is also of interest to a wider area of polymer science, namely synthesis of macromolecules.

A new area of polymer chemistry has appeared in the last few years, reactions between macromolecules in the melt and in solution, and main achievements in the study of such intermacromolecular reactions and their products are reported here.

In our opinion such a physicochemical approach provides the basis for modern and, more importantly, for future chemistry and the technology of processes of chemical modification of polymers. Such important and popular technological processes as the preparation of esters and other derivatives of cellulose, the production of polyvinyl alcohol, polyvinylbutyral and other polyacetals, chlorinated polyvinyl chloride, chloropolyethylene, chloropolypropylene and chlorosulfonated polyolefines, polyamidoacids, polyenes from polyvinyl chloride, polyacrylonitrile and from polyvinyl alcohol, the formation of three-dimensional networks for various polymeric binders, reactions in blends during processing and other processes are directly connected with specific features of the chemical behavior of polymeric species.

The preparation and usage of polymer reagents and polymer catalysts and the chemistry of macromolecular stabilizers and polymeric drugs—this is an area where a comprehensive knowledge of how functional groups of the chain and macromolecules as a whole behave is of extreme importance. We should not forget also that chemical reactions on polymers are efficient experimental pathways to synthesize new materials when direct polymerization or polycondensation is not possible. Since chemical modification of existing polymers as a means of preparing materials with improved properties will remain as one of the main pathways to new materials in the coming decade, it may be assumed that further development of the theory and experimental techniques of the study of macromolecular reactions will have an enhanced validity.

Dealing essentially with the first attempt at such a generalization in the area of chemical transformations of polymers, the authors are well aware that it is far from perfect, the area is far from being completed and there are still many more unclear problems than established facts and relationships. In accordance with the chosen approach, attention is mainly concentrated on reactions that proceed without variation of the length of the main polymer chain, and we deliberately excluded data relating to polymer degradation (these problems have been sufficiently well covered in the literature), to crosslinking in thermosetting oligomers and to network formation mechanisms. The latter problem, although undoubtedly belonging to macromolecular reactions, in our opinion is so specific and important that it deserves special consideration.

On the request of the authors Chapter 7 was written by Prof. Ivan M. Papisov and Dr Andrei A. Litmanovich, Section 1 of Chapter 8 by Dr Marina P. Filatova and Section 2 of Chapter 8 by Prof. Vladimir G. Zaikin. The authors are very grateful to these known specialists in corresponding areas for their contribution to this book. We would like also to mention gratefully Dr Boris A. Korolev whose technical assistance in shaping all the manuscript was extremely important.

CHAPTER 1

Properties of the Products of Macromolecular Reactions as Functions of the Units Distribution and Compositional Heterogeneity

When studying reactions involving macromolecules it should be borne in mind that if the reaction does not proceed to 100% conversion (which is very seldom obtained), chemically transformed units always remain in the same chain, and for the same average content may be differently distributed along the chain (we exclude here degradation reactions). Naturally the combination of chemical and physical properties of the modified or partially transformed polymer depends on its chemical composition, but of no less importance may be the distribution of reacted and unreacted units for the same overall composition. Furthermore, the statistical character of the process of chemical transformation itself always leads to compositional nonhomogeneity in the products of polymer-analogous and intramolecular reactions, even without an account of the molecular-mass distribution. Such nonhomogeneity is related to the fact that not all macromolecules participate equally in the reaction, and a sample with a given average chemical composition in reality consists of macromolecules differently affected by a given chemical reaction.

Hence the distribution of units in the chain of an isolated macromolecule and the composition nonhomogeneity of the polymer as a whole are factors affecting such important polymer properties as solubility, melting point, crystallinity, specific gravity, tensile strength and others for the same average chemical composition. This is true for all polymeric samples no matter how they were obtained: by copolymerization of two or more monomers or by the chemical reaction of units in already formed macromolecules. Thus solubility, melt viscosity, specific gravity, tensile strength of block copolymers of ethylene and propylene depend on the length of particular blocks in the chain for the same composition, i.e. on the character of distribution of both types of units [1] (Fig. 1.1).

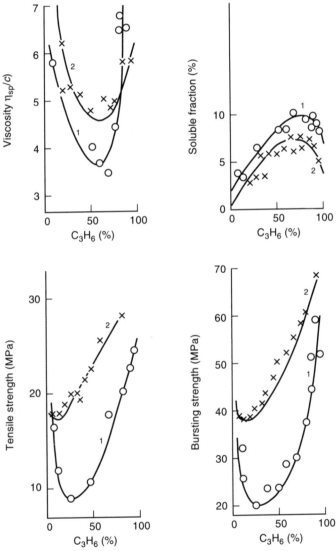

Fig. 1.1 Dependence of physicomechanical properties of ethylene–propylene block copolymers on the mass fraction of propylene for 6 (1) and 24 (2) blocks in the chain. (Reproduced by permission of Hüthtig & Wepf Verlag, Basel, Switzerland, from *Die Makromol. Chem.*, 1961, **44**, 347)

Merker *et al.* [2] noticed the differences in dependences of melting points of siloxane block copolymers:

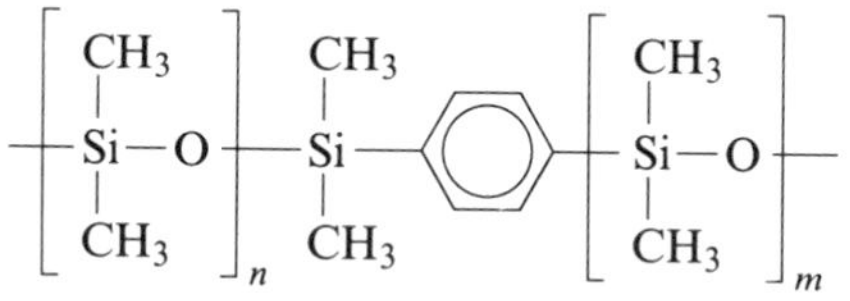

on the fraction of dimethylsiloxane units in the case where they are distributed along the chain as blocks of various length.

The type of units distribution in chains of copolymers of butyl methacrylate with vinyl chloride affects its glass transition temperature [3]. Similar effects were established and thoroughly studied by Johnston for copolymers of other alkyl methacrylates with vinyl chloride [4] and for methylstyrene–acrylonitrile copolymers [5,6]. Since copolymers of alkyl methacrylates with vinyl chloride are widely used for preparation of artificial leather, the problems of their glass transition temperature and frost resistance are of great importance.

The effect of units distribution on the glass transition temperature of copolymers was studied by many authors [7–12]. A comprehensive review of this problem was published by Johnston in 1976 [13], where numerous copolymers of various compositions were examined: styrene–methyl methacrylate, styrene–acrylonitrile, methyl methacrylate–acrylonitrile, ternary α-methylstyrene–methyl methacrylate–acrylonitrile copolymers and various vinylidene chloride copolymers. Many binary and ternary copolymers were shown to have glass transition temperatures lower or higher than that predicted by the ordinary copolymerization scheme. Most of these deviations may be explained by the character of the units distribution. The approach of Johnston was developed later by Havlicek *et al.* [14]. These authors compared their theoretical predictions with experimental data for styrene–n-octyl methacrylate copolymers and obtained the satisfactory coincidence.

A theoretical interpretation of the dependence of the glass transition temperature on units distribution was proposed by Tonelli [15–19]. He suggested that experimentally observed deviations from Fox's equation

$$\frac{1}{T_g^{cop}} = \frac{N_A}{T_g^A} + \frac{N_B}{T_g^B}$$

(where T_g^A, T_g^B, T_g^{cop} are glass transition temperatures of A and B homopolymers and AB copolymer respectively, N_A and N_B are the molar fractions of A and B units in the copolymer) are related to variations of copolymer chain flexibility. Calculation of conformational entropies of all diads for styrene–acrylonitrile copolymer (S–AN) shows that, for isotactic polymer,

$$S_{S,AN} < \tfrac{1}{2}(S_{S,S} + S_{AN,AN})$$

$$S_{rand} > S_{alter}$$

(where $S_{S,S}$, $S_{AN,AN}$ and $S_{S,AN}$ are entropies of the respective diads, S_{rand} and S_{alter} are the total entropies of random and alternating copolymers). For syndiotactic copolymer,

$$S_{S,AN} > \tfrac{1}{2}(S_{S,S} + S_{AN,AN})$$

$$S_{stat} < S_{alter}$$

In other words, styrene–acrylonitrile isodiads are less flexible and syndiodiads are more flexible than averaged diads; therefore the copolymer with alternating units will be less flexible than a random one, if it is isotactic, and more flexible, if it is syndiotactic. The experimentally observed deviations of glass transition temperatures from the value predicted by Fox's equation do not contradict the entropy ratios presented above. Indeed, T_g of isotactic styrene–acrylonitrile copolymer having random units distribution is higher, while for the syndiotactic copolymer it is lower than follows from Fox's equation and for the random copolymer (isotactic) it is lower than for the alternating one. Similar calculations for α-methylstyrene–acrylonitrile and α-methylstyrene–methyl methacrylate copolymers [15] give results that agree with experiment:

$$S_{\alpha-S,AN} > \tfrac{1}{2}(S_{\alpha-S,\alpha-S} + S_{AN,AN})$$

$$S_{\alpha-S,MMA} < \tfrac{1}{2}(S_{\alpha-S,\alpha-S} + S_{MMA,MMA})$$

(for syndiotactic polymers).

Although this approach allows only qualitative conclusions to be made about the entropy nature of the T_g dependence on the units distribution, it is of considerable interest, being practically the first attempt to study such effects on a molecular level. Tonelli successfully explained the T_g variation by molecular interactions in chains due to the different units distributions [17] and concluded

Table 1.1. Deviations from Fox's equation for various copolymers

Copolymer	Deviation
Styrene–butadiene	Without deviation
Acrylonitrile–methyl methacrylate	Negative, with maximum
Acrylonitrile–vinyl chloride	Negative
Vinylidene chloride–methyl acrylate	Positive, with maximum
Vinylidene chloride–ethyl acrylate	Positive, with maximum
Styrene–methyl acrylate	Positive
Styrene–butyl acrylate	Negative
Styrene–acrylic acid	Positive, with maximum
Styrene–methyl methacrylate	Negative, with maximum
Methyl methacrylate–methyl acrylate	Positive
Vinyl chloride–vinyl acetate	Negative

(Reproduced by permission of John Wiley & Sons Ltd. from *J. Polymer Sci.*, 1970, **30**, p. 573)

that in some systems (e.g. styrene–methyl methacrylate copolymers) the main factor was the intramolecular interaction affecting not only T_g but also properties of copolymer solutions [18,19].

As an illustration one can consider data obtained by Barton [9] on the deviation of T_g from values predicted by Fox's equation for copolymers having various chemical compositions (Table 1.1). The character of the deviation of the experimentally observed T_g dependence on copolymer composition from the Fox dependence may serve as a qualitative indication of a tendency to the alternation of units in the chain or to their block-type distribution.

To illustrate the effect of units distribution let us consider the data on mechanical properties of butadiene–nitrile rubbers obtained by random copolymerization of butadiene and acrylonitrile according to the method used in the industry for more than 40 years. According to the mechanism of alternating copolymerization developed by Furukawa *et al.* in the late 1960s and early 1970s [20] and taking into consideration the corresponding data the change of the character of the units distribution at a molar ratio close to 1:1 leads to such a sharp change of physicomechanical properties of a polymer that it can be considered as a transition to a new type of rubber. Some properties of products of sulphur vulcanization of butadiene–acrylonitrile copolymers (molar ratio of 52:48) are presented in the Table 1.2.

It was shown by Rahalkar [21] that the friction coefficient of a styrene–butadiene copolymer also depends on the units distribution in the chain. Not only mechanical properties of copolymers are determined by the character of the units distribution in the chain. As shown by Podesva and Kratochvil [22], the selective sorption from solution for methyl methacrylate–styrene copolymers also essentially depends on the chain structure. Comparison of the selective sorption of 2,2,3,3-tetrafluoropropanol from its mixture with propylene by methyl methacrylate–styrene copolymer with the alternating structure, with random units distribution and by the mixture of homopolymers shows that there is a linear dependence between the sorption coefficient and probability of a diad of different units.

Another example of the dependence of physical properties of copolymers on the units distribution is the capacity of polyurethanes based on polyethylene

Table 1.2.

Property	Alternating copolymer	Random copolymer
Modulus at 100% elongation (kg/cm^2)	43	10.5
Tensile strength (kg/cm^2)	23.7	19.4
Elongation at rupture (%)	400	210

Macromolecular Reactions

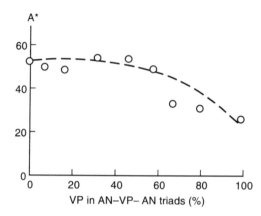

Fig. 1.2 Dependence of the normalized absorbance of reactive dye on the fraction of isolated vinylpyridine units. (Reproduced by permission of John Wiley & Sons Ltd from *J. Polymer Sci. Chem.*, 1982, **20**, 1658)

adipate and polyhydroxydiethylene adipate to crystallize [23]. The first polymer is crystalline, while the second one is amorphous. The crystallinity of a copolymer depends both on the ratio of comonomer units and on their distribution in the chain, increasing with an increase in the degree of regularity.

In studies by Randall the relation of the 'run number' R (the number of blocks of units per 100 units of the chain) with the density of low-density polyethylene [24] and capacity to crystallization of ethylene–1-butene, ethylene–1-hexene and ethylene–1-octene copolymers [25] was found.

Harris and Gilbert [26] have demonstrated the relationship between sequence distribution and acrylonitrile–4-vinylpyridine (AN–VP) copolymers properties such as dyeability, easiness to hydrolysis and T_g. Thus the dyeability of 'reactive red 3' dye is determined only by VP content, and in VP–VP–VP triads 4-VP is more reactive to dye than in AN–VP–AN triads (Fig. 1.2). On the contrary, hydrolysis of AN units proceeds better in AN–AN–AN triads than in other ones (Fig. 1.3)

Nicely *et al.* [27] studied the effect of sequence distribution on a phase diagram of copolymers made from poly(ethylene terephtalate) and *p*-acetoxybenzoic acid. It was shown that an acid unit (PHB) has a slightly greater than random chance of being bonded to another PHB, which leads to significant deviations from randomness at higher PHB levels in copolymers. The formation of the liquid crystalline phase in the melt requires significant amounts of sequences of four or more PHB units to initiate the liquid crystalline formation.

Galvin [28] demonstrated the effect of sequence distribution on miscibility in polymer blends. It has been shown in this work that for the styrene–methyl methacrylate system the ternary phase diagram for the mixtures of block, random or alternating copolymer with the two homopolymers are different. The effect of

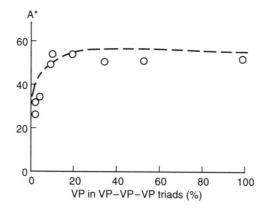

Fig. 1.3 Dependence of normalized absorbance reactive dye on the fraction of vinyl-pyridine triads. (Reproduced by permission of John Wiley & Sons Ltd from *J. Polymer Sci. Chem.*, 1982, **20**, 1658)

the block length in urethane copolymers on dielectric and NMR (nuclear magnetic resonance) relaxation is described by Lipatov *et al.* [29] and on UV absorption by Garcia-Rubio [30].

Generally speaking the differences in properties between random, alternating and block copolymers of the same composition are well known to chemists and is an example of the phenomenon under consideration. When different units are moieties of the same chemical but different steric structure, as in the case of *cis*- and *trans*-polydiene chains, not only the ratio of *cis* and *trans* units but also the distribution of these units in the chain affect the frost resistance and other physicochemial properties of vulcanizates on the base of polybutadiene and polyisoprene [31].

The effect of the functional group distribution in the chain on the chemical behavior of polymeric reagents is especially strongly manifested in macro-molecular catalysis, a subject that has been greatly advanced by Overberger and co-workers [32,33]. As an example the hydrolysis of 1-acetylbenzotriazole catalyzed by copolymer of acrylamide with 4,5-vinylimidazole and *N*-acryloyl-histidine can be discussed [34]. For these processes the change of the reaction rate as a function of the vinylimidazole content in the chain follows the same dependence as the change of the Im–Im diads fraction as a function of composition. In other words, two adjacent catalytically active groups are required for efficient hydrolysis of the low-molecular substrate. It is noteworthy that when the imidazole residues are not attached to the backbone rigidly, but via a 'spacer' (acryloylhistidine), this effect is not observed; presumably the imidazole cycles situated apart from each other along the chain come into the act here.

The examples presented above concern the copolymerization products in which the units distribution is governed by the statistics of the polymer chain growth according to the end or penultimate Markov model or by direct synthesis of alternating or block copolymers. Obviously the same holds for products of intramolecular and polymer-analogous reactions. The number of such systems for which analysis of units distribution has been carried out is relatively small as yet, and researchers confronted with this effect could not give it a quantitative interpretation, as for example in the case of anomalies of viscometric behavior of solutions of polyvinyl alcohol prepared by hydrolysis of polyvinyl acetate. Nevertheless, it was shown that the distribution of vinyl acetate and vinyl alcohol units in chains of this copolymer obtained by partial saponification of polyvinyl acetate affects the emulgating properties of the system used for emulsion polymerization of vinyl acetate [35]. Moreover, these copolymers with different units distributions permit the polyvinyl acetate dispersions of various stabilities due to different decreases in the surface tension to be obtained.

The process of iodine complex formation of partly alcoholized polyvinyl acetate (PVA) polymers is sensitive to the route of polyvinyl acetate formation [36]. The PVA samples can be prepared by saponification with alkali (1), base-catalyzed *trans* esterification with methanolic methoxide (2) and acid-catalyzed equilibration with acetic acid and hydrochloric acid mixture (3). The process of complex formation depends upon the content of long acetate block sequences which are preferably formed for the base-catalyzed reactions 1 and 2, than for reaction 3 (Fig. 1.4). The melting points of polymers prepared by each of these three routes increase smoothly as their acetate content decreases (Fig. 1.5). The melting point is diminished by residual acetate substitution—the more so as this is more random. A relatively high melting point for a given acetate content is

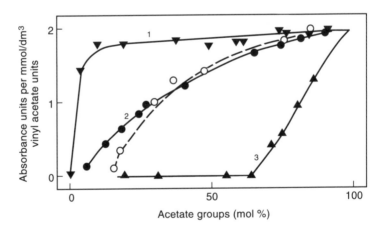

Fig. 1.4 Sensitivity of red complex formation between partly hydrolyzed polyvinyl acetate and iodine. Polymers prepared by saponification (1), methoxide-catalyzed *trans* esterification (2) and acid-catalyzed equilibration (3). (Reproduced from *Polymer*, 1979, **20**, 1492, by permission of the publishers, Butterworth Heinemann Ltd ©)

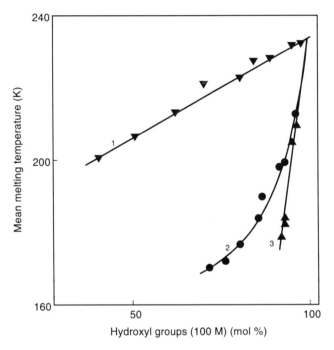

Fig. 1.5 Mean melting temperatures for partly hydrolyzed polyvinyl acetate polymers prepared by saponification (1), methoxide-catalyzed *trans* esterification (2) and acid-catalyzed equilibrium (3). (Reproduced from *Polymer*, 1979, **20**, 1492, by permission of the publishers, Butterworth Heinemann Ltd ©)

thus achieved only by the products of the saponification reaction which retain a very high degree of hydroxyl group block structure.

Properties of aqueous solutions of vinyl alcohol–vinyl acetate copolymers also depend on conditions of polyvinyl acetate hydrolysis, i.e. presumably on the distribution of residual acetate groups [37–40]. Therefore polyvinyl alcohol solutions display completely different stability (formation of associates causing turbidity) in time depending on the method of their preparation—by reacetylation of completely hydrolyzed polyvinyl acetate up to 5% content of acetate groups or by alkaline hydrolysis of polyvinyl acetate up to the same 5% content of acetate groups. Optical anisotropy and character of the structure formation are also different in these solutions. It was shown [41] that the cloud point of copolymer solutions depends more on the (AcAc)/(OHOH) ratio, i.e. on the diads distribution, than on the relative ratio of comonomers. The mixed solvent under study (water with NaF, NaCl, NaBr, NaI, Na_2SO_4 additives) is better for samples containing long OH sequences.

Sorokin *et al.* [42] have demonstrated that the side biological effects of polyvinyl alcohol specimens used in medicine, for instance as plasma expander, depend on the distribution of residual acetate groups. The presence of isolated acetate units as well as of block sequences, even in small amounts (5%), results in higher toxicity of specimens.

The product of the reaction of polyvinyl alcohol with vinyl sulfoxides and vinyl sulfones in water or DMSO in the presence of NaOH [43]:

$$-CH_2-CH- \ + \ CH_2=CHSOR \ \xrightarrow[\substack{H_2O \\ (DMSO)}]{NaOH}$$

$$\begin{array}{c} | \\ OH \end{array}$$

$$\left(CH_2-CH\right)_n CH_2-CH-$$
$$\qquad\qquad | \qquad\qquad\quad |$$
$$\qquad\qquad OH \qquad\quad OCH_2CH_2SOR$$

and

$$-CH_2-CH- \ + \ CH_2=CHSO_2R \ \xrightarrow[\substack{H_2O \\ (DMSO)}]{NaOH}$$

$$\begin{array}{c} | \\ OH \end{array}$$

$$\left(CH_2-CH\right)_n CH_2-CH-$$
$$\qquad\qquad | \qquad\qquad\quad |$$
$$\qquad\qquad OH \qquad\quad OCH_2CH_2SO_2R$$

(where $R = CH_3$, C_2H_5, *tert*-C_4H_9, C_6H_5) is a membrane material having the permeability dependent not only on the general number of sulfoxide or sulfone groups but also on their distribution along the chain.

Truong *et al.* [44] compared the effect of carboxylate groups distribution on potentiometric titration of acrylamide–acrylic acid copolymers for products of copolymerization and of alkaline hydrolysis of polyacrylamide, proceeding with autoretardation and characterized by the alternating distribution of units in hydrolysis products. ΔpK_a as a function of the ionization degree and composition is different for two cases of copolymer preparation.

Guzman [45] compared the effect of CHCl groups distribution on the melting points of two types of samples with the same chlorine content in chlorinated polyethylene and in copolymer of ethylene with vinyl chloride. Heterogeneous chlorination of polyethylene results in the uneven distribution of CHCl groups in the chain and the melting point is higher than that calculated according to the Flory equation for the random ethylene–vinyl chloride copolymer.

Takahashi *et al.* [46] studied the solubility behavior of *O*-methyl cellulose (MC) in water in terms of the distribution of substituents along the cellulose chain and in the anhydroglucose unit. For this purpose three different type of MC samples were prepared by (a) methylation of cellulose acetate prepared from cellulose triacetate followed by deacetylation, (b) methylation of cellulose acetate prepared by direct acetylation of cellulose in mixed solvent followed by deacetylation and (c) methylation of cellulose with dimethyl sulfate in the same solvent. Their solubility in water was compared with that of MC samples prepared by the alkali cellulose process (heterogeneous reaction including com-

Table 1.3.

Chlorine content (%)	In CCl$_4$ solution	In aqueous suspension
8	69	79
28	20	65
40	20	69
50	40	77

mercial products). It was found that water-soluble MC samples prepared by the heterogeneous process exhibit a thermally reversible sol-gel transition in aqueous solution, but all the MC samples prepared by the homogeneous reactions show a normal phase separation in aqueous solution. This result confirms that the highly substituted glucose sequences presented in commercial MC act as 'cross-linking loci' on warming. The distribution of substituents in the anhydroglucose unit also affect the solubility of MC.

The composition nonhomogeneity of copolymers is also a factor affecting properties of materials. Thus the mechanical properties of styrene–methyl methacrylate and vinyl chloride–methyl acrylate copolymers vary with the change of composition nonhomogeneity of copolymers [47].

The softening temperature of partially chlorinated polyethylene samples depends both on the distribution of chloromethylene groups in the chain and on the composition heterogeneity of the product. Thus for samples chlorinated in CCl$_4$ solutions, where the homogeneity is higher, the softening temperature is lower than in the product obtained in suspension [48] (see Table 1.3). Chloropolyethylene samples with a higher composition heterogeneity behave as more rigid materials at the same chlorine content.

The effect of the composition heterogeneity on properties of copolymerization products can be illustrated by the butyl acrylate–methacrylic acid system [49, 50]. Compositionally homogeneous samples exhibit an extremal dependence of ultimate stress in films as a function of the composition, whereas compositionally nonhomogeneous copolymers do not display such a dependence (Fig. 1.6).

Other mechanical properties such as elasticity viscosity η_{el}, flow viscosity η_{fl} and modulus of elasticity E_{el} for samples of the similar composition (29 mol % of methacrylic acid units) also differ practically by an order of magnitude for compositionally homogeneous and nonhomogeneous products, as shown in Table 1.4. The same difference in properties of homogeneous and nonhomogeneous samples was observed for solutions of these copolymers in acetone (Fig. 1.7).

The authors explain such behavior of copolymers by the formation of a network of intermolecular hydrogen bonds (via methacrylic acid units) of the higher density for the homogeneous sample. It should be noted that apparently in all cases of considerable composition nonhomogeneity of a copolymer its properties will be the result of too extensive averaging over properties of individual

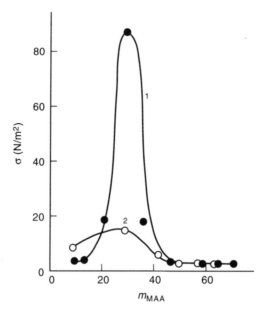

Fig. 1.6 Dependence of the tensile stress on the methacrylic acid content in copolymer for composition-homogeneous (1) and composition-heterogeneous (2) samples. (Reproduced by permission of Nauka, Moscow, Russia, from *Dokl. Akad. Nauk. SSSR*, 1975, **223**, 396)

Table 1.4.

	$[\eta]_{25^\circ}$	$\eta_{el} \times 10^{-3}$	$\eta_{fl} \times 10^{-3}$	$E_{el}(kg/cm^2)$
Compositionally homogeneous copolymer	0.8	4.95	5.7	82.0
Compositionally nonhomogeneous copolymer	0.8	0.61	0.6	22.8

(Reproduced by permission of Nauka, Moscow, Russia, from *Vysokomol. Soed. B*, 1979, **21**, 23)

fractions. Estimation of the degree of nonhomogeneity or its prediction when it is difficult or impossible to check it experimentally is a very important problem.

Existing experimental methods used to determine the composition non-homogeneity of copolymers as well as the theoretical approaches to its prediction from kinetics and mechanism of macromolecular reactions will be described in the following chapters. Since in the analysis of chemical reactions in polymers the problem of the distribution of transformed and untransformed units along the

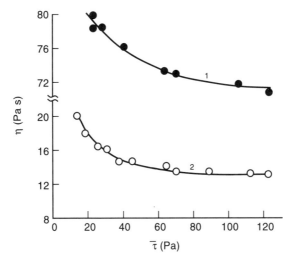

Fig. 1.7 Dependence of viscosity of 50% solutions in acetone on the shear stress for composition-homogeneous (1) and composition-heterogeneous (2) samples containing 29% of methacrylic acid units. (Reproduced by permission of Nauka, Moscow, Russia, from *Dokl. Akad. Nauk. SSSR*, 1975, **223**, 396)

chain arises, this problem as well as thermodynamics and kinetics is one of the most important in the field of macromolecular reactions.

REFERENCES

1. Bier, G., Lehmann, G. and Leugering, H. J., *Makromol. Chem.*, 1961, **44**, 347–61.
2. Merker, P. L., Scott, M. J. and Haberland, G., *Polymer Preprints ACS*, 1962, **3**, 301–6.
3. Johnston, N., *Polymer Preprints ACS*, 1969, **10**, 608–14.
4. Johnston, N., *J. Macromol. Sci. Chem.*, 1973, 7(2), 531–45.
5. Johnston, N., *Macromolecules*, 1973, **6**, 453–6.
6. Johnston, N., *Appl. Polymer Symp.*, 1974, **25**, 19–26.
7. Beevers, R. B. and White, E. F. T., *Trans. Faraday Soc.*, 1960, **56**, 1529–42.
8. Hirooka, M. and Kato, T., *J. Polymer Sci. Part B*, 1974, **12**, 31–4.
9. Barton, J. M., *J. Polymer Sci. Part C*, 1970, **30**, 573.
10. Wessling, R. A., Dickens, E. L., Kurowsky, S. R. and Gibbs, D. S., *Polymer Preprints ACS*, 1973, **14**, 283.
11. Illers, K. H., *Kolloid Z.*, 1963, **190**, 16.
12. Staubli, A., Mathiowitz, E. and Langer, R., *Macromolecules*, 1991, **24**, 2291–8.
13. Johnston, N., *J. Macromol. Sci. Rev. Macromol. Chem.*, 1976, **C14**(2), 215–50.
14. Havlicek, I., Giros, J., Podesva, J. and Krouz, J., *Polymer Bull.*, 1981, **4**, 9–16.
15. Tonelli, A. E., *Macromolecules*, 1974, **7**, 632–4.
16. Tonelli, A. E., *Macromolecules*, 1975, **8**, 544.
17. Tonelli, A. E., *Polymer Preprints ACS*, 1975, **16**(2).
18. Tonelli, A. E., *Macromolecules*, 1977, **10**, 153–7.
19. Tonelli, A. E., *Macromolecules*, 1977, **10**, 633–5.

20. Furukawa, J., Kobayashi, E. and Iseda, Y., *Bull. Inst. Chem. Res. Kyoto Univ.*, 1969, **47**, 222–38.
21. Rahalkar, R. R., *Polymer*, 1990, **31**, 1028–31.
22. Podesva, J. and Kratochvil, P., *sb Prednasek 'MAKROTEST 1973'*, 1973, **2**, 125–33.
23. Yagfarov, M. Sh. and Ogonina, N. S., in *Nekotorye Problemy Organicheskoi Khimii (Some Problems of Organic Chemistry)*, Ed. A. N. Vereshchagin (Proceedings of Scientific Session of the Inst. of Org. Phys. Chem. AN SSSR), 1972, pp. 97–100.
24. Randall, J. C. and Hsieh, E. T., in *NMR and Macromolecules*, ACS Symposium Series 247, 1984, Chap. 9.
25. Randall, J. C. and Ruff, C. J., *Macromolecules*, 1988, **21**, 3446–54.
26. Harris, S. H. and Gilbert, R. D., *J. Polymer Sci. Chem.*, 1982, **20**, 1658–68.
27. Nicely, V. A., Dongherty, J. T. and Renfro, L. W., *Macromolecules*, 1987, **20**, 573–7.
28. Galvin, M. E., *Macromolecules*, 1991, **24**, 6354–6.
29. Lipatov, SYu., Gritsenko, T. M. and Fabulyak, F. G., *Dokl. AN SSSR*, 1977, **234**, 375–8.
30. Garcia-Rubio, J. H., *J. Appl. Polymer Sci.*, 1982, **27**, 2043–51.
31. Marei, A. I., *Kauchi i Rezina*, 1965, **7**, 1–9.
32. Overberger, C. G., *et al.*, *J. Am. Chem. Soc.*, 1965, **87**, 296–301, 3270–2, 4310–13; *Ann. NY Acad. Sci.*, 1969, **155**, 431.
33. Overberger, C. G. and Guterl, A. C., *J. Polymer Sci. Symposia*, 1978, **62**, 13–28.
34. Takeishi, M., Hara, O., Niino, S. and Hayama, S., *Macromolecules*, 1979, **12**, 531–4.
35. Shrinyan, B. T., Mnatsakanov, S. S., Gromov, V. V., Perlova, T. I. and Ivanchev, S. S., *Vysokomol. Soed. A*, 1975, **17**, 182–6.
36. Ahmed, I. and Pritchard, J. G., *Polymer*, 1979, **20**, 1492–6.
37. Hayashi, K., Nakane, T. and Motoyama, T., *Kobunshi Kagaku*, 1965, **22**, 358–64.
38. Sakurada, I., *Kobunshi*, 1968, **17**, 21–6.
39. Toynshina, K., *Kobunshi*, 1968, **17**, 190–8.
40. Sorokin, AYa., Meiya, N. V., Rozenberg, M. E., Tolmacheva, T. P. and Broitman, A.Ya., *Plast. massy*, 1970, **11**, 11–12.
41. Eagland, D. and Crowther, N. J., *Europ. Polymer J.*, 1991, **27**, 299–302.
42. Sorokin, A. Ya., Kuznetsova, V. A., Domnicheva, N. A., Broitman, A. Ya. and Budtov, V. A., *Vysokomol. Soed B*, 1978, **20**, 104–6.
43. Imai, K., Shiomi, T., Tezuka, Y. and Tsukahara, T., *Polymer J.*, 1991, **23**, 1105–9.
44. Truong, N. D., Medjahdi, G., Sarazin, D. and Francois, J., *Polymer Bull.*, 1990, **24**, 101–6.
45. Guzman, J., *Revista Plasticos Modernos*, 1977, **256**, 533–8.
46. Takahashi, S. I., Fujimoto, T., Miyamoto, T. and Inagaki, H., *J. Polymer Sci. Chem.*, 1987, **25**, 987–94.
47. Nielsen, N., *J. Am. Chem. Soc.*, 1953, **75**, 1435–9.
48. Renfrew, A. and Morgan, P., *Polythene*, London Hife and Sons Ltd, 1957.
49. Slavnitskaya, N. N., Semchikov, Yu. D. and Ryabov, S. A., *Vysokomol. Soed B*, 1979, **21**, 23–5.
50. Ryabov, S. A., Slavnitskaya, N. N., Zvereva, E. V., Thshurov, A. F. and Semchikov, Yu. D., *Dokl. AN SSSR*, 1980, **253**, 118–21.

CHAPTER 2

Peculiarities of Macromolecular Reactions

2.1 GENERAL REMARKS

Chemical transformation in polymer chains is not merely a field of chemistry concerning synthesis of new polymer species but is also a field closely related to the reactivity of macromolecules and their functional groups. When considering macromolecular reactions it is necessary to visualize clearly how a macromolecule behaves as a chemical identity and what specific features of its chemical behavior can appear and manifest themselves as compared to low-molecular substances of a similar structure. Here, essentially, we want to know to what extent a chemist-researcher is justified in applying the known concepts and relationships suitable for the reactions of low-molecular organic substances to macromolecular reactions. It is necessary to reveal the existing differences in these reactions and to find specific relationships (if they can be found) for chemical transformations of macromolecules in goal-oriented chemical modifications of materials and for controlling these processes.

In the development of given concepts about reactivity of functional groups of macromolecules historically two periods may be singled out. At the first stage, that is in the 1930s, the so-called principle of 'equal reactivity' advanced by Flory [1,2] played a definite role. This principle holds that the reactivity of a given functional group is independent of whether it is attached to a long chain of any length. The principle had been formulated for the analysis of polycondensation reactions and later was verified many times in studies of different reactions with polymers and played a very important role. Essentially, it was shown that macromolecular substances, like low-molecular ones, can undergo a number of chemical transformations touching their functional groups and monomeric units, which proceed following the same mechanism and even at the same rates. The developments made in the 1930s and 1940s in the field of polymer-analogous reactions and the synthesis of dozens of modified and new polymeric materials based on cellulose, natural rubber, polyvinyl acetate and others were based mainly on the known laws of organic chemistry of small molecules being applied to polymeric substances.

Alfrey [3] has formulated conditions when one might expect equal reactivity of functional groups both in the polymer and its low-molecular analog in a reaction, in which the second reactant is *a priori* the low-molecular substance:

(a) The reaction proceeds in a homogeneous liquid medium and all initial substances, intermediate and end reaction products are soluble in this medium.
(b) In each elementary step of the reaction not more than one type of functional group of a polymer participates, and the molecules of all other reactants are small and mobile.
(c) When selecting a low-molecular model one should take into account the steric hindrances caused by the polymeric chain itself.

An analysis of the conditions when this principle holds is available in the works of Alfrey [3] and Platé [4–6]. Of importance is the problem concerning time factors, which affect the rate of a reaction due to diffusion hindrances caused by the motion of small particles in the solution of macromolecules of increased viscosity.

Since it is well known that the propagation rate constant during radical polymerization is practically independent of the degree of conversion of a monomer in the process in bulk, it should be assumed that macroviscosity of the system will not in general affect the rates of reactions between macromolecules and small particles. Changes in diffusion mobility can be compensated to some extent by reacting particles staying close to each other for longer periods.

The reasons mentioned above are based on experimental data listed in Table 2.1. It is evident from these data that a number of kinetic parameters of chemical reactions with different mechanisms are indeed practically identical for polymers and their low-molecular analogs [4, 5]. Thus, many reactions (the list given in Table 2.1 is not complete) provided and provide serious theoretical support to the fact that the reaction between functional groups or monomeric units of a macromolecule and other low-molecular (sometimes high-molecular) reactions proceeds following those kinetic laws, which are observed in corresponding reactions completely at the low-molecular level.

However, with the increase in the number of polymeric species studied by chemists from the viewpoint of their chemical transformations and modification, the information relating to the fact that reactivity of functional groups of macromolecules differs from that of low-molecular analogs also increased. The reason for this difference is the chain nature of the reactant with all ensuing configurational, conformational and other effects.

At the second stage of development of concept about the chemical behavior of polymers and macromolecular reactions, the works of Smets, Morawetz and

Table 2.1.

Polymer and its low-molecular analog (bibliography)	Reaction type	Rate constant (1 mol/s)	E_{act} (kJ/mol)	Pre-exponential factor $A \times 10^{-3}$
Poly-4-vinylpyridine	Menshutkin	29.4×10^3	65.8	—
Pyridine [7]	reaction	23.1×10^3	67.2	—
Polyacrylamide	Alkaline amide	—	57.3	5.63
Isobutyramide [8]	hydrolysis	—	59.5	5.56
Poly-N-vinylpyrroli-done	Lactam hyd-rolysis	—	104.7	11.7
N-isopropyl-2-methylpyrrolidone [9]		—	113.1	12.3
Hydroperoxide of atactic polypropylene	Initiation of radical poly-	1.96×10^{-7}	92.2	—
tert-Butylhydro-peroxide [10]	merization of styrene	2.11×10^{-7}	100.6	—
Polyethylene glycol	Polyesterifica-tion of tere-	2.5×10^{-4}	—	—
Diethylene glycol [11]	phthalic acid chloride	2.7×10^{-4}	—	—

Sakurada, related to the study of polymer-analogous reactions, played an important role. Based on these works, as well as on the investigations made by Fuoss, Harwood, Klesper, Platé, Litmanovich and a number of other chemists, the main differences in the chemical behavior of macromolecular substances compared with corresponding low-molecular analogs can be formulated.

Before discussing these differences one should mention that some types of macromolecular reactions do not generally have direct similarities with low-molecular substances. These reactions are caused merely by the presence of a sufficiently long chain of uniform units. Into the category of such reactions are placed, for example, the processes of chain radical depolymerization (polymetha-crylates, polystyrene), the reactions of chain degradation of polyoxymethylene and other heterochain polymers proceeding according to the ionic mechanism. Taking into consideration the fact that these and similar processes have been considered in detail by Fettes [12], Grassie [13], Madorsky [14] and Yenikolopyan [15] we shall not dwell on them in our discussion. These processes have been mentioned mainly to give the reader a complete concept of all types of reactions.

Into the category of such reactions are also placed well-known processes of intramolecular cylization and the formation of systems of conjugated bonds along the chain as takes place in the case of polyacrylonitrile, polyvinyl alcohol and polyvinyl chloride during their thermal treatment. Such reactions proceed as cooperative ones. Thus the formation of even one double bond in the chain results in an increase in the rate of dehydrochlorination of polyvinyl chloride by about two orders compared to the rate of thermal elimination of HCl from monochlorohydrocarbons. If there are several conjugated double bonds this effect increases even more [16, 17].

This group of reactions is quite typical for macromolecular substances and the chemistry of such processes has formed the basis of an entirely new branch of the chemistry of polymers with conjugated bonds, which exhibit a number of unique physical properties, such as conductivity, semiconductor properties, catalytic and thermostabilizing properties and others. For greater details the reader may refer to the monographs of Davydov, Krentsel and Berlin [18, 19], as well as to other books describing the synthesis of such polymers.

From the viewpoint of a comparison with reactions of low-molecular substances the following effects may be singled out as chracterizing the possble differences and the specific nature of the chemical behavior of functional groups of macromolecules:

(a) Chain effect and effect of surrounding medium. The variation in the rate, direction and the degree of conversion of the reaction owing to the presence of a neighboring unit of the same or other chemical nature and to the steric hindrances caused by the chain can be considered as a chain effect. On the other hand, the interaction of solvent molecules with a macromolecule may play a much more important role than in the case of an analogous interaction with small molecules, and such features should also be taken into consideration, like a demonstrations of the chain effect. Effects resulting from the variation of the local concentration and orientation of reacting groups in the immediate vicinity of the macromolecule in solution compared to the average volume concentration and the variation associated with it in the reaction rate are also an example of the chain effect.

(b) The neighboring group effect is manifested as the variation of the kinetics and mechanism of the reaction with the low-molecular reactant due to different surroundings of the given functional group or unit at the beginning and at the end of the reaction. As a result of this effect the variation in reactivity of functional groups with conversion arises.

(c) The configurational or stereochemical effect is related to the dependence of reactivity of functional groups on microtacticity of the chain, i.e. on the spatial configuration of monotypic units.

(d) The conformational effect is produced due to the intial conformation of a macromolecule and its variation in the given medium during the reaction period.

(e) The supermolecular effect and the effect of the polymer matrix are related to possible association and aggregation of reacting particles in solutions and in the solid phase and lead to composition nonhomogeneity and variation in the reaction rate.

(f) Effects related to electrostatic interaction of a charged macromolecule with the reacting particles. Such an interaction can vary with the degree of conversion, resulting in the variation of macromolecule conformation and the reaction rate.

2.2 CHAIN EFFECT AND EFFECT OF SURROUNDING MEDIUM

2.2.1 STERIC HINDRANCES AND REACTIONS RELATED TO THE PRESENCE OF THE SAME FUNCTIONAL GROUPS

Let us look at some particular reactions that take place in polymers and are caused by the presence of neighboring groups. The effect of the chain, polymer matrix and the functional groups located in the neighborhood can result in an increase of reactivity of the given group and the given reagent (in the form of substrate or catalyst) or can change its stability in time and under the action of temperature. A detailed review of this topic is available in reference [21]. The low-molecular acylation reagents of the R–O–C–O–C–R′ type are known to be quite unstable, whereas the polymeric aromatic derivatives prove to be quite stable compounds and successfully enter into acylation reaction of simple and complex amines [22]. Upon storage these substances decompose with time, decomposition being caused by intramolecular interaction of the neighboring group as per the scheme shown:

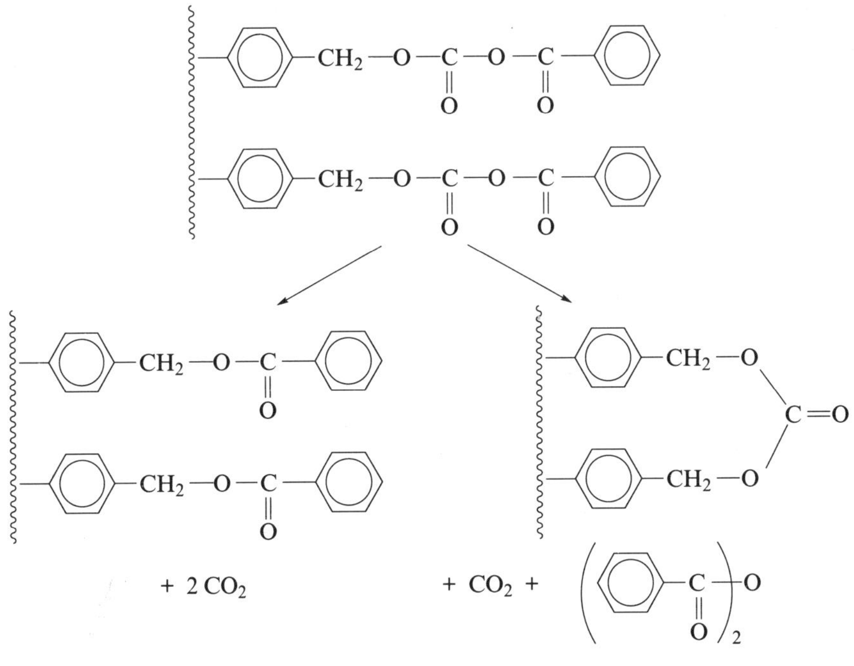

The ratio of different products of degradation and the role of interaction of neighboring groups have been studied in detail by Martin *et al.* [23]:

A decrease in the reactivity of the group attached to the polymer chain and as a consequence of this an increase in the lifetime of the intermediate active compound of benzyne (dehydrobenzene) derivative has been reported by Jayalekshmy and Mazur [24]. Oxidation of the polymeric derivative of 1-aminobenzotriazole attached to the polystyrene matrix results in the formation of 'benzyne' which can be quantitatively identified due to its addition and dimerization products.

The authors identify the retardation effect in the reaction in the polymeric matrix as 'pseudodilution'. The decrease in the mobility of the functional groups linked to the chain due to the formation of clusters and simultaneous increase of their relative concentration at a given point has also been described by Crosby and Kato [25] by discussing the oxidation of 1,7-heptanediol with polymeric thioanisole chloride.

The 'dilution' effect of the macromolecular chain can also be demonstrated by considering the reaction with chloromethylated polystyrene studied by Kraus and Patchornik [26]. If an ester is attached to such a chain capable of enolization, then the ester retains its rigidity and ability to further chemical transformations with the formation of a particular product, while in the free state such an ester instantaneously undergoes ester condensation:

Such behavior of a polymer is characteristic for a not very large number of functional groups in the chain. For densely located functional groups a contrary phenomenon, 'the concentration effect', is observed. As a result such reactions can proceed with a good yield of particular products, which is not possible on the low-molecular level. For example, from chloromethylated polystyrene carrying both enolated and nonenolated ester groups it is easy to synthesize nonsymmetrical ketones [27]:

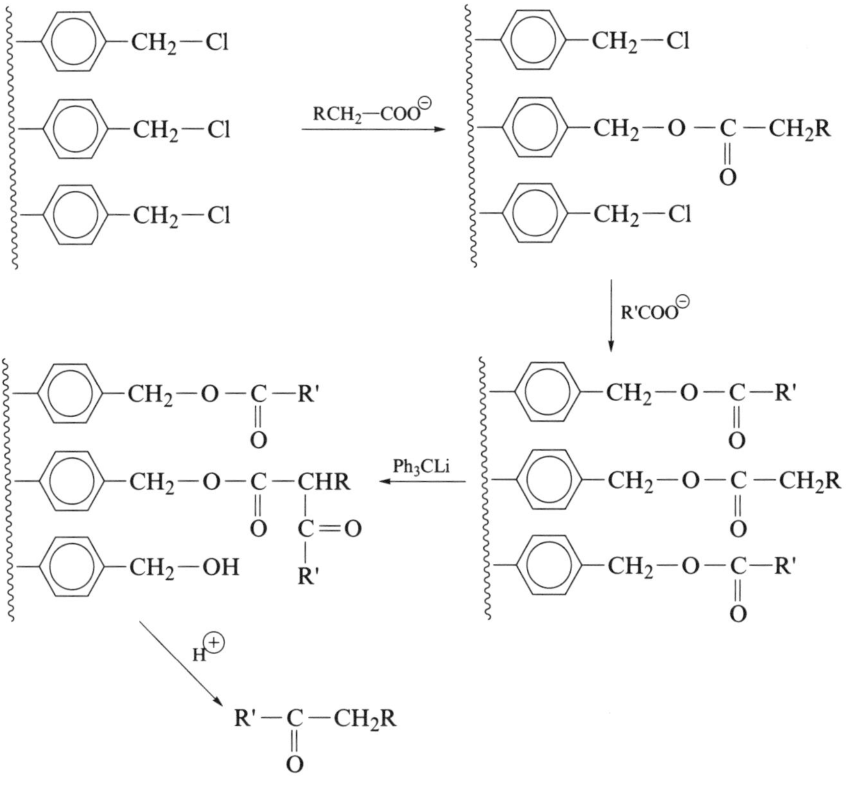

More detailed information about the behavior of polymeric reagents can be found in the reviews of Manecke and Reuter [28] and Leznoff [29].

The classical reaction of reduction of a nitro group into an amino one in aromatic compounds proceeds differently with the formation of different intermediate products [30] in the case of polystyrene and its low-molecular analog, ethyl benzene. Under similar conditions the reduction of the aromatic nitro group in an alkaline medium proceeds with the formation of nitroso- and hydroxylamino-derivatives for a polymer, whereas in the case of low-molecular aromatic compounds azoxy-, azo- and hydrazo- compounds are formed.

The chain rigidity and impossibility of achieving the necessary spatial arrangement of aromatic nuclei for an 'interunit' reaction or for an intermacromolecular one with the participation of two nitro groups belonging to different molecules apparently result in such a pathway of the reaction which, in general, is not typical for low-molecular nitroaromatic compounds:

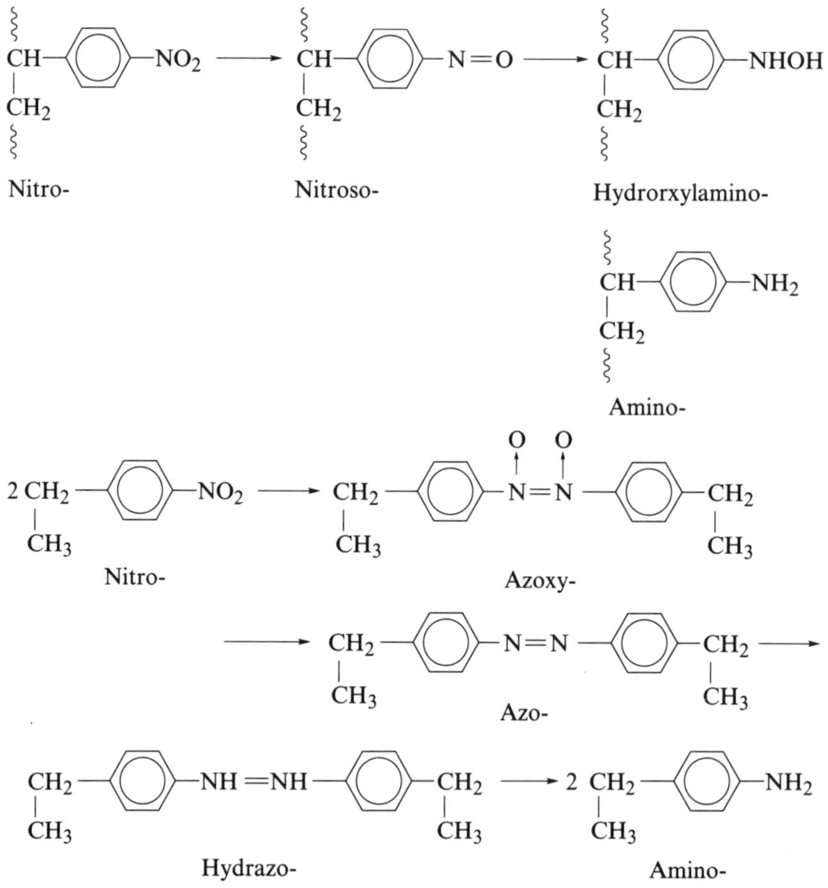

Nitro-   Nitroso-   Hydrorxylamino-

Amino-

Nitro-   Azoxy-

Azo-

Hydrazo-   CH₃   Amino-

Similar results were obtained by Yaroslavsky *et al.* [31] when it was found that the reaction of *N*-bromopolymaleimide with cumene in the presence of benzoyl peroxide did not result in the formation of 2-bromo-2-phenylpropane and 1,2-dibromo-2-phenylpropane, being a product of the analogous reaction for *N*-bromosuccinimide. This unusual behavior of a polymeric reagent is apparently related to the formation of a polar micromedium within the nonpolar solvent owing to the interaction of closely located maleimide groups. This leads to instantaneous dehydrobromination of 2-bromo-2-phenyl propane formed at the beginning. A similar phenomenon has been described by the same authors [32] for chlorination of alkylbenzenes using N-chloropolymaleimide in the absence of free-radical initiators.

In reference [33] the reaction concerning the elimination of hydrogen by poly-4-vinyl benzophenone from tetrahydrofuran in the presence of diphenylamine in benzene solution was studied. The rate of this reaction was found to be two times higher than in the case of the low-molecular analog, benzophenone [33]. The reason for this is that in a macromolecular photosensitizer the efficiency of the reaction increases owing to the migration of the photoexcited state along the chain as along the unidimensional crystal. If the distance between chromophores is small enough, then just the polymeric reagent acts better than the low-molecular one.

The study of the mechanism of the elimination of hydrogen by radicals formed as a result of decay of cumene hydroperoxide from different saturated macromolecular and low-molecular hydrocarbons shows that if for n-heptane the efficiency of the reaction amounts to 11 relative units, then for polyethylene it is equal to 3 and for polyisobutylene to 2 [34]. The activation energy of this process for polystyrene is 8 kcal/mol higher than for its analog, isopropylbenzene. A rise in the energetic barrier and a decrease in the efficiency of radical reactions with the participation of macromolecules are apparently related with steric hindrances and the shielding effect of the chain.

A large amount of experimental data on kinetic features of radical reactions in solid polymers is available [35]. Here we describe only a few of them. The rate constant of the reaction of abstraction of hydrogen atoms by alkyl radicals on passing from low-molecular substances in the gaseous phase to high-molecular compounds decreases by one–three orders (1.6×10^{-18} cm^3/s for n-heptane, 10^{-20} cm^3/s for polyethylene, 10^{-21} cm^3/s for polypropylene). For the reaction between a macroradical and polyethylene this value further decreases by 4–6 orders of magnitude.

It is shown above (Table 2.1) that the Menshutkin reaction, quaternization of pyridine by alkyl halides, has practically the same parameters when carried out at macromolecular and low-molecular levels. In reference [36] this fact was confirmed for quaternization of copolymers of styrene with 4-vinylpyridine and styrene with 2-methyl-5-vinylpyridine using butyl and octyl bromides. However, it is not the case for copolymers of styrene with 2-vinylpyridine and homopolymer of 2-methyl-5-vinylpyridine. Here the initial rate constant decreases respectively by 3 and 10 times compared with low-molecular models, and in addition to that the retardation of the process is observed. The reasons are steric hindrances and the chain rigidity of the macromolecular reagent.

Similar steric hindrances in reactions of polyvinyl alcohol, polyacrylic acid and amidoxime polyacrylonitrile with organotin and organosilicon compounds and different chloroacyl derivatives compared with corresponding low-molecular analogs were observed in the works of Carraher [37]. Drobnik *et al.* [38], in the course of studying the reaction of poly-N-2-hydroxypropyl methacrylamide with chromogenic substrata for chymotrypsin linked by the amino group to the terminal carboxylic group of the side chain, also mentioned the effect of steric hindrances. It was shown in this work that the longer the spacer was the easier the cleavage of the enzyme susceptibly bound proceeded. This process is possible only for the spacer length more than six carbon atoms. Imai *et al.* [39] observed the steric hindrances in the reaction of polyvinyl alcohol with vinyl sulfoxides $CH_2=CH—SO—R$. For $R=CH_3$ and C_2H_5 the reaction proceeds easier than for more bulky *tert*-C_4H_9. Arranz and Sanchez-Chaves [40] have shown that for acetylated dextrans prepared by the reaction with acetyl chloride the reactivity of individual secondary hydroxyl groups decreases in the order C-2 > C-4 > C-3:

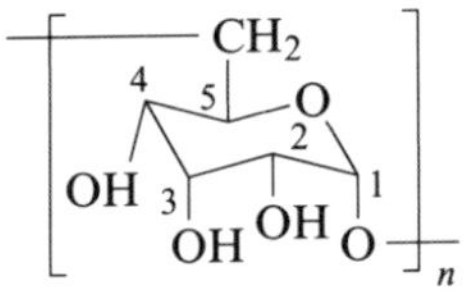

For those modified dextrans prepared with acetic anhydride the ease of acetylation was C-2 ∼ C-3 > C-4. These results were also explained by steric hindrances.

The Arndt–Eistert reaction between polymethacryl chloride and diazomethane studied by Smets and co-workers [41] is an example of the complete change of the reaction path resulting in the formation of other end products than those formed in the case of low-molecular analogs. If the reaction were to proceed following the scheme typical for low-molecular compounds then diazoketone formed at the beginning should rearrange into polyisopropenylcarboxylic acid:

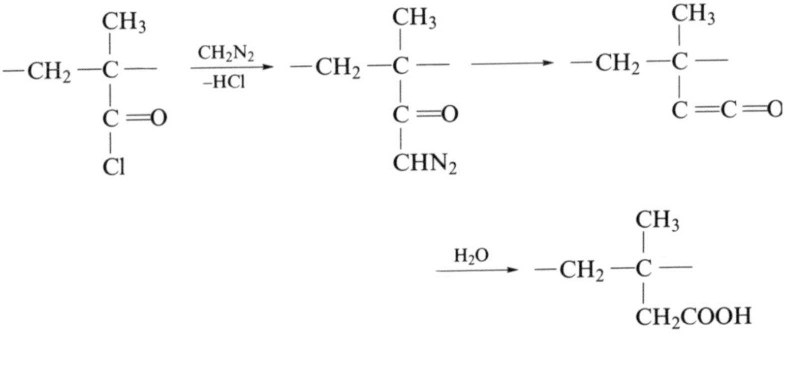

However, this product is not formed. As a result of the reaction between intermediate ketone groups and neighboring unreacted acid chloride group β-ketoketone cycles are formed, which upon subsequent hydrolysis and decarboxylation give five-member cyclic ketone:

$$
\begin{array}{cc}
\quad\;CH_3 \qquad\quad CH_3 \\
-CH_2-C-CH_2-C- \\
\quad\;\; | \qquad\qquad\quad | \\
\quad CH=C=O \quad C=O \\
\qquad\qquad\qquad\qquad | \\
\qquad\qquad\qquad\quad Cl
\end{array}
\longrightarrow
\begin{array}{cc}
\quad\; CH_3 \qquad\quad CH_3 \\
-CH_2-C-CH_2-C- \\
\quad\;\; | \qquad\qquad\quad | \\
\qquad C\!\!-\!\!-\!\!-\!\!-\!\!-\!\!C=O \\
\qquad \| \\
\qquad C=O
\end{array}
\xrightarrow{H_2O}
$$

$$
\begin{array}{cc}
\quad\; CH_3 \qquad\quad CH_3 \\
-CH_2-C-CH_2-C \\
\quad\;\; | \qquad\qquad\quad | \\
\qquad CH\!\!-\!\!-\!\!-\!\!-\!\!-\!\!C=O \\
\qquad | \\
\qquad COOH
\end{array}
\xrightarrow{-CO_2}
\begin{array}{cc}
\quad\; CH_3 \qquad\quad CH_3 \\
-CH_2-C-CH_2-C- \\
\quad\;\; | \qquad\qquad\quad | \\
\qquad CH_2\!\!-\!\!-\!\!-\!\!-\!\!-\!\!C=O
\end{array}
$$

It is known that hydrolysis of alkyl sulfonates proceeds with the rupture of the bond between the alkyl group and oxygen, but if this reaction is carried out for partially β-substituted polyvinylsulfonic esters, then the cyclic polymer is formed instead of polyvinylamine [42]:

$$
\begin{array}{ccc}
-CH_2-CH-CH_2-CH-CH_2-CH- \\
\quad\;\; | \qquad\qquad\; | \qquad\qquad | \\
\quad OSO_2R \qquad OSO_2R \qquad OH
\end{array}
\xrightarrow{R'_3N}
$$

$$
\begin{array}{ccc}
\qquad\qquad\qquad O \\
\qquad\qquad\quad / \;\; \backslash \\
-CH_2-CH \qquad CH- \\
\qquad\quad | \qquad\qquad | \\
\qquad\; CH_2 \qquad CH_2 \qquad + RSO_3H \\
\qquad\qquad \backslash \quad / \\
\qquad\qquad\; CH \\
\qquad\qquad\;\; | \\
\qquad\qquad\; HR'_3
\end{array}
$$

A number of reactions with the participation of neighboring units which result in nontypical structures from the viewpoint of the chemistry of low-molecular analogs simulating the chain unit are described by Morawetz [43]. These reactions which include the formation of cycles can be used for the analytical control of the composition and structure of copolymers. Some interesting data concerning this problem have been generalized by Harwood [44] in the review dedicated specially to the problems of chemical modification of polymers for analytical purposes. We shall touch on this problem in more detail in Chapter 8.

The formation of cycles in the course of intra- and intermolecular transformations can change the rigidity of the system and create steric hindrances so much that the reaction is broken off and does not reach 100% conversion. It should be noted that it is apparently typical for macromolecular reactions when non-changed functional groups 'cut off' from the possible interaction due to steric reasons are formed. This phenomenon is quite often encountered in macromolecule–macromolecule type reactions, but it can be illustrated by the reaction studied by Platé and co-workers [4, 45]. It was found that reactions of complex formation in solution between Ni^{2+} ions and two polymers of diketone type proceed differently and do not reach 100% conversion despite the fact that the corresponding values of equilibrium constants are sufficiently high (of the order of 10^8). Thus the polymer obtained by polymerization of methylvinyl ketone with simultaneous acetylation and containing complexing groups in each monomer unit (I) combines only with about 50% Ni^{2+} of its possible amount referred to all available groups, while the similar polymer (II) obtained by ester condensation of polybutyl methacrylate with benzophenone and containing complexing groups not in every monomer unit reacts with all available Ni^{2+}:

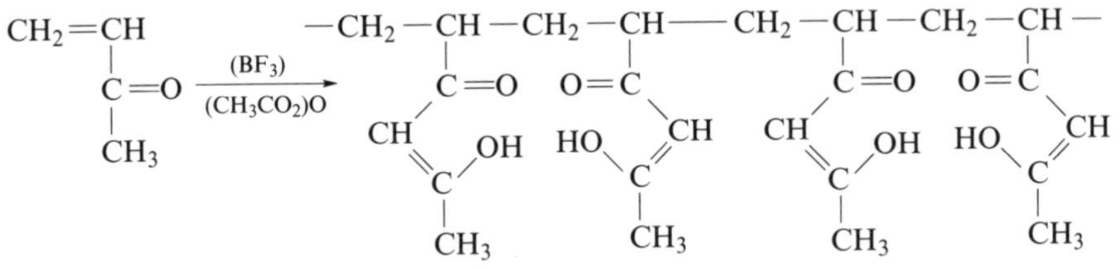

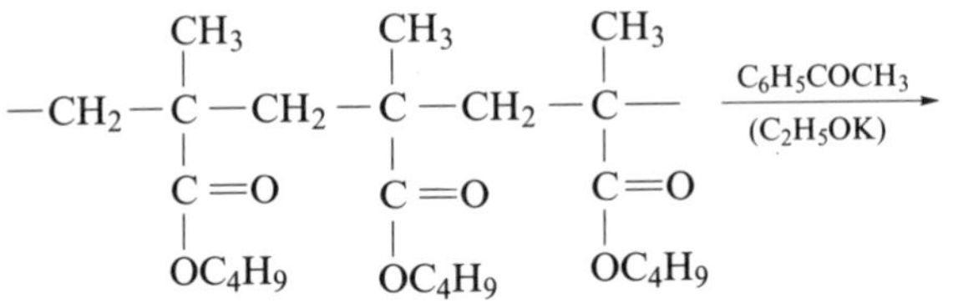

I

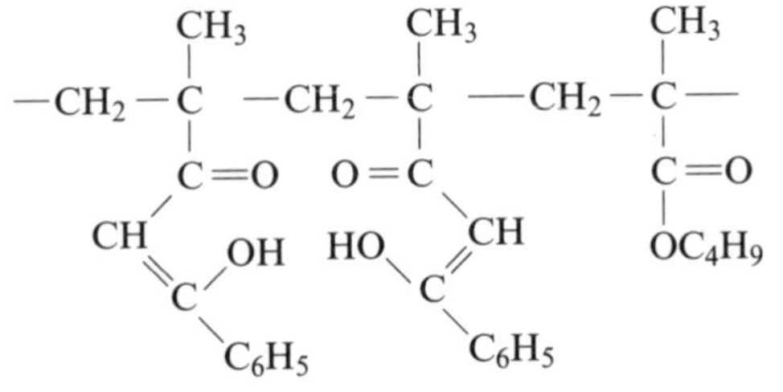

II

Apparently a very rigid structure which arises upon formation of chelate groups in neighboring units hinders the reaction of diketone groups in its vicinity. These diketone groups seem to be blocked. On the contrary, the presence of an 'inert diluent' in the chain results in more efficient participation of complexing groups in intra- and intermolecular reactions of chelate formation.

Only 50% reactive groups of polyacrylonitrile amidoxime react with dicarboxylic acid dichloride, whereas at the low-molecular level the reaction proceeds easily up to the 100% conversion [37]. Carraher believes that the retardation and low conversion in this reaction are related to steric hindrances, preventing the necessary arrangement of functional groups, especially if a part of the cycles has already been formed.

Another case occurs when a chemical transformation of several functional groups in the chain stimulates another reaction with neighboring groups. Such an effect is observed for intramolecular transformations of polyacryl and polymethacryl aldehydes [46]. The authors have found and proved that the reduction of aldehyde groups of polymethacryl aldehyde to hydroxyl ones up to 10–15% under the action of alkaline reagents (metal alcoholates) stimulates the accompanying reaction of ester condensation with adjacent aldehyde groups. This fact is explained by the catalytic action of the alcoholate formed in the reaction and corresponds to aldehyde on the subsequent Tishchenko reaction when the intermediate high-molecular products of the lactone structure are formed [47].

2.2.2 EFFECT OF SOLVENT POLARITY

The chain effect is closely related to the solvent effect as any reactive group of a polymer has in its microenvironment both groups of the same macromolecule and solvent molecules. Because of the competition of these two types of groups the solvent effect for polymers might be much weaker than for small molecules [48].

This problem was described in detail by Galin [49, 50]. The polarity of the microenvironment of some reaction sites is a complex function of the nature of its vicinal units, of the concentration and of the polymer–solvent interactions. In reference [49] tautomerism of keto-β-heterocycles, $COCH_2Het$, is considered:

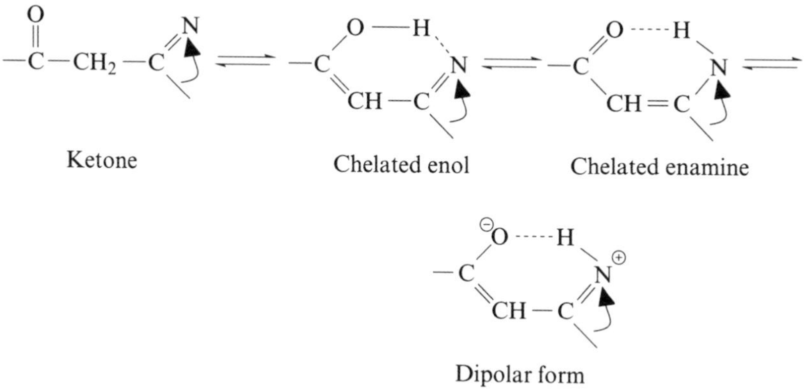

Ketone   Chelated enol   Chelated enamine

Dipolar form

Tautomerism of polymers is quite sensitive to the neighboring group effect and to microenvironment polarity in solution. The dependence of the tautomerism constant on the solvent polarity has been studied for *t*-butyl-2-picolyl ketone and corresponding PMMA copolymers. Although the solvent polarity effect for the low-molecular model is much greater, some effect is also observed for copolymers. Based on these results the quantitative approach to an evaluation of the solvent polarity effect was later proposed [50]. The local polarity of some chromophore-labeled polymers in various solvents has been measured by the solvatochromy method.

2.2.3 VARIATION OF LOCAL CONCENTRATION OF REACTING GROUPS IN THE PRESENCE OF MACROMOLECULES

The rate of the reaction proceeding in the homogeneous solution depends on the concentrations of reacting groups in accordance with the homophase reaction kinetics. However, for reactions where a polymer acts as a macromolecular catalyst the rate is often higher than that with participation of the low-molecular analog and higher than that calculated from the average concentrations of reagents. An example is the hydrolysis of the ester bond in organic molecules catalyzed by polystyrene sulfonic acid [51]. It has been found that in the wide range of acid concentrations (substrate–ethyl acetate) and at different temperatures the hydrolysis rate in the presence of polyacid is essentially higher than in the system involving the low-molecular analog (toluene sulfonic acid), and the difference increases with the dilution of the system. As the activation energy for both processes is found to be practically equal, one can assume that an increase in the hydrolysis rate is the result of a rise in the stationary concentration of protonated ethyl acetate due to an increase in the concentration responsible for hydrolysis of $[H_3O]^+$ ions in vicinity of polymer chains.

It is clear that the greater the dilution of an acid the more is the difference between the local concentration of $[H_3O]^+$ ions near the chain and their average concentration in solution, and the greater is an increase in the rate in comparison with the low-molecular catalyst. The very dilute solution of the polyacid behaves as a heterogeneous liquid with detached droplets of the concentrated sulfuric acid in which the hydrolysis proceeds at a very high rate.

The factors affecting the conformational properties of a polymer catalyst, i.e. variation of the dielectric constant of a medium, and addition of the neutral electrolyte, also affect the efficiency of the polymer catalyst, whereas in the case of the low-molecular catalyst such external factors do not affect the rate of hydrolysis [52, 53]. Similar effects related to an increase in the catalytic activity of polyethylene sulfonic acid as compared with H_2SO_4 in hydrolysis of polypeptides have been observed by Kern and co-workers [54, 55] and by Sakurada and co-workers [56, 57] in studies of hydrolysis of esters by polymer sulfonic acids and HCl.

Enzymatic catalysis is the limiting case of homogeneous catalysis when the role of specific adsorption of substrate by a catalyst is very important. Kabanov and co-workers [53], studying properties of synthetic macromolecular analogs of enzymes, have shown that polyethyleneimine arylated by benzyl chloride acts as a nucleophilic catalyst with hydrophobic fragments of the chain and is 100–1000 times more active in splitting ester bonds in *p*-nitrophenyl acetate and *p*-nitrophenyl butyrate than the corresponding low-molecular analog. A similar effect was observed for the reaction of diisocyanates with alcohols catalyzed by polyvinylpyridine and its copolymers with styrene and methyl methacrylate compared with catalytic action of pyridine. One of the reasons for this is the increased sorption of substrate in hydrophobic domains of the polymer, resulting in much higher local concentrations of substrate.

The increased affinity of substances to macromolecular catalysts is responsible for the appearance of the 'polymeric' effect in the reaction rate, while different solubilities of different substrates can result in some selectivity of the action of the polymer catalyst. This phenomenon was observed in many works concerning the synthetic models of biocatalysts [57–59].

An example of the 'concentrational' effect in catalysis with participation of macromolecules is the acceleration effect in saponification of diallylphthalyc copolymer. In this reaction the $—OH^-$ and $—COO^-$ groups formed in the chain due to external hydrolysis make the copolymer into a more hydrophilic one; as a result the local concentration of the catalytic OH^- ions near the chain increases [60].

Some works concerning catalysis by polymer derivatives of imidazole were reviewed by Overberger and Guterl [61] who described in detail the role of the catalyst–substrate hydrophobic interactions, promoting in some cases the substrate hydrolysis as compared with the low-molecular catalyst of the analogous structure.

Thus in certain cases a macromolecule in solution can be considered to be a 'molecular reactor' acting as a nucleus of the new phase and selectively collecting the molecules of low-molecular substances from the multicomponent system. One can assume that this phenomenon will always be observed in studies of macromolecular reactions in solution. An analysis of this effect applicable to reactions between a polymer and low-molecular reagents has been given by Morawetz [62, 63] where some experimental methods of the study of reactivity of functional groups of macromolecules are reviewed. It has also been found that this effect is not confined to simple microconcentration of a reagent from the volume and its enrichment around macromolecules. As shown in a number of works dedicated to catalysis in macromolecular systems [52, 53], the specific sorption of a low-molecular substrate takes place.

It has been found [64–66] that the molecules of comb-like polymers, e.g. of polyhexadecyl acrylate, are capable of 'organizing' the solvent molecules around them. As a result this solvent (e.g. toluene) in the form of solvate 'coat' around the polymer loses its mobility (the relaxation time of its molecules increases up to 10^{-9} s compared with 10^{-12} s for ordinary liquid toluene). If aliphatic alcohols, such as n-octyl and n-decyl alcohols, are used as solvents, then thermotropic gels

Table 2.2. Kinetic parameters of H–D exchange reactions in polymers and their low-molecular analogs

Polymer and its low-molecular analog	Solvent–catalyst	$k_{ef}(s^{-1})$ at 25 °C	E (J/mol)	lg A
Polyisopropylpropion-amide	D_2O–DCl	5×10^{-5}	24	10.3
N-Isopropylpropion-amide [70]	D_2O–KOD	2.5×10^{-3}	84	12.0
Polyaminomethacryl-2-lysine	D_2O–DCl	6.6×10^{-5}	80	9.8
Aminoisobutyryl lysine [71]	D_2O–KOD	2.5×10^{-3}	67	9.1
Polymethylpropion-amide [72]	D_2O–KOD	4×10^{-4}	76	9.7
N-Methylacetamide	D_2O–KOD	1×10^{-2}	72	10.4

are formed at rather small concentrations of a polymer (0.1–1%) and a new phase of the solvent appears that has a different structure and thermodynamic properties compared with the ordinary liquid alcohol. The latter appear due to the 'organizing' role of macromolecules of a comb-like polymer.

Such an 'organizing' role is characteristic not only of comb-like polymers. This exemplifies only a rather general phenomenon known as the 'short-range orientational order' in polymers [67, 68]. Because of the shape anisotropy of solvent and polymer molecules the energy of their interaction depends on their mutual disposition. As a result some orientations of solvent molecules in the vicinity of a polymer chain are more favorable. If we substitute solvent for reactant we can see that a chemical reaction can proceed in the medium of such a mobile but oriented solvent owing to the polymer chain effect.

The 'increased' local concentration of functional groups attached to the polymer chain can result not only in the acceleration but also in the retardation of the reaction as compared with the low-molecular analogs [69–72]. This is evident from the hydrogen–deuterium (H–D) exchange data shown in Table 2.2. Compared with low-molecular analogs the decrease in the H–D exchange rate constant by about two orders of magnitude in polymeric substances is related according to Scarpa *et al.* [70], with a fall in the dissociation constant of water in the vicinity of amide groups of a macromolecular particle compared with a more 'uniform' solution of the low-molecular substance.

2.3 NEIGHBORING GROUP EFFECT

To the number of effects that bring about a change in the reaction mechanism of intramolecular or polymer-analogous reactions can be grouped the so-called 'neighboring group effect'. This term usually denotes a change in the reactivity of the given functional group or unit under the action of an already reacted group which is located adjacent to the given one. This effect is observed in a number of cases of hydrolysis, cyclization, quaternization, epoxydation and many other reactions, representing very distinctly the effect of the polymer nature of a reagent on the kinetics and mechanism of macromolecular reactions compared with reactions of low-molecular compounds.

The reacted chain units or functional groups B in the course of the reaction, which can be represented as

$$\sim AAAAAA \cdots AAA \cdots AAAAA \sim$$
$$\downarrow$$
$$\sim ABAAAA \cdots BAA \cdots ABABA \sim$$
$$\downarrow$$
$$\sim BBBBBB \cdots BBB \cdots BBBBB$$

can increase or decrease the reactivity of adjacent groups A that have not reacted so far or can remain unchanged. If this happens (there are many reactions that proceed with the 'neighboring group effect'), then a principally new aspect, being not characteristic for the chemistry of monofunctional low-molecular substances, is observed, when the reactivity of a given group or monomer unit is a function of the degree of conversion or the duration of the reaction. This indicates that in the course of the reaction one deals with a reagent of a variable chemical activity. This fact results in the appearance of a number of mainly new problems related to the description of the kinetics of such a process, of the distribution of reacted and unreacted groups in the chain and of the compositional heterogeneity.

A detailed analysis of the present state of the art and of future developments that may take place in this field is given in subsequent chapters of this monograph. Here we mention only that the experimentally observed increases in the rate of the elementary reaction by 10^3–10^4 times due to the appearance of the neighboring reacted group are not so rare, as well as the cases of the process inhibition by 10 times. This once again provides support to the idea of the specific nature of the chemical behavior of large molecules, which should be taken into account both in developing the theory of macromolecular reactions and in practical modification and transformation of polymers.

Let us discuss briefly some well-known examples of the manifestation of the neighboring group effect. Apparently this effect was first observed by Morawetz [73, 74] when studying the hydrolysis rate of the copolymer of methacrylic acid and *p*-nitrophenyl methacrylate. The substance was found to be hydrolyzed at a rate higher by several orders than a low-molecular analog, *p*-nitrophenyl ester of isobutyric acid. As shown in references [43] and [75], this happens due to the change in the reaction mechanism: hydrolysis of ester groups in a copolymer proceeds mainly not due to the action of external OH ions (as one could expect from the data for the low-molecular model) but under the action of adjacent ionized carboxyl groups that attack a carbonyl of the ester group:

It is quite natural for the low-molecular ester which simulates only one unit of the chain not to have an adjacent carboxyl group, and this effect is not observed there. The reaction proceeds in the same way as the usual mechanism of acid hydrolysis. It was shown later on [76] that in the 3–6 pH range the hydrolysis rate was proportional to the degree of ionization of adjacent carboxyl groups.

The fact that it is the adjacent carboxyl group that produces the effect is proved by autocatalytic acceleration of alkaline hydrolysis of polyethyl acrylate. The acceleration effect is not observed in copolymers of ethyl acrylate with butadiene, as the monomer units of ethyl acrylate (and hence of the acid) are separated by inert units [77].

Activation of the given group at the cost of already reacted groups can be observed in the chain degradation reaction. Thus, pyrolysis of poly-*tert*-butyl acrylate proceeds with the formation (isolation) of isobutylene with an almost constant rate up to 10% conversion. Then the rate constant increases by 40 times, apparently due to the activation of the ester group by the adjacent formed carboxyl group [78]:

The reacted adjacent group can produce an inhibiting effect on the reactivity of the given group. An illustration of that is given by the data on alkaline hydrolysis of polymethacrylamide which can not be completed due to blocking of an amide group by two ionized carboxyl groups [79]:

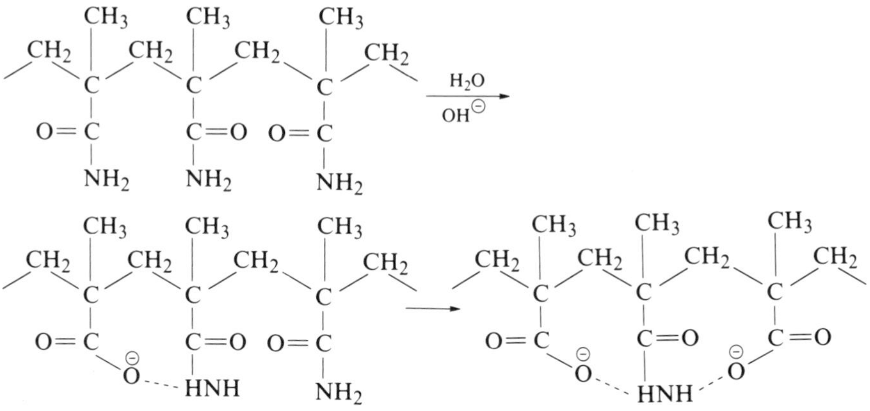

The inhibiting effect of the adjacent reacted group was observed in the study of the chlorination process of polyethylene, quaternization of polyvinyl pyridine [80] and a number of other hydrolysis, solvolysis and oxidation reactions. We have described here the reactions that are 'classical' from the viewpoint of the neighboring group effect. These and more recent results are discussed in detail in Chapter 4.

2.4 CONFIGURATIONAL EFFECT

In vinyl polymers each triad of units can exist in at least one of the three spatial configurations:

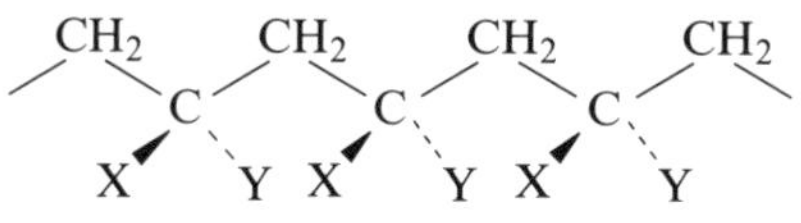

Isotactic triad (*meso* dyads)

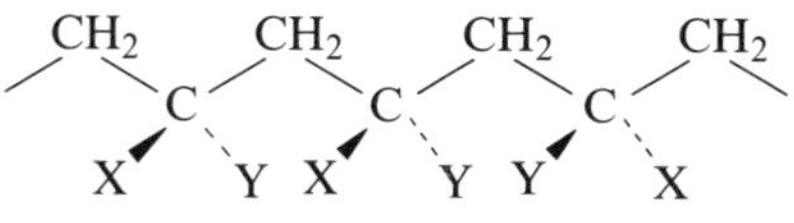

Heterotactic triad

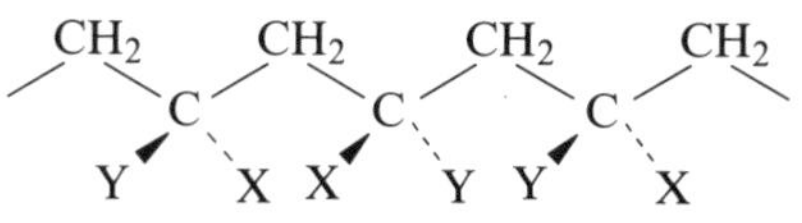

Syndiotactic triad (racemic dyads)

Generally speaking, the reactivity of the group X can vary depending on whether this group is situated in the center of the iso-, hetero- or syndiotactic triad.

All of the three spatial groups considered above are typical of polymers formed by the 'head-to-tail' addition of monomer units. The 'head-to-head' and 'tail-to-tail' combinations of units are also known, as well as the formation of disordered structures containing arbitrarily arranged groups of both types.

Another type of isomerism involves the formation of *cis* and *trans* isomers, e.g. in the course of polymerization of diene hydrocarbons. The *cis* and *trans* isomers can also exhibit different reactivities.

The effect of stereoisomerism on the kinetics of polymer-analogous reactions was first experimentally established for hydrolysis of the copolymer of methacrylic acid with *p*-nitrophenyl methacrylate [81, 82]. It was shown that one part of ester groups (up to 20%) was hydrolyzed with a rate exceeding almost 10 times the rate of hydrolysis of another part of ester groups. A similar effect was observed in the case of solvolysis of copolymers of methacrylic acid and substituted phenyl esters of acrylic acid [83]. De Loecker and Smets showed [83–85] that hydrolysis of isotactic copolymers of methacrylic acid and its various esters proceeded at a higher rate. Similar results were obtained by Glavis [86] who showed that in heterophase hydrolysis in alkaline medium isotactic polymethyl methacrylate was hydrolyzed at a higher rate and up to a higher degree of conversion than the specimens of another microstructure. Chapman [87] has found that isotactic poly-N, N-dimethylacrylamide is hydrolyzed 6–7 times faster than the atactic one. The rate of hydrolysis of isotactic copolymer of acrylic acid and methyl acrylate is 3–5 times higher than for the atactic one [88].

Cyclization of polyacrylic acid with the formation of polyanhydride [89] is an example of the stereochemical effect on the mechanism of the intramolecular reaction. On treating syndiotactic polyacrylic acid with thionyl chloride no anhydride units are formed, while dehydration of isotactic polyacrylic acid proceeds easily with the formation of corresponding anhydride.

The stereochemical effect in various reactions of polyvinyl chloride such as dehydrochlorination, chlorination, ionic elimination, nucleophilic substitution and others has been studied by Millan and co-workers [90–99]. Thus for dehydrochlorination at elevated temperatures [90–93] it was shown that at 175–185 °C the rate of the reaction was proportional to the relative content of syndiotactic triads in a polymer. The average length of the resulting polyene sequences increases with an increase in the degree of syndiotacticity, although there are reasons to assume the more complicated situation in this case. For nucleophilic substitution on polyvinyl chloride with sodium thiophenate in cyclohexanone solution [95] the central chlorine in the isotactic triad is more reactive than in the heterotactic one.

In the reaction of polyvinyl alcohol with vinyl sulfoxides and vinyl sulfones the reactivity of OH groups situated in different triads changes in the range iso > hetero > syndio [100]. Ketalization of polyvinyl alcohol [101] proceeds presumably in isotactic triads where the energetically favorable equatorial– equatorial conformation can be maintained in the produced six-membered 1,3-dioxane ring skeleton. The reactivity of the central unit in isotactic triads of polyvinyl chloroformate is higher than in syndiotactic triads [102].

These and many other examples describe the effect of microtacticity on the reactivity of macromolecules only qualitatively. Therefore here we will consider the experimental data permitting a quantitative evaluation of the configurational effect in macromolecular reactions.

The most illustrative method of estimation of the effect of stereoisomerism on the macromolecule reactivity is a comparison of the reaction rate constants for samples of polymers of different microstructure. Such a study of kinetic constants was made in the works of Fujii *et al.* [103, 104], dedicated to the investigation of reactions of polyvinyl alcohol and its derivatives.

Having determined that the rate of hydrolysis of polyvinylacetals depends on the microstructure of parent polyvinyl alcohol [103], the authors made an attempt to estimate quantitatively the effect of stereoisomerism on the rate of hydrolysis of polyvinyl acetate [104]. The effect was found to be small and became noticeable only in the case where the length of isotactic sequences exceeds a certain critical value. The accelerating effect of the neighboring groups in this case is not very distinct (Fig. 2.1) and the rate constant determined from the initial part of the kinetic curve has the same value as for syndiotactic and atactic specimens.

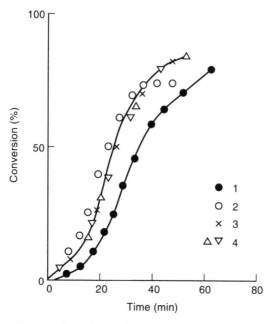

Fig. 2.1 Kinetics of hydrolysis for isotactic (1), stereoblock (2), atactic (3) and syndiotactic (4) polyvinyl acetate samples. (Reproduced by permission of John Wiley & Sons Ltd from *J. Polymer Sci., B*, 1963, **1**, 687)

Table 2.3.

PVAc	k_0(l/mol min)	m^*
Isotactic	0.07	110
Stereoblock	0.13	110
Atactic	0.07	170

Reproduced by permission of John Wiley & Sons
Ltd from *J. Polymer Sci., B*, 1963, **1**, 687.

The rate constants of the hydrolysis of polyvinyl acetate specimens of different microstructures are given in Table 2.3 [104]. Here m^* is a dimensionless parameter which characterizes the degree of acceleration and appears in the empirical equation suggested by Sakurada [105] to describe the autocatalytic polymer-analogous reaction:

$$\frac{dX}{dt} = k(A - X)(B - X)$$

In this equation $k = k_0(1 + m^*X/A)$, A is the initial concentration of reactive groups, B is the concentration of a low-molecular reagent and X is the concentration of reactive groups at time t.

Important results were obtained by Fujii *et al.* [104] when studying the reversible reaction of acetalation of polyvinyl alcohol and its low-molecular analogs 2,4-pentanediols of different spatial configurations:

$$
\begin{array}{cc}
\begin{array}{c}
CH_3 \\
| \\
H-C^*-OH \\
| \\
CH_2 \\
| \\
H-C^*-OH \\
| \\
CH_3
\end{array}
&
\begin{array}{c}
CH_3 \\
| \\
H-C^*-OH \\
| \\
CH_2 \\
| \\
HO-C^*-H \\
| \\
CH_3
\end{array} \\
\textit{meso}\text{-2,4-Pentanediol (1)} & \textit{d,l}\text{-2,4-Pentanediol (2)}
\end{array}
$$

It is assumed that (1) is a model of the isotactic dyad in the chain of polyvinyl alcohol and (2) is a model of the syndiotactic dyad.

The equilibrium constants of the acetylation reaction of polyvinyl alcohol specimens of different microstructures and of low-molecular models are given below:

	$K \times 10^{-3}$
Isotactic polyvinyl alcohol	3.01
Atactic polyvinyl alcohol	1.14
Syndiotactic polyvinyl alcohol	0.796
meso-2,4-Pentanediol	8.70
d,l-2,4-Pentanediol	0.37

It is evident from these data that the difference between the K values for low-molecular stereoisomers is very large but for the polyvinyl alcohol specimens K has intermediate values. This seems to be quite natural, as both iso- and syndiotactic polymers are not made entirely of corresponding dyads but are characterized only by the prevailing amount of either series. If we assume that the found values of the equilibrium constant for the low-molecular models correspond to the K values for pure iso- and syndiotactic specimens, then the amount of iso- and syndiotactic dyads in a polymer under study (I and S) can be calculated. As a result, for isotactic polyvinyl alcohol $I = 0.746$ and $S = 0.254$, for atactic polyvinyl alcohol $I = 0.470$ and $S = 0.530$ and for syndiotactic polyvinyl alcohol $I = 0.329$ and $S = 0.671$ were found. Thus studies of low-molecular models can provide valuable information about the microstructure of polymers (for details see Chapter 4).

The same reactions of the hydrolysis of polyvinyl acetate and acetylation of polyvinyl alcohol were studied by Sakurada [106] who obtained k values for polyvinyl acetate specimens having different microstructures very close to those obtained by Fujii *et al.* For isotactic specimens $k_0 = 0.14$, for atactic $k_0 = 0.21$–0.23 and for polymers enriched with 'head-to-head' units $k_0 = 0.50 \, 1/\text{mol} \, \text{min}$. The values of equilibrium constants K for the acetylation reaction practically coincide with Fujii's ones: $K_{iso} = 3.1 \times 10^3$, $K_{atactic} = 1.2$–1.4×10^3. For 'head-to-head' samples $K = 17$.

Sakurada calculated the kinetic constants for the reaction of alkaline hydrolysis of polymethyl acrylate of different microstructure and compared them with the hydrolysis rate constant for methylisobutyrate which can serve as a low-molecular model:

	$k_0(\text{l/mol min})$
Atactic polymethyl acrylate	0.09
Syndiotactic polymethyl acrylate	1.7

$$CH_3-CH-CH_3$$
$$|$$
$$CO$$
$$|$$
$$OCH_3 \qquad 1.2\text{--}2.8$$
Methylisobutyrate

As can be seen from these data, the k_0 value for the syndiotactic specimen is close to the k_0 value for the model. This fact is rather interesting, although there are good reasons to select bi- or trifunctional low-molecular compounds as corresponding models.

For the reaction between polyvinyl alcohol and iodine Matsuzawa et al. [107] found the relationship between the mechanism of the reaction and the chain configuration. A blue color complex is formed as a result of this reaction, the color intensity increases with an increase in the degree of syndiotacticity of polyvinyl alcohol and reaches its maximum for 34.2–39.3% of syndiotactic triads. The authors [107] assumed that for the formation of a complex the helical conformation of polyvinyl alcohol molecules is necessary. Polyiodine chains penetrate into such helixes, forming a complex. The fraction of molecules present in helical conformation being stabilized by hydrogen bonds rises with an increase in the content of syndiotactic configurations and the color becomes more intensive. When the content of syndiotactic triads exceeds 34.2–39.3% the formation of microgel prevents the further growth of the complex.

Of course it is widely known that stereoisomerism of chains affects all physicochemical properties of polymers (T_g, T_m, mechanical properties, etc.). Loebl and O'Neil [108] have shown also that polymethacrylic acid of isotactic structure is weaker than the atactic one (by about 0.3 pK). Dissociation of the isotactic polymethacrylic acid proceeds with the zero thermal effect $\Delta H_{dis} = 0$, while for the atactic acid $(\Delta H_{dis})_{\alpha = 0.5} \approx 700$ cal.

As shown in reference [109], polymethyl methacrylate of different microstructure behaves differently in the course of thermal depolymerization. Syndiotactic and isotactic specimens of the same molecular mass have different thermal stabilities, and for syndiotactic polymethyl methacrylate the rate of depolymerization proves to be slightly less than that for the isotactic specimen.

In the examples described above the values of rate constants are evaluated using the kinetic method for sterically relatively pure models specially synthesized syndiotactic and isotactic samples. In such a case one question always remains unanswered, i.e. are other effects manifested simultaneously with that one? If it is unknown, the quantitative data obtained by the kinetic method for estimating the stereochemical effect cannot always prove to be correct, although in most cases they are quite reliable.

A more precise experimental approach to gain reliable results has been proposed by Harwood [110, 111], who named it the 'double radiometric titration method'. This method was applied to estimate the rate constants of acid hydrolysis of polymethyl methacrylate of different structure. In this approach, a polymer containing a radioactive tracer in each monomer unit and having groups of the same chemical structure, but reacting with different rates, is subjected to the polymer-analogous reaction accompanied by splitting out labeled groups. At the intermediate stage of the process, due to the different reactivity of functional groups a polymer loses more fast reacting groups than slow reacting ones. If we denote the fast reacting groups as F, slow reacting ones as S and reacted ones as X the process can be represented schematically as follows:

$$\overset{*}{F}-\overset{*}{S}-\overset{*}{F}-\overset{*}{F}-\overset{*}{S}-\overset{*}{S}-\overset{*}{S}-\overset{*}{F}-\overset{*}{S}-\overset{*}{F}-\overset{*}{F}-\overset{*}{S}-\overset{*}{F}-\overset{*}{S}$$

$$\downarrow$$

$$X-\overset{*}{S}-\overset{*}{F}-X-\overset{*}{S}-\overset{*}{S}-\overset{*}{S}-X-\overset{*}{S}-\overset{*}{F}-X-\overset{*}{S}-X-\overset{*}{S}$$

If this last polymer is transformed again into the initial one, but without the tracer reagent, then the radioactivity of the obtained sample will be less than that of the initial one. This difference will depend on the difference in reactivities between 'fast' and 'slow' groups and on the degree of conversion, when the process is stopped and starts to go in the opposite direction. The structure of such a polymer will look like this:

$$F-\overset{*}{S}-\overset{*}{F}-F-\overset{*}{S}-\overset{*}{S}-\overset{*}{S}-F-\overset{*}{S}-\overset{*}{F}-F-\overset{*}{S}-F-\overset{*}{S}$$

By repeating this cycle many times, by using different samples of initial polymers (of different microstructure), by carrying out the reaction up to different degrees of conversion and by measuring the radioactivity of initial and 'reconstructed' polymer it is possible to obtain information both about the fraction of functional groups of different reactivity and about the relative reactivity of these groups.

Polymethyl methacrylate is a convenient compound for such an approach as it is easy for it to carry out the reactions in the sequence described above, including diazomethylation of carboxyl groups:

$$\begin{array}{c} CH_3-\overset{*}{C}-COO\overset{*}{C}H_3 \\ | \\ CH_3-\overset{*}{C}-COO\overset{*}{C}H_3 \end{array} \xrightarrow{H^{\oplus}} \begin{array}{c} CH_3-\overset{*}{C}-COOH \\ | \\ CH_3-\overset{*}{C}-COO\overset{*}{C}H_3 \end{array} \xrightarrow{CH_2N_2}$$

$$\begin{array}{c} CH_3-\overset{*}{C}-COOCH_3 \\ | \\ CH_3-\overset{*}{C}-COO\overset{*}{C}H_3 \end{array}$$

On the base of methyl-C^{14}-methacrylate the specimens of isotactic (74% of *meso* dyads), syndiotactic (94% *dl* dyads) and atactic (radical) polymethyl methacrylate were synthesized. The results of the study of their hydrolysis in the presence of H_2SO_4 and remethylation following the described scheme [111] show that isotactic ester groups are 16 times more active than heterotactic ones and 43 times more active than syndiotactic ones. This method proves to be more sensitive than the usual kinetic method. Thus it can be used to detect the less reactive groups in the samples having a degree of isotacticity equal to 52 and 74%; due to the large difference in reactivities of isotactic and other units this problem can not be solved by the kinetic method.

Chemical transformations associated with *cis–trans* isomerization and cycliz- ation of polybutadiene and polyisoprene represent an independent area of intra- and intermolecular reactions related to chain stereoisomerism. A more detailed discussion of this topic can be found in the papers of Dolgoplosk and Kropacheva and co-workers [112, 113]. In reference [112] it was found that poly-1,4-*cis*-butadiene in the presence of some low-molecular radicals (the phenyl sulfide radical C_6H_5—S· is the most active among them) are capable of chain isomerization with the formation of *trans* structures.

The reaction is assumed to proceed with the formation of π complexes with the double bond; it can be seen from Fig. 2.2 that it is an equilibrium reaction. The final product is an equilibrium mixture of *cis* and *trans* structures. It is interesting to note that the kinetic length of such a chain reaction is unusually large; a single phenyl sulfide radical can cause isomerization of up to 300 monomer units of polybutadiene.

In addition to the reactions with radical reagents that cause isomerization, polydienes are capable of another type of intramolecular rearrangement proceed-

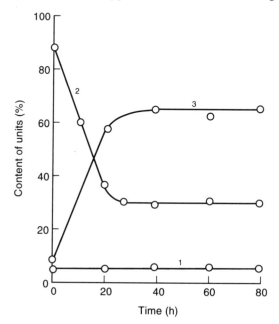

Fig. 2.2 Contents of units of different structures in polybutadiene treated by diphenyl sulfide: 1,2-units (1), *cis*-1,4-units (2), *trans*-1,4-units (3) (benzene, 170 °C, rubber concen- tration 2% (mass), diphenyl sulfide concentration 3%). (Reproduced by permission of Nauka, Moscow, Russia, from *Vysokomol. Soed. B*, 1969, **11**, 717)

ing under selective action of catalytic systems based, for example, on tungsten and molybdenum chlorides and resulting in the formation of cyclic groups [113]. Such a rearrangement can be accompanied by *cis–trans* isomerization. In principle similar phenomena are also observed for 1,4-*cis*-polyisoprene when the intramolecular cyclization proceeds, under the action of cationic systems, resulting in a sharp decrease in unsaturation and viscosity of polymer solutions at constant molecular mass [114].

Macromolecular reactions of isomerization and redistribution of double bonds in chains can be used as a method of directed change of the stereochemical and chemical structure of synthesized chains. On the other hand, these reactions proceed spontaneously in the same systems where polybutadiene and polyisoprene are synthesized and can change the stereoregularity of macromolecules being formed.

Another example of reactions that change the stereochemical structure of the main chain is the asymmetrical reduction of double carbon–carbon and carbon–nitrogen bonds in the polyketo- and polyaminoethers [115]. These reactions result in the formation of optically active polymers due to the asymmetric induction when forming the asymmetric center.

Isomerism of the 'head-to-tail' and 'head-to-head' type affects some chemical reactions and can be used as an analytical tool. Thus the exchange reaction between polyvinyl alcohol and aluminum isopropylate proceeds with inhibition and involves not more than 4–12% of the total amount of OH groups [116]. The authors relate this phenomenon to the fact that only neighboring OH groups situating in the 'head-to-head' monomer units react as their reactivity is much higher than that of the usual OH groups of polyvinyl alcohol.

2.5 CONFORMATIONAL EFFECT

Two possibilities should be distinguished here. First, if for carrying out a reaction it is necessary to bring within a definite distance the functional groups distributed far apart in the chain, then the result, i.e. the reaction will proceed or not, will depend on whether the conformation necessary for the reaction has been achieved or not. If it has been achieved, then how large is the probability of this event and what is the lifetime of such a favorable conformation? Such effects with variations in reaction rates of tens of thousands and a million times are typical of enzymatic reactions when the conformational contribution plays a significant role.

Second, the very process of chemical transformation of a macromolecule of one type and structure into a molecule of another type and structure in accordance with the given reaction is definitely related to the variation in shape of the macromolecule in solution as a change in chemical composition, intra- and intermolecular interactions, potential barriers of rotation, etc., takes place. The chain conformation which provides access of the reagent to all units in the beginning of the process cannot be formed, for example, at later stages, and the reaction slows down. Opposite cases are also possible, when the reaction rate increases due to uncoiling of the chain in the given medium in the course of the reaction.

Such conformational effects in the clear form are very seldom observed. Usually they are masked by other effects. However, it can be affirmed with confidence that the conformational contribution to observed chemical behavior of the chain will be significant.

Definite conformation and the degree of extension of a macromolecular coil affects, on the one hand, the rate of transport of the low-molecular reagent to reactive groups of a polymer and, on the other hand, the equilibrium concentration of this reagent near reactive groups. Finally, if the time of the conformational rearrangement and reaction time are of the same order then it will also affect the mechanism and the rate of reaction.

Interesting experimental data have been obtained by Turska in the course of the study of the conformational effect in polycondensation [117] and hydrolysis [118] reactions. A comparison of kinetic data on alkaline hydrolysis of polyvinyl acetate with different degrees of polymerization (9 and 8950) and of low-molecular model compounds (ethyl acetate and 1,3-diacetoxybutane) shows that at the initial stages of the reaction the rate of hydrolysis is much lower than that of low-molecular esters. The kinetic curves are presented in Fig. 2.3 (where a and b are the concentrations of NaOH and ester groups at time t and a_0 and b_0 are the same concentrations for $t = 0$). In such plots for a second-order reaction a linear dependence should be observed with a slope equal to the kinetic constant k. As can be seen from Fig. 2.3, such linear kinetics are observed only for low-molecular models and for polymers after 20% conversion. For polymers at lower conversions hydrolysis proceeds at a lower rate, and the retardation effect is more pronounced for the sample with a higher degree of polymerization. This retardation cannot be related to the neighboring groups effect, as the appearance of hydroxyl groups in the immediate vicinity of ester groups should result in an acceleration of the reaction.

Turska and Jantas [118] believe that this effect can be related to steric hindrances for an oligomer (PVA-9) and to a decrease in the accessibility of ester groups for the hydrolyzing agent due to coiling of the macromolecule for

---

Fig. 2.3 (a) Kinetics of hydrolysis of polyvinyl acetate of the degree of polymerization $P = 8950$ (1, 1′) and 9 (2), of ethyl acetate (3, 3′) and 1,3-diacetoxy butane (4, 4′) in acetone/water solvent: 70:30 volume % (1, 2, 3, 4) and 75:25 volume % (1′, 3′, 4′) at 20 °C; $k_1 = 1.72$, $k_2 = 1.78$, $k_3 = 1.76$, $k_4 = 1.88$, $k'_1 = 2.19$ l mol/min. (b) Kinetics of hydrolysis of polyvinyl acetate (1) and its copolymers with vinyl alcohol: 5.3 (2), 11.8 (3), 20.6 (4), 33.4% (5) hydroxyl groups (acetone/water, 70:30 volume %, 20 °C). (Reproduced by permission of John Wiley & Sons Ltd from *J. Polymer Sci. C*, 1974, **47**, 359)

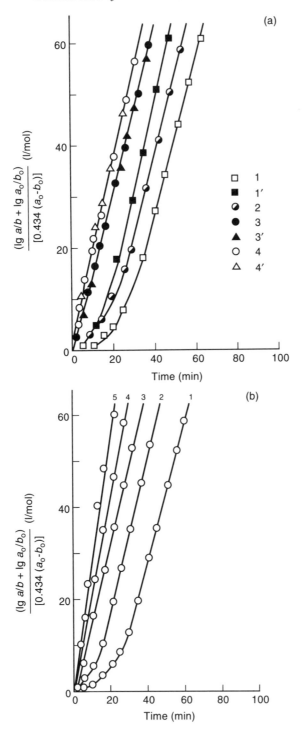

a polymer (PVA-8950). This assumption is confirmed by a comparison of kinetic data obtained in different solvents (Fig. 2.3a). Improving the solvent quality results in a decrease in the inhibition effect. The comparison of kinetic curves of hydrolysis of copolymers of vinyl acetate with vinyl alcohol (Fig. 2.3b) shows that an increase in the number of hydroxyl groups (accompanied by an increase in the affinity of the solvent to the polymer) is accompanied by an increase in the rate of hydrolysis, and the kinetic curves become linear when the concentration of alcohol groups in a copolymer becomes equal to 20%. The rate constants calculated from the slope of linear parts of kinetic curves are found to be close to those obtained for low-molecular models (Fig. 2.3a). It is evident that at 20% concentration of hydroxyl groups the macromolecular coils unfold so much that all functional groups are equally accessible for the hydrolyzing agent.

This reaction of hydrolysis of polyvinyl acetate was studied by Pichot *et al.* [119] using HCl as a catalyst in the THF/CH_3OH mixed solvent. In this case, as in the case of alkaline hydrolysis, the initial rate constant in the poor solvent (CH_3OH) was found to be lower than in the good one (THF). The reaction of acidic hydrolysis proceeds with autoacceleration being a result of the attack on the carbonyl group with $CH_3C^+(OH)(OCH_3)$ ion formed on adjacent reacted groups [119]. In the case of the unfolded chain conformation such ions quickly diffuse into solution, while in the case of the coiled conformation these ions are held back in the immediate neighborhood of the chain and do not accelerate the reaction.

The analogous manifestation of the conformational effect was observed by Boucher *et al.* [120] in the quaternization of poly-4-vinylpyridine. In the course of this reaction, because of the electrostatic repulsion between reacted functional groups the chain is unfolded. On the other hand, with conversion the solvent becomes poorer. As a result the viscosity increases at first and then decreases. An addition of salt promotes chain coiling and results in retardation of the reaction.

Huang *et al.* [121] studying epoxidation of polybutadiene and styrene–butadiene triblock copolymers with monoperoxyphthalic acid observed an increase in the rate of the reaction when adding chloroform to the dioxane solvent. Such an effect is a result of the improvement of a solvent. Both in dioxane and in the mixed solvent the coil dimensions increase in the course of the reaction.

Aminolysis of side-chain esters of poly-γ-alkyl glutamate [122] is greatly affected by the conformation of macromolecules The reaction proceeds much faster in random-coil parts of the polymer chain, while it is very slow in the helical segment. In the presence of helicogenic solvents (dimethyl formamide, dimethyl amine) the retardation of the reaction is observed, related both to the effect of helical conformation and to its stabilization by the solvent solvatation.

Irzhak, Yenikolopyan and co-workers [123–125] studying the ratio between intra- and intermolecular reactions in the course of processes of networks formation detected the conformation effect on this ratio. Thus for the reaction of polyvinylbutyral crosslinking with diisocyanates [124] it has been found that the yield of a sol fraction depends on the nature of a solvent, increasing on passing from a good solvent (dioxane) to the poor one (pyridine). This fact may be ascribed to an increase in the probability of intramolecular crosslinking as compared with an intermolecular one with an increase in the degree of coiling in a poor solvent.

Very demonstrative are the data obtained by Varshavskii and co-workers [126] on the kinetics of tritium–hydrogen exchange in polyriboadenylic acid (poly-A) and its low-molecular analog adenosine-5-monophosphate. The stability of helix conformation of poly-A chains is achieved mainly by the stacking action of adenylic bases. The ^1H—^3H exchange rate in C—H bonds of adenylic fragments decreases with an increase in the bases fraction as compared with the exchange rate in adenosine monophosphate. Involvement of adenylic groups in the stacking interaction and the presence of the fixed helix conformation of the chain results in a decrease in the reaction rate.

For the reaction of photodegradation of polybutyl methacrylate Yermilov [127] has found the dependence of the number of unreacted butoxyl groups determined by the reaction with hydroiodic acid on conditions of the polymer preparation and processing. The dependence of the 'determined' number of butoxyl groups on the solvent indicates that their incomplete detection is related to different conformational states, one of which may be unfavorable for the reaction with HI.

One more example of the conformational effect is the different catalytic activity of natural catalyst chymotrypsin in hydrolysis of the ester bond in nitrophenyl ester molecules [128]. In the native form chymotrypsin hydrolyzes the ester bond with a sufficiently high rate. After denaturation when the chemical sequence of units is retained but the conformation of the molecule is changed, the rate of hydrolysis decreases by a million times. The reason for this phenomenon is that in the native conformation of α-chymotrypsine two of its amino acid groups, histidine and serine, are located close to each other. As a result the catalytic complex including OH groups and imidazolyl rings is formed, providing the fast two-stage reaction. The histidine and serine residues are separated by hundreds of units along the chain, and when this conformation is destroyed the catalytic activity disappears.

2.6   ROLE OF SOLID POLYMER MATRIX AND SUPERMOLECULAR EFFECTS

If on passing from a good to a poor solvent the shielding of functional groups as a result of the change of a conformation can affect the rate of a macromolecular reaction, then still more pronounced effects can be assumed in the case when supermolecular formations in the system arise or for a reaction proceeding generally in the solid polymer matrix. This is confirmed, for example, by the dependence of the rate of cellulose reactions on its morphology. Such a dependence is explained first by the variation in the availability of functional groups for diffusion on changing over from one type of supermolecular structure to another. Thus the extent to which the acetylation reaction proceeds depends on the humidity of the initial product [129, 130]. Water does not participate in the reaction, but affects the morphology of cellulose by increasing the number of available hydroxyl groups. The rate of cellulose hydrolysis depends on features of its morphological structure [131, 132]. The thermal stability of cellulose materials is related to the degree of crystallinity, crystallite size and packing density [133]. A detailed discussion of these examples which are related mainly to diffusion processes can be found in reference [12].

In the studies of thermal dehydration of polyvinyl alcohol in the oriented and nonoriented states it has been found [134] that in the IR spectrum of the nonoriented sample the 1141 cm^{-1} absorption band characterizing the degree of structural order of the polymer at the supermolecular level disappears when water elimination starts. In oriented samples that have a very distinct texture these structures at the same temperature 'live' much longer and are stable at much higher temperatures.

Vol'f and Meos [135] have shown that the number of double bonds which are accumulated in polyvinyl alcohol fibers as a result of dehydration depends on the

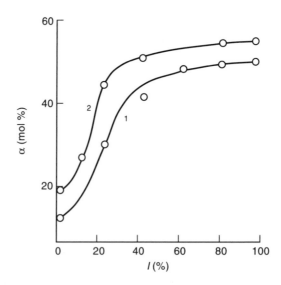

Fig. 2.4 Content of double bonds in polyvinyl alcohol fiber versus the degree of drawing during the dehydration (thermal action, 20 s, 240 °C) at acidity of the sample 0.14 (1) and 0.18 g KOH/g (2). (Reproduced by permission of Khimiya, Moscow, Russia, from *Uspekhi khimii i tekhnologii polimerov* (*Advances in Polymer Chemistry and Technology*), Khimiya, Moscow, 1971, p. 68)

degree of fiber stretching, and at a certain degree of stretching the dehydration efficiency for some standard conditions increases abruptly (Fig. 2.4). Apparently in the course of stretching a change of conformation of polyvinyl alcohol molecules takes place and oriented structures are formed, promoting water formation.

It is possible that the mechanism of reactions involving macromolecules may also change due to the different physical structure of the substance of the same composition. Apparently this is a reason for an observed phenomenon [136] when upon $\Delta\gamma$-irradiation of isotactic crystalline polypropylene as a result of chain migration of formed hydrogen atoms and double bonds from initial alkyl radicals the allyl and conjugated double bonds are formed that impart to polypropylene the radiation resistance. In amorphous polypropylene, due to the high chain mobility, the alkyl radicals recombine mainly with each other and the stabilization effect is less pronounced.

The structure–physical effects associated with the chemical behavior of a polymer can also be manifested in solutions. As an example one can consider the reaction of polyethylene chlorination studied by Shibaev *et al.* [137]. This reaction is discussed in detail in Chapters 4 and 6. Here we mention only that the existence of associates of polyethylene macromolecules [138] results in the formation of two types of products, which are the strongly chlorinated macromolecules (which are formed as a result of chlorination of molecules on the surface of associates and at the molecular level after their decay) and the slightly chlorinated macromolecules situated within associates into which chlorine hardly enters.

Important results concerning interrelation of the polymer morphology and reactivity have been obtained by Wunderlich [139] who showed that some chemical reactions could proceed at the surface of macromolecular crystals without affecting the chains included in the crystalline lattice. It proceeds as if 'etching' of the noncrystalline phase takes place, which is accompanied by an increase in the degree of crystallinity of a sample. This phenomenon has been illustrated by acidic hydrolysis of cellulose and polyethylene terephthalate, resulting in a decrease in the molecular mass down to a certain critical value corresponding to the length of a chain constituting the crystal. A similar effect is observed when treating polyethylene with fuming nitric acid [140]. This reaction also proves to be sufficiently selective and to stop at the crystal surface. Such reactions which break the backbone of macromolecules in the amorphous phase and do not affect practically the crystalline phase can be used to obtain oligomers with high crystallinity and apparently with quite narrow molecular mass distribution.

A spectacular example of the dependence of reaction kinetics on supermolecular structures is thermooxidative degradation of polypropylene studied in detail by Shlyapnikov and co-workers [141–145]. The discussion of degradation processes is beyond the scope of this book, but the supermolecular effects manifested in polypropylene oxidation are of interest for macromolecular reactions in general. It has been shown in reference [141] that thermooxidative degradation proceeds mainly in amorphous regions. On comparing the kinetics of reactions that proceed in samples of various crystalline structure it has been found that polypropylene having large spherulites is oxidized at a lower rate than that having small spherulites [142]. The reaction is also sensitive to sample thickness: degradation in the surface layers of thick samples (300 μm) proceeds up to higher conversions than in a thin (30 μm) oriented film being oxidized under similar conditions [143]. The rate of oxidation also depends on the degree of elongation of oriented samples [144]. In addition an increase in the reaction rate in this case is accompanied by a rise in crystallinity, and as a result the inhibition of the reaction should be observed. It has been assumed in reference [145] that in oriented polypropylene the reaction kinetics varies due to the change of conformation of chains constituting the amorphous domains and due to their conformational mobility. One of the possible explanations of the dependence of the oxidation rate on the degree of oxidation proposed by Rapoport and Miller [145] is that the presence of nonequilibrium extended conformations results in a decrease in the rate of formation of polypropylene hydroperoxide, being a product on which the branching of the kinetic chain takes place.

Comprehensive and systematized data on features of reactions proceeding in solid polymers including degradation, oxidation and stabilization reactions are available in works of Emanuel' and Zaikov [35, 146]. In solid polymers the most ordered crystalline domains are characterized by the minimal reactivity. The structural defects, for example dislocations, result in greater permeability for small molecules and increase the reactivity. The maximal reactivity is observed in amorphous domains with a low chain packing density.

Similar results were obtained by Guzman *et al.* [147] in the course of the study of chlorination of polyethylene single crystals. It has been shown that melting temperatures and heats of fusion change after melting and recrystallization of PE crystals. Even at a high degree of chlorination the chemical attack takes place in a preferential and selective way on the amorphous interface of the single crystals.

Gibson and Bailey [148] observed two stages in the reaction of sulfonation of polystyrene—a fast stage for the reaction on the surface and a slow one for the reaction in the volume.

The intensity of molecular motions significantly affects the kinetics of chemical reactions in solid polymers. It is related not only to the structure rearrangements rate but also to the diffusion rate of low-molecular compounds in a polymer matrix. This is a reason for the unusually high values of Arrhenius parameters— the reaction rate constant and the activation energy. The effect of the physical structure of a polymer is manifested in the experimentally observed 'step-like' kinetics of reactions in the solid polymer phase [35, 149]. A number of papers are dedicated to the effect of molecular mobility on the decrease in the rate of macroradical decay on passing from the amorphous to the crystalline phase

[150, 151] to the decrease in the rate constant of degradation in oriented samples compared with nonoriented ones [152]. An addition of low-molecular compounds (plasticizers) which increase the molecular mobility results in a rise in the rate of radical reactions in polymers [35, 146].

In studying the diverse reactions in solid polymers—oxidation of radical photo- and thermal ionization, decay of radicals, postpolymerization and others—it has been found that the 'kinetic stop' of a reaction occurs due to the accumulation of active centers and their consumption (in the course of the reaction) in structurally nonhomogeneous regions of a polymer where the most favorable conditions for the reaction are created [146, 153, 154].

In a study of the effect of molecular mobility on the change of the singlet oxygen suppression rate constants, suppression being carried out by many suppressors, the governing factor has been found to be the difference between the experimental temperature and glass transition point of a given polymer. Polybutadiene, polystyrene and copolymer of butyl methacrylate with methacrylic acid behave as polymer matrices in which this reaction proceeds [155, 156].

In reference [157] an example is discussed showing the difference in the structure of macroradicals according to the conformation of chain fragments permitted by one or other type of supermolecular packing. It has been shown that upon low-temperature mechanochemical or radiation treatment of amorphized polypropylene and polybutene-1 the signals in their ESR spectra appear to correspond to free radicals in which the unpaired electron is located on the tertiary carbon atom of the chain. After annealing of irradiated crystalline samples of these polymers at 300 K a quite different type of ESR spectrum is observed. The form of this spectrum is a result of the presence of macroradicals located in crystalline regions of a polymer. The radicals located in amorphous and crystalline regions have different conformations and unlike spectra. The difference in conformations of the same 'middle' macroradicals is related to different effects of surrounding molecules on radicals in the crystalline lattice or in the amorphous matrix.

Radtzig [157] has considered in detail the following questions. What are the reasons for the difference in conformations of these radicals? What are the permissible chain fragment rotations around the methylene groups? How do conformational angles vary in this case? From the analysis of these results it can be concluded that the reactivity of macroradicals showing different ESR spectra may also be different This is manifested in different reactivities.

The solid-phase polycyclization of polyhydrazides can serve as an interesting example of the effect of supermolecular structures on the kinetics of macromolecular reactions. A detailed analysis of polycyclization is published in the works of Korshak, Berestneva and co-workers [158–162].

The formation of cycles in polyhydrazide chains is possible only in the case of the *cis* configuration of hydrazide fragments. However, from the viewpoint of intramolecular interactions the most favorable is the *trans* form, which owing to the packing effects becomes still more favorable in the solid state. Therefore the formation of a cycle requires a turn around the N—N bond, which is possible only at temperatures above the glass transition point. With the formation of cycles the chain becomes more rigid, the glass transition temperature rises and at the point close to the temperature at which the cyclization is carried out the reaction practically stops due to vitrification of a polymer. For this reason a high degree of transformation can be achieved only at very high temperatures, when the polymer degradation process begins. The very complicated process of polycyclization becomes more complicated if the original polyhydrazide has an oriented or crystalline structure [160], as in this case the conformational transitions are still further complicated. As a result the rate of polycyclization decreases and the reaction cannot proceed to higher degrees of conversion.

The latter reaction is of significant interest for the reason that it illustrates simultaneously the manifestation of both conformational and supermolecular effects, the parallel action of which also affects the other. An analogous example of simultaneous action of conformational and configurational effects can be the formation of the complex of polyvinyl alcohol with iodine [107], which has been considered above.

The degree of crystallinity of polymers which affects most of their reactions is determined by stereoregularity; i.e. the possible supermolecular effects are also related to configurational ones. Thus the specific effects typical for macro-molecular reactions show up in combination more frequently than in the 'pure' form.

2.7 ELECTROSTATIC EFFECTS

A tentative analysis of this phenomenon is given by Morawetz [128]. The electrostatic free energy necessary to remove the protons from the charged polyanion (or hydroxyl ions from the polycation) can affect the shift of the ionization equilibrium in the reaction of a polymer with a low-molecular reagent accompanied by the appearance of charges in the chain. The latter will change the rate of the reaction compared with the reaction of a monofunctional analog for the same pH value. A classical example of such a reaction is the hydrolysis of 3-nitro-4-acetoxybenzene sulfonate by the polymer reagent poly-4-vinylpyridine and its low-molecular analog 4-picoline [58].

Strictly speaking, nonionized pyridine nuclei, which are nucleophilic reagents, act as a catalyst. The reaction, however, is promoted by electrostatic attraction of the negatively charged anion of substrate to the polycation of partially ionized polyvinyl pyridine. As seen from Fig. 2.5, the curve of the dependence of the rate on the fraction of ionized pyridine groups has a maximum corresponding to 75% of nonionized groups. As the high positive charge of the chain inhibits its ionization the number of reactive pyridine groups at low pH values turns out to be 60–100 times more than for the low-molecular analog 4-picoline. For these reasons, i.e. the small number of active nonionized groups and the absence of the electrostatic attraction of the substrate to the catalyst, the low-molecular analog of polyvinyl pyridine acts with a lower efficiency:

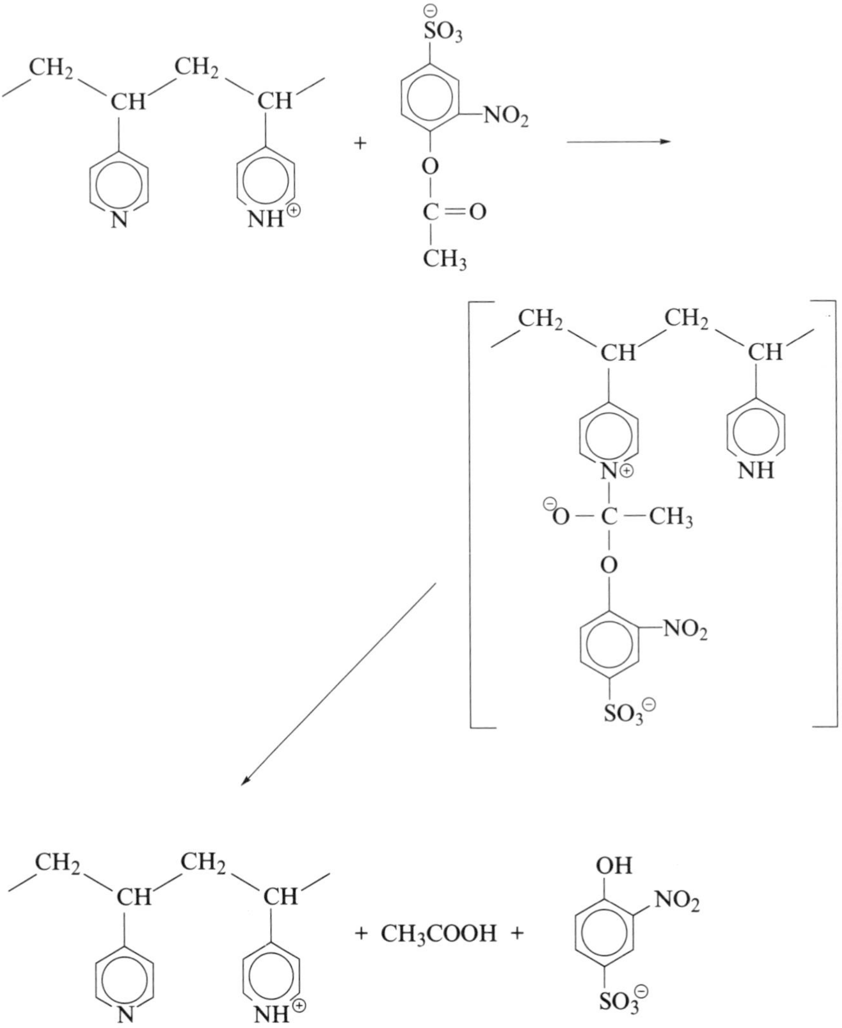

The 'polymeric' effect is observed even in the case of the equal degree of ionization of the macromolecular and low-molecular catalysts. One of the possible explanations of this fact [163, 164] is the dependence of the low-molecular reagent microconcentration near the macromolecule on the degree of ionization of functional groups, in particular if the reaction proceeds with participation of uncharged organic substrate in a mixed solvent. In other words, two polymeric

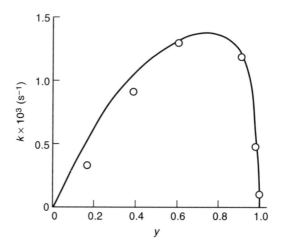

Fig. 2.5 Constant of solvolysis of 3-nitro-4-acetoxybenzene sulfonate in 0.01 N solution of poly-4-vinylpyridine versus the fraction of pyridine groups in basic form (y) (50% ethyl alcohol, ionic strength 0.04, 36.8 °C). (Reprinted with permission from *J. Am. Chem. Soc.*, 1962, **84**, 114. Copyright 1962 American Chemical Society)

effects are manifested simultaneously, resulting in a rise in the effective rate constant of the reaction.

Electrostatic effects may result in a significant decrease in the reaction rate, as observed for the reaction between acetamide and benzamide derivatives of lithium and polymethyl acrylate [165]. The substitution of ester groups by amide ones proceeds via formation of anions in side groups, and due to the electrostatic repulsion the reaction is inhibited and finally stops at 60% conversion. In this case the isolated ester groups situated between two already substituted amide groups prove to be blocked and do not participate in the reaction.

The role of electrostatic effects is very high in reactions catalyzed by polyelectrolytes. Detailed treatment of these phenomena can be found in Chapter 7 of this monograph.

An attempt to give a quantitative account of the electrostatic interactions in macromolecular reactions was made by Kachalsky and Feitelson [166] for the hydrolysis of pectins catalyzed by OH⁻ ions. The observed rate constant of this reaction K can be written as

$$K = K_0 \exp\left[-\frac{\Delta G_{el}^*(\alpha)}{kT} \right]$$

where K_0 is the rate constant for uncharged groups (in this case it corresponds to the initial rate of the process), α is the degree of ionization of macromolecules and ΔG_{el}^* is the free electrostatic activation energy corresponding to the work of the transfer of hydroxyl ions to ester groups against the repulsion forces of the negatively charged polyanion. It is also assumed that for most systems the ΔG_{el}^* value coincides with the free energy of ionization, which means that the effective rate constant is a function of the ionization constant and depends on the degree of conversion. Here again we are dealing with a sufficiently complicated dependence, when it is difficult to differentiate the electrostatic effect from the conformational ones and the effects related with the change of local concentrations of reagents.

2.8 CONCLUSIONS

We can say that all the variety of macromolecular effects has the same origin—the chain structure of polymer molecules. Therefore the quantitative description of these effects requires a general approach: one should separate the effects and consider them subsequently. Below we describe the known approaches to evaluation of macromolecular effects, beginning with the neighboring group effect, which is the most inherent for polymer substances as it reflects the quite general property of chemical compounds—the mutual influence of groups of atoms in molecules.

REFERENCES

1. Flory, P., *J. Am. Chem. Soc.*, 1939, **61**, 3334–40.
2. Flory, P., *Principles of Polymer Chemistry*, Interscience, New York, 1953.
3. Alfrey, T., in *Chemical Reactions of Polymers*, Ed. E. Fettes, Interscience, New York, 1964.
4. Platé, N. A., in *Kinetika i Mekhanizm Obrazovaniya i Prevrashcheniya Makromolekul* (*Kinetics and Mechanism of Formation and Transformation of Macromolecules*), Khimiya, Moscow, 1968.
5. Platé, N. A. in *Uspekhi Khimii i Tekhnologii Polimerov* (*Achievements of Chemistry and Technology of Polymers*), Khimiya, Moscow, 1971.
6. Platé, N. A., Litmanovich, A. D. and Noah, O. V., *Makromolekulyarnye Reaktsii* (*Macromolecular Reactions*), Khimiya, Moscow, 1977.
7. Coleman, B. and Fuoss, R., *J. Am. Chem. Soc.*, 1955, **77**, 5472–6.
8. Smets, G., *Makromol. Chem.*, 1954, **34**, 190–7.
9. Conix, A. and Smets, G., *J. Polymer Sci.*, 1955, **15**, 221–9.
10. Wallace, R. and Hedley, K., *J. Polymer Sci.*, 1966, **A4**, 71–9.
11. Tiger, R P., Yevreinov, V. V. and Entelis, S. G., *Vysokomol. Soed*, 1966, **8**, 2000–11.
12. Fettes, E. (Ed.), *Chemical Reactions of Polymers*, Interscience, New York, 1964.
13. Grassi, N. and Scott, G., *Polymer Degradation and Stabilization*, Cambridge University Press, Cambridge, 1985.
14. Madorsky, K. L., *Thermal Degradation of Polymers*, Academic Press, 1964.
15. Yenikolopyan, N. S. and Vol'fson, S. A., *Khimiya i Tekhnologiya Poliformal'degida* (*Chemistry and Technology of Polyformaldehyde*), Khimiya, Moscow, 1968.
16. Minsker, K. S. and Fedoseeva, G. T. *Destruktsiya i Stabilizatsiya Polivinilkhlorida* (*Degradation and Stabilization of Polyvinyl Chloride*), Khimiya, Moscow, 1972.

17. Mayer, Z., Overeigner, B. and Lim, D., *J. Polymer Sci. Part C*, 1971, **33**, 289.
18. *Khimiya Polisopryazhennykh Sistem* (*Chemistry of Polyconjugated Systems*), Khimiya, Moscow, 1972.
19. Dvorko, G. F. (Ed.), *Organicheskie Poluprovodniki* (*Organic Semiconductors*), Mir, Moscow, 1965.
20. Morishima, J., Fujisawa, K. and Nozakura, S., *J. Polymer Sci. Polymer Letters Ed.*, 1976, **14**, 467–9.
21. Daly, M., Main Lecture at the International Symposium on Macromolecular Chemistry, Mainz, 1979.
22. Martin, G. E., Shambhu, M. B. and Digenis, G. A., *J. Pharm. Sci.*, 1978, **67**, 110–14.
23. Martin, G. E., Shambhu, M. B., Shakir, S. R. and Digenis, G. A., *J. Org. Chem.*, 1978, **43**, 4571–8.
24. Jayalekshmy, P. and Mazur, S., *J. Am. Chem. Soc.*, 1976, **98**, 6710; 1979, **101**, 677.
25. Crosby, G. A. and Kato, M., *J. Am. Chem. Soc.*, 1977, **99**, 278.
26. Kraus, M. A. and Patchornik, A., *J. Am. Chem. Soc.*, 1970, **92**, 7587.
27. Kraus, M. A. and Patchornik, A., *J. Am. Chem. Soc.*, 1971 **93**, 7335.
28. Manecke, G. and Reuter, P., *J. Polymer Sci. Polymer Symp.*, 1978, **62**, 227–50.
29. Leznoff, C., *Chem. Soc. Rev.*, 1974, **3**, 65.
30. Yergozhin, Ye., Dissertation, Alma-Ata, 1974.
31. Yaroslavsky, C., Patchornik, A. and Katchalski, E., *Tetrahedron Letters*, 1970, 3629.
32. Yaroslavsky, C. and Katchalski, E. *Tetrahedron Letters*, 1972, 5173.
33. Kamachi, M., Kikuta, Y. and Nozakura, Sh., *Polymer J.*, 1979, **11**, 273–7.
34. Rado, R. and Shimunkova, D., paper presented at the International Symposium on *Macromolecular Chemistry*, Brussels, 1967.
35. Emanuel', N. M. and Zaikov, G. Ye., *Vysokomol. Soed*, 1975, **A17**, 2122–32.
36. Loucheux, C., in *Reactions on Polymers*, Proceedings of NATO Advanced Study Symposium, Ed. J. Moore, Reidel Publishing Co., Dordrecht, 1973, pp. 102–25.
37. Carraher, Ch., in *Reactions on Polymers*, Proceedings of NATO Advanced Study Symposium, Ed. J. Moore, Reidel Publishing Co., Dordrecht, 1973, pp. 126–48.
38. Drobnik, J., Kopecek, J., Labsky, J., Rejmanova, P., Exner, J., Saudek, V. and Kalal, J., *Makromol. Chem.*, 1976, **177**, 2833–44.
39. Imai, K., Shiomi, T., Tezuka, Y., Takahashi, K. and Satoh, M., *J. Macromol. Sci. Chem.*, 1985, **A22**, 1359–69.
40. Arranz, F. and Sanchez-Chaves, M., *Polymer*, 1988, **29**, 507–14.
41. Rondon, S., Smets, G. and de Wilde-Delvaux, M., *J. Polymer Sci.*, 1957, **24**, 261–8.
42. Reynolds, D. and Renyon, W., *J. Am. Chem. Soc.*, 1950, **72**, 1584–90.
43. Morawetz, H., in *Chemical Reactions of Polymers*, Ed. E. Fettes, Interscience, New York, 1964.
44. Harwood, J., *Angew. Makromolek. Chem.*, 1968, **4/5**, 279–85.
45. Davydova, S. L., Barabanov, V. A., Platé, N. A. and Kargin, V. A., *Vysokomol. Soed*, 1968, **A10**, 1004–13.
46. Andreeva, I. V., Koton, M. M., Madorskaya, L. Ya., Turbina, A. I., Pokrovskii, E. I. and Lyubimova, G. V., *Vysokomol. Soed*, 1972, **A14**, 1565–9.
47. Andreeva, I. V., Koton, M. M., Madorskaya, L. Ya., Turbina, A. I. and Pokrovskii, E. I., Proceedings of the Conference on *Sintez, Structura i Svoistva Polimerov* (*Synthesis, Structure and Properties of Polymers*), Nauka, Leningrad, 1970, pp. 18–22.
48. Morawetz, H., *Pure Appl. Chem.*, 1979, **51**, 2307–11.
49. Galin, J. C. in *Modification of Polymers*, Ed. Ch. Carraher, ACS Symposium Series 121, 1980, pp. 119–38.
50. Velasquez, D. L. and Galin, J. C., *Macromolecules*, 1986, **19**, 1096–109.

51. Kargin, V. A., Kabanov, V. A. and Kargina, O. V., *Dokl. AN SSSR*, 1963, **153**, 845–7; 1965, **161**, 1131–3; 1966, **170**, 1130–2.
52. Kabanov, V. A., Main Lecture at the 3rd Symposium on *Physiological Polymers and Macromolecular Models of Biopolymers*, Zinatne, Riga, 1973, p. 97.
53. Massukh, I., Petrovskaya, V. A., Pshezhetskii, V. S., Kabanov, V. A. and Kargin, V. A., *Vysokomol. Soed*, 1967, **A9**, 899–905; **B9**, 834–7.
54. Kern, W. and Sherhag, E., *Makromol. Chem.*, 1958, **28**, 209–23.
55. Kern, W., Herold, W. and Sherhag, E., *Makromol. Chem.*, 1956, **17**, 231–48.
56. Sakurada, I. and Ono, T., *Makromol. Chem.*, 1966, **94**, 280–99.
57. Sakurada, I. and Sequiguchi, Y., *Makromol. Chem.*, 1963, **81**, 243–57.
58. Leitsinger, R. and Saverlide, T., *J. Am. Chem. Soc.*, 1962, **84**, 114–15, 3122–7.
59. Overberger, C. G., *et al.*, *J. Am. Chem. Soc.*, 1965, **87**, 296–301, 3270–2, 4310–13; *Ann. NY Acad. Sci.*, 1969, **155**, 431.
60. Imai, I., Matsumoto, A. and Oiwa, M., *J. Polymer Sci. Polymer Chem. Ed.*, 1979, **17**, 821–33.
61. Overberger, C. G. and Guterl, A. C., *J. Polymer Sci. Symposia*, 1978, **62**, 13–28.
62. Morawetz, H., *J. Polymer Sci. Symposia*, 1978, **62**, 271–82.
63. Morawetz, H., *J. Macromol. Sci. Chem.*, 1979, **A13**, 311–20.
64. Anufrieva, E. V., Pautov, V. D., Freidzon, Ya. S., Shibaev, V. P. and Platé, N. A., *Vysokomol. Soed*, 1975, **A17**, 586–93.
65. Platé, N. A., Shibaev, V. P. and Tal'roze, R. V., in *Uspekhi Khimii i Tekhnologii Polimerov* (*Achievments of Chemistry and Technology of Polymers*), Khimiya, Moscow, 1973, pp. 127–74.
66. Platé, N. A. and Shibaev, V. P., *J. Polymer Sci. Reviews*, 1974, **8**, 117–253.
67. Frisman, E. V. and Dadivanian, A. K., *J. Polymer Sci. Symposia*, 1967, **16**, 1001–9.
68. Dadivanian, A. K. and Agranova, S. A., *Vysokomol. Soed*, 1980, **A22**, 1499–502.
69. Moiseev, Yu. V. and Zaikov, G. Ye., *Khimitcheskaya Stoikost' Polimerov v Agressivnykh Sredakh* (*Chemical Stability of Polymers in Hostile Media*), Khimiya, Moscow, 1979, p. 76.
70. Scarpa, J. S., Mueller, D. D. and Klotz, J. M., *J. Am. Chem. Soc.*, 1967, **89**, 6024.
71. Kakuda, Y., Perry, N. and Muller, D. D., *J. Am. Chem. Soc.*, 1971, **93**, 5992.
72. Hvidt, J. and Corret, R., *J. Am. Chem. Soc.*, 1970, **92**, 5546.
73. Morawetz, H. and Zimmering, P., *J. Phys. Chem.*, 1954, **58**, 753–7.
74. Zimmering, P., Westhead, E. and Morawetz, H., *Biochim. Biophys. Acta*, 1957, **25**, 376–9.
75. Bender, M. and Neveu, M., *J. Am. Chem. Soc.*, 1958, **80**, 5388–92.
76. Gaetjens, E. and Morawetz, H., *J. Am. Chem. Soc.*, 1960, **82**, 5328–35.
77. Cooper, W., *Chem. and Ind.*, 1958, 263–5.
78. Schaefgen, J. and Serasohn, I., *J. Polymer Sci.*, 1962, **58**, 1049–58.
79. Pinner, S., *J. Polymer Sci.*, 1953, **10**, 379–87.
80. Platé, N. A. and Litmanovich, A. D., *Vysokomol. Soed*, 1972, **A14**, 2503–17.
81. Morawetz, H. and Gaetjens, E., *J. Polymer Sci.*, 1958, **32**, 526–34.
82. Gaetjens, E. and Morawetz, H., *J. Am. Chem. Soc.*, 1961, **83**, 1738–41.
83. De Loecker, W. and Smets, G., *J. Polymer Sci.*, 1959, **40**, 203–12.
84. Smets, G. and De Loecker, W., *J. Polymer Sci.*, 1960, **45**, 375–83.
85. Smets, G. and De Loecker, W., *J. Polymer Sci.*, 1959, **41**, 461–7.
86. Glavis, F. J., *J. Polymer Sci.*, 1959, **36**, 547–58.
87. Chapman, C. B., *J. Polymer Sci.*, 1960, **45**, 237–46.
88. Smets, G. and Van Humbeek, W., *J. Polymer Sci.*, 1963, **A1**, 1228–41.
89. Mercier, J. and Smets, G., *J. Polymer Sci.*, 1963, **A1**, 1491–505.

90. Millan, J., Madruga, E. L., Bert, M. and Guyot, A., *J. Polymer Sci.*, 1973, **A1** (11), 3299–308.
91. Millan, J., Carranza, M., Bert, M. and Guzman, J., *J. Polymer Sci.*, 1973, **C42**, 1411–18.
92. Millan, J., Madruga, E. L. and Martinez, G., *Angew Makromol. Chem.*, 1975, **45**, 177–84.
93. Martinez, G., Mijangos, C. and Millan, J., *Makromol. Chem.*, 1979, **180**, 2937–45.
94. Millan, J., Martinez, G. and Mijangos, C., *Rev. Plast. Mod.*, 1980, **39**, 61–9.
95. Martinez, G., Mijangos, C. and Millan, J., *J. Macromol. Sci.*, 1982, **A17**, 1129–48.
96. Millan, J., Martinez, G. and Mijangos, C., *J. Polymer Sci. Chem. Ed.*, 1985, **23**, 1072–9.
97. Martinez, G., Mijangos, C. and Millan, J., *Polymer Bull.*, 1985, **13**, 151–62.
98. Martinez, G., Mijangos, C. and Millan, J., *Makromol. Chem. Rap. Com.*, 1986, **7**, 415–18.
99. Millan, J., Martinez, G., Mijangos, C. and Gomez-Daza, M., *Makromol. Chem.*, 1989, **190**, 223–31.
100. Imai, K., Shiomi, T., Tezuka, I. and Tsukahara, T., *Polymer J.*, 1991, **23**, 1105–9.
101. Kanamura, N., *J. Appl. Polymer Sci.*, 1987, **33**, 2065–73.
102. Jameison, F. A., Schilling, F. C. and Tonelli, A. E., *Macromolecules*, 1986, **19**, 2168–74.
103. Fujii, K., Ukida, J. and Matsumoto, M., *Makromol. Chem.*, 1963, **65**, 86–90.
104. Fujii, K., Ukida, J. and Matsumoto, M., *J. Polymer Sci.*, 1963, **B1**, 687–92, 693–6.
105. Sakurada, I., *Ann. Rev. Synthetic Fiber Soc. Japan*, 1943, **1–1**, 192–213.
106. Sakurada, I., *Pure Appl. Chem.*, 1968, **16**, 263–83.
107. Matsuzawa, S., Jamaura, K. and Noguchi, H., *Makromol. Chem.*, 1974, **175**, 31–41.
108. Loebl, E. M. and O'Neil, J. J., *J. Polymer Sci.*, 1960, **45**, 538–40.
109. Platé, N. A. and Shibaev, V. P., paper presented at the International Symposium on *Macromolecular Chemistry*, Budapest, 1969, p. 10/9.
110. Harwood, H. J. and Robertson, A. B., paper presented at the International Symposium on *Macromolecular Chemistry*, Budapest, 1969, p. 11/31.
111. Robertson, A. B. and Harwood, H. J., *ACS Polymer Preprints*, 1971, **12**, 620–31.
112. Dolgoplosk, B. A., Kop'eva, I. A., Tinyakova, Ye. N., Red'kina, L. I. and Zavadovskaya, E. N., *Vysokomol. Soed*, 1967, **A9**, 645–52; Kop'eva, I. A., Tinyakova, Ye. N. and Dolgoplosk, B. A., *Vysokomol. Soed*, 1969, **B11**, 717–20.
113. Kropatcheva, Ye. N., Sterenzat, D. Ye., Patrushin, Yu. A. and Dolgoplosk, B. A., *Dokl. AN SSSR*, 1972, **206**, 878–83; Kropatcheva, Ye. N., Dolgoplosk, B. A., Sterenzat, D. Ye. and Patrushin, Yu. A., *Dokl. AN SSSR*, 1970, **195**, 1388–92.
114. Yermakova, I. I., Kropatcheva, Ye. N., Kol'tsov, A. I. and Dolgoplosk, B. A., *Vysokomol. Soed*, 1969, **A11**, 1639–45.
115. Platé, N. A., Davydova, S. L. and Alieva, Ye. D., *Vysokomol. Soed*, 1973, **A15**, 1856–67; Platé, N. A., Davydova, S. L. and Popova, T. R., *Dokl AN SSSR*, 1973, **210**, 267–71.
116. Hippe, R. and Germen, A., *Europ. Polymer J.*, 1978, **14**, 845–7.
117. Turska, E., *J. Pract. Chem.*, 1971, **313**, 387–96.
118. Turska, E. and Jantas, R., *J. Polymer Sci. Symposia*, 1974, **47**, 359–68.
119. Pichot, C., Guillot, J. and Guyot, A., *J. Macromol Sci. Chem.*, 1974, **A8**, 1073–86.
120. Boucher E. A., Khosravi-Babadi, E. and Mollett, C. C., *J. Chem. Soc. Faraday Trans. 1*, 1979, **75**, 1728–35.
121. Huang, W. K., Hsiue, G. H. and Hou, W. H., *J. Polymer Sci. Pol. Chem.*, 1988, **26**, 1867–83.
122. Noskova, D., Kotva, R. and Rypacek, F., *Polymer*, 1988, **29**, 2072–5.

123. Irzhak, V. I., Kuzub, L. I. and Yenikolopyan, N. S., in *Sintez i Fiziko-Khimiya Polimerov* (*Synthesis and Physical Chemistry of Polymers*), 1973, **12**, 36–40.
124. Irzhak, V. I., Kuzub, L. I. and Yenikolopyan, N. S., *Dokl. AN SSSR*, 1974, **214**, 1340–2.
125. Kuzub, L. I., Irzhak, V. I., Bogdanova, L. M. and Yenikolopyan, N. S., *Vysokomol. Soed*, 1974, **B16**, 431–3; Raspopova, Ye. N., Bogdanova, L. M., Irzhak, V. I. and Yenikolopyan, N. S., *Vysokomol. Soed*, 1974, **B16**, 434–6.
126. Maslova, R. N., Lesnik, Ye. A. and Varshavskii, Ya. M., *Molek. Biol.*, 1969, **3**, 728–33.
127. Yermolov, S. P., *Vysokomol. Soed*, 1976, **B18**, 496–9.
128. Morawetz, H., *Macromolecules in Solution*, Interscience, 1964.
129. Happey, F., *J. Textil. Inst.*, 1950, **41**, 381.
130. Urquhart, A. R., *Textil. Res. J.*, 1958, **28**, 159–66.
131. Krassig, H., *Makromol. Chem.*, 1958, **26**, 17–31.
132. Nickerson, R. F. and Habrle, J. A., *Ind. Eng. Chem.*, 1947, **39**, 1507–14.
133. Waller, R. S., Bass, K. C. and Rosenveare, W. E., *Ind. Eng. Chem.*, 1948, **40**, 138–49.
134. Distlyar, R. S., Sotnikov, P. S. and Korshunova, Ye. I., *Dokl. AN SSSR*, 1964, **156**, 652–5.
135. Vol'f, L. A. and Meos, A. I., in *Uspekhi Khimii i Tekhnologii Polimerov* (*Achievements of Chemistry and Technology of Polymers*), Khimiya, Moscow, 1971, pp. 68–9.
136. Slovokhotova, N. A., Il'itcheva, Z. F., Vasil'yev, L. A. and Kargin, V. A., *Vysokomol. Soed*, 1964, **6**, 608–19.
137. Shibaev, V. P., Platé, N. A., Grushina, R. K. and Kargin, V. A., *Vysokomol. Soed*, 1964, **6**, 231–9.
138. Nikonorova, N. I., Bakeev, N. F., Fakirov, S. Kh. and Kargin, V. A., *Vysokomol. Soed*, 1969, **A11**, 2197–203.
139. Wunderlich, B., in *Reactions of Polymers*, Proceedings of NATO Advanced Study Symposium, Ed. J. Moore, Reidel Publ. Co., Dordrecht, 1973, pp. 395–411.
140. Palmer, R. P. and Cobbold, A. J., *Makromol. Chem.*, 1964, **74**, 175–88.
141. Korduner, N. Ye., Bogaevskaya, T. A., Gromov, B. A., Miller, V. B. and Shlyapnikov, Yu. A., *Vysokomol. Soed*, 1970, **A12**, 693–9.
142. Bogaevskaya, T. A., Miller, V. B., Gromov, B. A., Monakhova, T. V. and Shlyapnikov, Yu. A., *Vysokomol. Soed*, 1972, **A14**, 1552–8.
143. Rapoport, N. Ya. and Shlyapnikov, Yu. A., *Vysokomol. Soed*, 1975, **A17**, 738–44.
144. Rapoport, N. Ya., Berulova, S. I. and Kovarskii, A. L., Musaelyan, I. N., Yershov, Yu. A. and Miller, V. B., *Vysokomol. Soed*, 1975, **A17**, 2521–7.
145. Rapoport, N. Ya. and Miller, V. B., *Vysokomol. Soed*, 1976, **A18**, 2343–7.
146. Emanuel', N. M., *Vysokomol. Soed*, 1979, **A21**, 2624–49.
147. Guzman, J., Fatou, J. G. and Perena, J. M., *Makromol. Chem.*, 1980, **181**, 1051–8.
148. Gibson, H. W. and Bailey, F. C., *Macromolecules*, 1980, **13**, 34–41.
149. Pudov, V. S. and Butchatchenko, A. L., *Kinetika I Kataliz*, 1974, **15**, 1110–18.
150. Lebedev, Ya. S., Tsvetkov, Yu. D. and Voevodskii, V. V., *Kinetika i Kataliz*, 1960, **1**, 496–501.
151. Shimada, S., Maeda, M., Hori, Y. and Klashiwabara, H., *Polymer*, 1977, **18**, 19–26.
152. Butyagin, P. Yu., *Dokl. AN SSSR*, 1961, **140**, 145–9.
153. Savinov, Ye. N., Anisimov, V. M. and Karpukhin, O. N., *Dokl. AN SSSR*, 1977, **233**, 164–7.
154. Rainov, M. M., *Vysokomol. Soed*, 1976, **A18**, 2022–32.
155. Bystritskaya, Ye. V. and Karpukhin, O. N., *Vysokomol. Soed*, 1976, **A18**, 1963–8.
156. Bystritskaya, Ye. V., Karpukhin, O. N. and Repina, T. S., *Dokl. AN SSSR*, 1978, **240**, 1380–3.

157. Radtzig, V. A., *Vysokomol. Soed*, 1975, **A17**, 154–62.
158. Korshak, V. V., Berestneva, G. L., Bragina, I. P. and Zhuravleva, I. V., *Dokl. AN SSSR*, 1973, **211**, 596–601.
159. Korshak, V. V., Khomutov, V. A., Berestneva, G. L. and Bragina, I. P., *Vysokomol. Soed*, 1973, **A15**, 2662–8.
160. Korshak, V. V., Berestneva, G. L., Bragina, I. P. and Yeremina, G. V., *Vysokomol. Soed*, 1974, **A16**, 1714–22.
161. Korshak, V. V., Berestneva, G. L., Bragina, I. P. and Astafiev, S. A., *J. Polymer Sci. Symposia*, 1974, **47**, 25–34.
162. Karyakin, N. V., Mochalov, A. N., Rabinovich, I. B., Kamelova, G. P., Berestneva, G. A., Astafiev, S. A. and Korshak, V. V., *Vysokomol. Soed*, 1975, **A17**, 1885–91.
163. Ladenheim, H. and Morawetz, H., *J. Am. Chem. Soc.*, 1959, **81**, 4860–3.
164. Arranz, F. and Mason, C., *Revista Plasticos Modernos*, 1975, **224**, 1–15.
165. Ise, N., *J. Polymer Sci. Symposia*, 1978, **62**, 205–26.
166. Katchalsky, A. and Feitelson, J., *J. Polymer Sci.*, 1954, **13**, 385.

CHAPTER 3

Neighboring Groups Effect: Theoretical Approaches

3.1 GENERAL REMARKS

As shown in Chapter 1, the chemical and physicomechanical properties of copolymers depend essentially on their chain structure and compositional heterogeneity. Therefore it is very important to know how to calculate the parameters of units distribution and composition heterogeneity on the basis of kinetic characteristics of a process.

In Chapter 2 we considered the characteristic polymeric effects that arise due to the chain structure of macromolecules. Generally the theory of polymer- analogous reactions should take all these effects into account. However, the consideration even of one of them is very difficult and becomes even more so when more than one of these effects are manifested because of their simultaneous and mutually dependent action.

In this chapter we consider the possibilities of including some characteristic polymeric effects in the quantitative description of macromolecular reactions, keeping in mind three main problems: the calculation of the kinetics of a process, of the sequence length distribution and of the composition heterogeneity of products. Some approaches to the solution of these problems are mentioned in monographs and reviews [1–4]. Here we try to describe this region as a whole.

We begin with the simplest case of the irreversible polymer-analogous reaction proceeding in conditions excluding the possibility of exhibition of any polymer effects listed above; i.e. in homogeneous conditions in the dilute solution with an excess of the low-molecular reactant, the polymer is stereoregular, and conformational, electrostatic and neighboring groups effects can be neglected. We denote the unreacted groups as A and the reacted ones as B.

The kinetics of such a reaction obeys the equation:

$$\frac{dP(A)}{dt} = -kP(A) \tag{3.1}$$

The solution of (3.1),

$$P(A) = P(A)_0 \exp(-kt) \tag{3.2}$$

gives the fraction of the unreacted groups $P(A)$ at any time t (k is the constant of the reaction $A \to B$).

The calculation of parameters of units distribution in the chain is also quite easy. At any intermediate moment of the reaction the product is a binary copolymer with a random distribution of A and B units. The probability of finding any sequence of units for such a polymer can easily be calculated from the probabilities of finding the A (or B) unit on a given site of the polymeric chain:

$$P(X_1, X_2, \ldots, X_n) = \prod_i P(X_i) \tag{3.3}$$

where $X = A$ or B with the condition

$$P(A) + P(B) = 1 \tag{3.4}$$

Thus the solution of the kinetic equation (3.2) gives the complete description of the sequence length distribution.

In this simplest case the solution of the kinetic equation also permits an evaluation of the composition heterogeneity of reaction products to be made. Because of the stochastic and independent nature of the $A \to B$ substitutions the process can be regarded as a Markov chain of zeroth order. From the general theory of regular Markov chains [5] it is known that in this case the composition distribution is a Bernullian one with the dispersion defined as

$$\lim_{n \to \infty} \frac{D_n}{n} = \frac{(1 - P_{A/A})P_{A/B}(1 - P_{A/B} + P_{A/A})}{(1 - P_{A/A} + P_{A/B})^3} \tag{3.5}$$

where n is the chain length, $P_{A/A}$ and $P_{A/B}$ are the Markov transitional probabilities to find A on the right (or on the left) of A or B respectively. For the Markov chain of the zeroth order $P_{A/A} = P_{A/B} = P(A)$ and

$$\lim_{n \to \infty} \frac{D_n}{n} = P(A)P(B) \tag{3.6}$$

Thus in the absence of any specific polymer effects all three problems of the description of the kinetics and statistics of a polymer-analogous reaction can easily be solved.

This scheme is valid for polymer-analogous reactions, i.e. for reactions between a functional group of the chain and the low-molecular reactant. Another type of macromolecular reaction is the intramolecular reaction, i.e. the reaction between functional groups of the same chain. If these groups are adjacent ones the kinetics and statistics of such reactions are described almost by the same mathematics as for polymer-analogous reactions.

The main feature of intramolecular reactions as compared with polymer-analogous ones is the pair transformations of functional groups. Therefore for reactions proceeding following the random mechanism there are always some isolated unreacted groups in the chain of the reaction product that cannot react as they are bordered by reacted groups. As an example one can consider chlorine elimination from polyvinyl chloride in the presence of zinc:

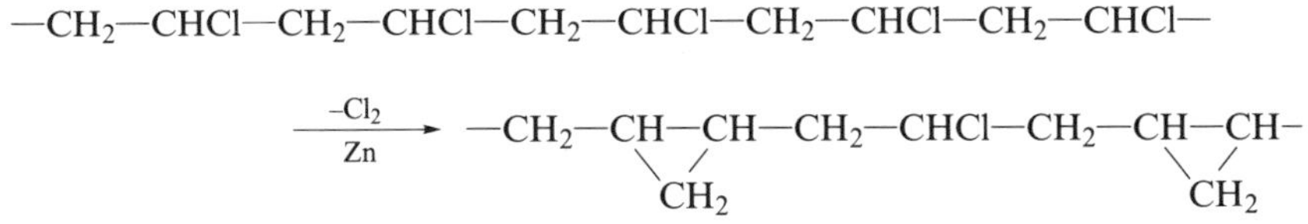

$$-CH_2-CHCl-CH_2-CHCl-CH_2-CHCl-CH_2-CHCl-CH_2-CHCl-$$

$$\xrightarrow[\text{Zn}]{-Cl_2} -CH_2-CH-CH-CH_2-CHCl-CH_2-CH-CH-$$
$$\qquad\qquad\qquad\qquad\searrow\;\nearrow\qquad\qquad\qquad\qquad\qquad\searrow\;\nearrow$$
$$\qquad\qquad\qquad\qquad CH_2\qquad\qquad\qquad\qquad\qquad\qquad CH_2$$

Another example is intramolecular aldol condensation of polymethyl-vinyl ketone:

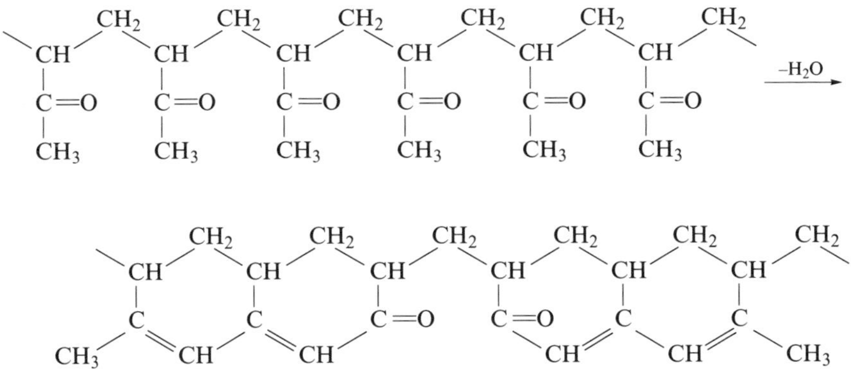

These schemes demonstrate the possibility of formation of isolated unreacted groups. The problem of the calculation of the number of such groups at the end of the reaction was firstly solved by Flory in 1939 [6].

Denoting the average number of unreacted groups in the chain run of n units as S_n Flory wrote the following obvious relationships

$$S_0 = 0$$
$$S_1 = 1$$
$$S_2 = 0$$
$$S_3 = 1$$

$$S_4 = \frac{2S_2 + 2S_1}{3} \qquad S_5 = \frac{2S_3 + 2S_2 + 2S_1}{4}$$

$$\cdots$$

$$S_n = \frac{2}{n-1}(S_1 + S_2 + \cdots + S_{n-2})$$

For new variables $\Delta_n = S_n - S_{n-1}$ one can write

$$\Delta_n = 1 - \frac{2}{1!} + \frac{4}{2!} - \frac{8}{3!} + \cdots \frac{(-2)^{n-1}}{(n-1)!}$$

For rather large n, $\Delta_n \to \Delta_\infty = 1/e^2$ and $S_n \approx n/e^2$; i.e. the fraction of isolated unreacted groups in rather long chains is roughly equal to 0.1353.

After Flory some authors also looked into the kinetics of intramolecular reactions [7–19]. Cohen and Reiss [7] used the method of multiplets. They defined a n-tuplet as a sequence of n unreacted units bordered by reacted or unreacted units and derived for the number of n-tuplets in the jth chain at the instant t $C_n^{(j)}(t)$ the following kinetic equation:

$$\frac{-dC_n^{(j)}}{dt} = k[(n-1)C_n^{(j)} + 2C_{n+1}^{(j)}]$$

where k is a constant of intramolecular reaction.

For the average value of $C_n^{(j)}$ (for a set of M chains),

$$\bar{C}_n(t) = M^{-1} \sum_{j=1}^{M} C_n^{(j)}(t)$$

the following kinetic equation is valid:

$$\frac{-d\bar{C}_n}{dt} = k[(n-1)\bar{C}_n + 2\bar{C}_{n+1}]$$

The solution of this system with the initial condition $\bar{C}_n(0) = N - n + 1$ (N is the number of units in a chain) results in

$$\bar{C}_n = \exp[-(n-1)kt] \sum_{s=0}^{N-n} (N - n - s + 1) \frac{(2e^{-kt} - 2)^s}{s!}$$

The probability of finding an unreacted n-tuplet is

$$P_n(t) = \frac{\bar{C}_n}{N - n + 1} = \exp[-(n-1)kt] \sum_{s=0}^{N-n} \left(1 - \frac{s}{N-n+1}\right) \frac{(2e^{-kt} - 2)^s}{s!}$$

If $N \to \infty$ one obtains, for all finite values of n,

$$P_n(t) = \exp[-(n-1)kt] \exp[-2(1 - e^{-kt})] \tag{3.7}$$

For $n = 1$,

$$P_1(t) = \exp[-2(1 - e^{-kt})] \tag{3.8}$$

Equation (3.8) is a final kinetic equation for the intramolecular reaction, while Eq. (3.7) describes the sequence length distribution of unreacted units. For $t \to \infty$ $P_1 \to 1/e^2$, corresponding to the classical result of Flory.

McQuiston and Lichtman [10] used another approach to the solution of this kinetic problem. They supposed that paired interactions in the polymer chain can be simulated by the process of random throwing of the 'dumb-bell' consisting of two units on the one-dimensional lattice of N cells. After m trials with the density of throwing on one unit per time unit $v = m/Nt$ the degree of filling θ is equal to

$$\theta(t) = 1 - \exp[-2(1 - e^{-vt})]$$

At $v = k$ this expression is identical to Eq. (3.8). The same result was obtained by Barron and Boucher [12].

All the methods described above relate to macromolecular reactions proceeding without any specific effects. More complicated cases occur when some of the polymer effects mentioned in Chapter 2 are manifested. The complete theoretical description of macromolecular reactions including all characteristic effects is a very difficult task. Therefore it seems reasonable to begin with a model taking into account only one of these effects, for example the neighboring groups effect.

3.2 KINETICS

Let us consider the following reaction model: units A are transformed into B units; the reaction is irreversible, of the first order; all the conditions are the same as described above, but the reactivity of the A units depends on the nature of the nearest neighbors. Examples of such reactions are presented in Chapter 2. We denote the rate constants of the transformation of A units having 0, 1 and 2 reacted neighbors B as k_0, k_1, k_2 respectively. These constants do not depend on the concentration of the reactants or on the degree of conversion.

In the mid 1960s several different approaches to the calculation of the kinetics of polymer-analogous reactions with the neighboring groups effect appeared in the literature [20–26]. The most accurate solution of this problem was proposed by McQuarrie *et al.* [26]. Two types of unreacted units sequence were introduced: a *j*-cluster being the sequence of *j* A units bordered by two B units and a *j*-tuplet being the sequence of *j* A units bordered either by A or B units. If $P(BA_jB)$ is the probability of finding a *j*-cluster in the chain of N units and $P(A_j)$ is the probability of finding a *j*-tuplet in the same chain, one can write

$$P(A_1) = P(BAB) + 2P(BA_2B) + 3P(BA_3B) + \cdots + NP(BA_NB)$$
$$P(A_2) = P(BA_2B) + 2P(BA_3B) + 3P(BA_4B) + \cdots + (N-1)P(BA_NB) \qquad (3.9)$$
$$\cdots$$
$$P(A_N) = P(BA_NB)$$

where N is the maximum length of the unreacted units sequence. From Eq. (3.9) one can obtain the inverse relationships:

$$P(BA_jB) = P(A_j) - 2P(A_{j+1}) + P(A_{j+2}) \qquad (3.10)$$

The change of j-tuplets with time (for $N \to \infty$) can be described by

$$\frac{dP(A_1)}{dt} = -k_2 P(BAB) - 2k_1 P(BA_2B) - 2k_1 P(BA_3B) - k_0 P(BA_3B)$$

$$- 2k_0 P(BA_4B) - 2k_1 P(BA_4B) - \cdots$$

$$= -k_0 P(A_3) - 2k_1[P(A_2) - P(A_3)] - k_2[P(A_1) - 2P(A_2) + P(A_3)]$$

$$\frac{dP(A_2)}{dt} = -2k_1 P(BA_2B) - 2k_1 P(BA_3B) - 2k_0 P(BA_3B) - 2k_1 P(BA_4B)$$

$$- 4k_0 P(BA_4B) - \cdots$$

$$= -2k_1 P[P(A_2) - P(A_3)] - 2k_0[P(A_3)]$$

$$\cdots$$

$$\frac{dP(A_j)}{dt} = -2k_1[P(A_j) - P(A_{j+1})] - k_0[(j-2)P(A_j) + 2P(A_{j+1})] \qquad (j \geq 2)$$

$$(3.11)$$

McQuarrie suggested searching the probabilities of j-tuplets in the form:

$$P(A_j) = \exp(-jk_0 t)\Psi(t) \qquad (j \geq 2) \qquad (3.12)$$

The validity of this assumption was proved by Mityushin [27]. It follows from (3.11) and (3.12) that

$$\frac{d\Psi}{dt} = 2(k_0 - k_1)[1 - \exp(-k_0 t)]\Psi(t) \qquad (3.13)$$

The solution of Eq. (3.13) for the initial product being a homopolymer, i.e. with the initial condition $P(A_j)_{t=0} = 1$, gives

$$\Psi(t) = \exp\left\{ 2(k_0 - k_1)\left[t - \frac{1 - \exp(-k_0 t)}{k_0} \right] \right\} \qquad (3.14)$$

$$P(A_j) = \exp\left\{ -jk_0 t + 2(k_0 - k_1)\left[t - \frac{1 - \exp(-k_0 t)}{k_0} \right] \right\}$$

This expression is a solution of the kinetic problem for all $j \geq 2$. Now we find $P(A_1)$:

$$\frac{dP(A_1)}{dt} + k_2 P(A_1) = 2(k_2 - k_1)P(A_2) + (2k_1 - k_0 - k_2)P(A_3) \qquad (3.15)$$

The solution of (3.15) is

$$P(A_1) = \exp\left\{ -k_2 t + 2(k_2 - k_1)\exp\left[\frac{2(k_1 - k_0)}{k_0}\right] \right.$$

$$\times \int \exp\left[(k_2 - 2k_1)t + \frac{2(k_0 - k_1)\exp(-k_0 t)}{k_0} \right] dt$$

$$+ (2k_1 - k_0 - k_2)\exp\frac{2(k_1 - k_0)}{k_0}$$

$$\left. \times \int \exp\left[(k_2 - k_0 - 2k_1)t + \frac{2(k_0 - k_1)}{k_0}\exp(-k_0 t) \right] dt + C \right\} \quad (3.16)$$

can be expressed by incomplete γ-functions:

$$\gamma(a; x) = \int_0^x u^{a-1} e^{-u} du$$

The final expression for $P(A_1)$ is

$$P(A_1) = e^{-k_2 t}\left(1 - \frac{2(k_2 - k_1)\exp\{[2(k_1 - k_0)]/k_0\}}{k_0\{[2(k_1 - k_0)/k_0\}^{(2k_1 - k_2)/k_0}} \left\{ \gamma\left[\frac{2k_1 - k_2}{k_0}, 2\left(\frac{k_1}{k_0} - 1\right)\right]\right. \right.$$

$$\left. - \gamma\left[\frac{2k_1 - k_2}{k_0}, \frac{2(k_1 - k_0)}{k_0}\right]\right\} - \frac{(2k_1 - k_0 - k_2)\exp\{[2(k_1 - k_0)]/k_0\}}{k_0\{[2(k_1 - k_0)]/k_0\}^{(2k_1 + k_0 - k_2)/k_0}}$$

$$\times \left\{ \gamma\left[\frac{2k_1 + k_0 - k_2}{k_0}, \frac{2(k_1 - k_0)\exp(-k_0 t)}{k_0}\right] - \gamma\left[\frac{2k_1 + k_0 - k_2}{k_0}, \frac{2(k_1 - k_0)}{k_0}\right]\right\} \right)$$

$$(3.17)$$

It should be pointed out that for the constants ratio $k_2 = 2k_1 - k_0$ (i.e. k_0, k_1, k_2 form the arithmetic progression) the equation for $dP(A_1)/dt$ changes to

$$\frac{dP(A_1)}{dt} = -2k_1[P(A_1) - P(A_2)] - k_0[2P(A_2) - P(A_1)] \qquad (3.15')$$

i.e. it is reduced to the particular case of Eq. (3.11), and $P(A_1)$ in this case is also defined by Eq. (3.14).

Later Ueda [28] proposed the simplified solution of Eq. (3.15). Introducing $s = \exp(-k_0 t)$, $\alpha = k_1/k_0$, $\beta = k_2/k_0$ this author wrote the following equations:

$$P(A_j) = s^{j+2\alpha-2} \exp[2(\alpha-1)(1-s)] \tag{3.18}$$

$$P(A_1) - P(BAB) = 2P(A_2) - P(A_3) = \sum_{i=2}^{N} iP(BA_iB)$$

$$= (2-s)s^{2\alpha}\exp[2(\alpha-1)(1-s)] \tag{3.19}$$

for $j \geq 2$

$$P(BA_jB) = P(A_j) - 2P(A_{j+1}) + P(A_{j+2}). \tag{3.20}$$

Differentiation of (3.20) with respect to t for $j = 1$ gives

$$\frac{dP(BAB)}{dt} = \frac{dP(A_1)}{dt} - 2\frac{dP(A_2)}{dt} + \frac{dP(A_3)}{dt}$$

Using Eqs. (3.10) and (3.11) one can obtain

$$-\frac{dP(BAB)}{dt} = 2k_0[P(A_3) - P(A_4)] + 2k_1[P(A_2) - 2P(A_3) + P(A_4)]$$

$$+ k_2[P(A_1) - 2P(A_2) + P(A_3)]$$

$$= 2k_0 \sum_{i=3}^{N} P(BA_iB) + 2k_1 P(BA_2B) - k_2 P(BAB) \tag{3.21}$$

The solution of Eq. (3.21) in terms of s, α and β is

$$P(BAB) = 2s^\beta \int_s^1 (1-x)[\alpha(1-x)+x]^{2\alpha-\beta-1} \exp[2(\alpha-1)(1-x)]dx \tag{3.22}$$

or

$$P(BAB) = (1-s)^2 s^{2\alpha-1} \exp[2(\alpha-1)(1-s)]$$

$$+ (2\alpha - \beta - 1)s^\beta \int_s^1 (1-x)^2 x^{2\alpha-\beta-2} \exp[2(\alpha-1)(1-x)]dx \tag{3.22'}$$

For $\beta = 2\alpha - 1$ (the constants form the arithmetic progression) the second term in Eq. (3.22') is equal to zero.

Thus $P(A_1)$ is determined as a function of time and of kinetic parameters of the process. As $P(A_1)$ is a mole fraction of unreacted units in the chain, Eqs. (3.16) and (3.22') describe the kinetics of the polymer-analogous reaction.

We also review here shortly other approaches to the solution of the kinetic problem [20–25]. The first successful attempt was made by Fuoss *et al.* [20]. Although there is no final kinetic equation in this work, the suggested method of solution is the same as that of McQuarrie *et al.* [26], and the result apparently should be the same.

The simplest and more convenient method of describing the kinetics of reactions proceeding with the neighboring groups effect was suggested by Keller [21]. Denoting the average fractions of unreacted units with 0, 1 and 2 reacted neighbors as N_0, N_1 and N_2 respectively, the author derived for them the following equations:

$$\frac{dN_0}{dt} = -(k_0 - 2\bar{k})N_0$$

$$\frac{dN_1}{dt} = -(k_1 + \bar{k})N_1 + \bar{k}N_0 \qquad \text{where } \bar{k} = \frac{k_0 N_0 + k_1 N_1}{N_0 + N_1} \qquad (3.23)$$

$$\frac{dN_2}{dt} = -k_2 N_2 + 2\bar{k}N_1$$

The solution of the equations (3.23) with the initial conditions $(N_0)_{t=0} = 1$, $(N_1)_{t=0} = (N_2)_{t=0} = 0$ gives

$$N_0(\tau) = \exp\left[-(2k+1)\tau - 2(k-1)\right]$$

$$N_1(\tau) = (e^{-\tau} - 1)\exp\left[-(2k+1)\tau - 2(k-1)(e^{-\tau} - 1)\right] \qquad (3.24)$$

$$N_2(\tau) = 2e^{-k'\tau} \int e^{k'\tau}[1 - 2k + (k-1)e^{-\tau} + ke^{\tau}]$$

where $\tau = k_0 t$, $k = k_1/k_0$, $k' = k_2/k_0$. The equations (3.23) are valid only when the initial product is a homopolymer [29]. The sum of expressions (3.24) is a mole fraction of unreacted units, i.e. it represents the solution of the kinetic equation.

This approach is not very accurate because of the assumption that the fractions of A units converting into B with rate constants k_0 and k_1 are determined by ratios $N_0/(N_0 + N_1)$ and $N_1/(N_0 + N_1)$ respectively. One can show that this means that the probability of finding A or B to the right (or to the left) of the AA diad does not depend on the nature of units situated on the another side of this diad. In fact the products of polymer-analogous reactions have such a property, but it was shown later by Mityushin [27].

The approach used by Alfrey and Lloyd [22] considers the change with time of the numbers of sequences of i unreacted units bordered by reacted ones N_i:

$$\frac{dN_1}{dt} = -k_2 N_1 + 2k_1 N_2 + 2k_0 \sum_{n=3}^{\infty} N_n$$

$$\frac{dN_2}{dt} = -2k_1 N_2 + 2k_1 N_3 + 2k_0 \sum_{n=4}^{\infty} N_n$$

...

$$\frac{dN_m}{dt} = -2k_1 N_m - (m-2)k_0 N_m + 2k_1 N_{m+1} + 2k_0 \sum_{n=m+n}^{\infty} N_n$$

This system includes m equations with $m+1$ unknown parameters and needs one more equation to be soluble. The authors postulated that $N_{n+1} = 2N_n - N_{n-1}$, supposing that this expression is more exact for larger n. Introducing some additional parameters one can derive and solve the kinetic equation.

Arends [23] used the stochastic approach to the kinetic problem. He introduced the following parameters: β, the probability of finding B to the left of A; γ_u, the probability of finding B to the right of AA; and γ_r, the probability of finding B to the right of BA. Probabilities of various sequences can be expressed through these probabilities and the fraction of reacted units, f. Because of the equivalence of the probabilities of the mirror-symmetric sequences AAB and BAA only two probabilities among three introduced are independent. Consideration of the evolution of probabilities of one, two or three units A with time permits the following expression to be derived:

$$\gamma_u = 1 - \exp(-k_0 t) \qquad (3.25)$$

With the aid of this expression one can solve the kinetic equation.

This approach is based on the assumption of the possibility of describing the sequence length distribution using only two independent parameters, namely the probability of finding B to the left of A and the probability of finding B to the right of AA. In fact that is valid only for the unreacted units distribution [29].

One more approach to the derivation of the kinetic equation was proposed by Lazare [25]. This author applied the Bose–Einstein statistics to calculate the distribution of N_0 objects on N_1 cells (N_0 is the number of A units in the centre of AAA triads and, N_1 is the number of A units in the centre of AAB or BAA triads). This approach is based on an assumption about the random character of the distribution of the sequences of unreacted units bordered by reacted ones.

It is interesting to note that despite the absence of any vigorous grounds for assumptions made in the approaches described above the final kinetic equations are identical to the exact kinetic equation of McQuarrie. The equivalence of results obtained in references [21] to [23] was shown by Keller [24]. The identity of these results with the McQuarrie equations [26] have been shown in references [2] and [29]. The final equations of the Lazare approach [25] are identical to Keller's equations (3.23), and therefore also coincide with exact ones.

We have presented here the complete solution of the kinetic problem for the model described above. Further development of the theory should include a complication of this model. One of the possible complications is to consider the reversible reactions [30–36].

If one takes into account the effect of the nature of the nearest neighbors on the reaction rate in this case as well, one has to introduce six rate constants describing the possible mutual transformations of A and B units according to the following scheme [30]:

$$
\begin{array}{ccc}
\text{A–A–A} & \text{B–A–A} & \text{B–A–B} \\
k_0 \downarrow \uparrow k_2' & k_1 \downarrow \uparrow k_1' & k_2 \downarrow \uparrow k_0' \\
\text{A–B–A} & \text{B–B–A} & \text{B–B–B}
\end{array}
$$

Following the McQuarrie approach one can write the kinetic equations for the probabilities of different sequences of units being analogous to Eq. (3.11). In this case, like the irreversible reaction, the equation for the j-tuplet probability contains the probability of the $(j + 1)$-tuplet. The final system of j equations with $j + 1$ unknown parameters can be solved only with some additional relationship between these parameters. For the irreversible reaction it follows from Eqs. (3.14) [27]:

$$P(A_{j+1}) = P(A_j)\exp(-k_0 t) \qquad (j \geq 2) \qquad (3.26)$$

Silberberg and Simha assumed [30, 33] that for the reversible reaction

$$P(YX_{j-1}B) = \frac{P(YX_{j-1})P(X_{j-1}B)}{P(X_{j-1})} \qquad (3.27)$$

(where X and Y are A or B). For irreversible reactions Eq. (3.27) is valid for $X = A$ and $j = 3$, thus Silberberg and Simha [30, 33] accepted $j = 3$ for reversible reactions as well. Kinetic calculations for short chains performed without any assumptions and with Eq. (3.27) give rather close results [33].

The most important point in the study of reversible processes is the analysis of the equilibrium state. For the polymer-analogous reaction with the neighboring groups effect the equilibrium conditions can be written as

$$k_0 P(AAA) = k'_2 P(ABA)$$
$$k_1 P(BAA) = k'_1 P(BBA) \qquad (3.28)$$
$$k_2 P(BAB) = k'_0 P(BBB)$$

Expressing the probabilities of various triads in terms of $P(A)$ and transitional probabilities $P_{A/B}$ (the probability of finding A on the right (or on the left) of B) and $P_{B/A}$ (the probability of finding B on the right (or on the left) of A) one can write [30]

$$k_0 P(A)P_{A/A}^2 = k'_2[1 - P(A)]P_{A/B}^2$$
$$k_2 P(A)P_{B/A}^2 = k'_0[1 - P(A)]P_{B/B}^2 \qquad (3.29)$$
$$k_1 P(A)P_{A/A}P_{B/A} = k'_1[1 - P(A)]P_{A/B}P_{B/B}$$

(where $P_{A/A} = 1 - P_{B/A}$, $P_{B/B} = 1 - P_{A/B}$). From this system one can derive

$$(k_0/k'_2)(k_2/k'_0) = (k_1/k'_1)^2 \qquad (3.30)$$

The same condition of the equilibrium state can be derived from thermodynamic considerations. Vainshtein *et al.* [36] supposed that because of the relation between the change of energy of the central and side units of a triad the analysis of a reversible reaction should include ten equilibrium constants corresponding to ten different pentades:

AAAAA	ABAAA	BBAAA	BAAAA	ABABA
$K_0 \Updownarrow$	$K_1 \Updownarrow$	$K_2 \Updownarrow$	$K_3 \Updownarrow$	$K_4 \Updownarrow$
AABAA	ABBAA	BBBAA	BABAA	ABBBA

BBABA	BBABB	BAABA	BAAAB	BBAAB
$K_5 \Updownarrow$	$K_6 \Updownarrow$	$K_7 \Updownarrow$	$K_8 \Updownarrow$	$K_9 \Updownarrow$
BBBBA	BBBBB	BABBA	BABAB	BBBAB

where K_i are the equilibrium constants of the reaction $A \to B$. Only four of the ten constants are independent, and the relation between them can be derived by expressing all of the constants in terms of free energies. Each unit can exist in one of the six energetic states with different values of the free energy:

AAA	BAA	BAB	ABA	BBA	BBB
F_1	F_2	F_3	F_4	F_5	F_6

The equilibrium constants are expressed in terms of free energies as follows:

$$- RT \ln K_0 = 3F_1 - F_4 - 2F_2$$
$$- RT \ln K_1 = F_4 + F_1 - 2F_5$$
$$- RT \ln K_2 = F_1 - F_6$$
$$- RT \ln K_3 = 2F_1 - F_3 - F_4$$
$$- RT \ln K_4 = F_3 + 2F_4 - 2F_5 - F_6 \qquad (3.31)$$
$$- RT \ln K_5 = F_3 + F_4 - 2F_6$$
$$- RT \ln K_6 = 2F_5 + F_3 - 3F_6$$
$$- RT \ln K_7 = F_4 + 2F_2 - 2F_5 - F_3$$
$$- RT \ln K_8 = 2F_2 + F_1 - F_4 - 2F_3$$
$$- RT \ln K_9 = 2F_2 - F_3 - F_6$$

As only the immediate neighbors affect the reactivity of the central unit of a triad, $K_0 = K_3 = K_8$, $K_1 = K_2 = K_7 = K_9$, $K_4 = K_5 = K_6$, and the following relations can be written:

$$F_3 + F_1 = 2F_2 \qquad F_4 + F_6 = 2F_5$$

From these relations it follows that

$$K_0 K_4 = K_1^2$$

which is identical to Eq. (3.30).

Krieger and Klesper studied the case of the reversible polymer-analogous reaction using the method of mathematical simulation [37–39]. They considered two cases:

(1) $K_1 = K_2 = K_3$

where $K_1 = k(AAA)/k(ABA)$

$\qquad K_2 = k(AAB)/k(ABB)$

$\qquad K_3 = k(BAB)/k(BBB)$

and

(2) $K_1 K_3 = K_2^2$

One more possible complication of the model of the reaction with the neighboring groups effect is to take into account the effect of more than two nearest neighbors. Krishnaswami and Vadav [40] considered the effect of four neighboring groups (two at each side) and using the Alfrey and Lloyd approach [22] obtained the kinetic equation with nine kinetic parameters. Serdyuk [41] showed the possibility of taking the effect of any number of neighboring units into account.

3.3 DISTRIBUTION OF UNITS

3.3.1 EXACT SOLUTION

Prior to starting the quantitative description of units distribution in the products of polymer-analogous reactions let us make one preliminary remark. The model of the reaction under discussion can be regarded as a Markov process with locally interacting components [27, 42]. The process is a Markov one for this model in the sense that the state of the chain at time $t + \Delta t$ depends stochastically on the state of the chain at time t, but not on the previous states. (This Markov property in time should not be confused with the Markov property in space, which is a characteristic of the copolymerization process.)

Some features of such processes can be used to describe the units distribution. One of these features is the independence to find any sequences of A and B units to both sides from the AA diad. This theorem is proved in reference [27]. We note that it follows from Eq. (3.12) that for all $j \geq 2$ for any t,

$$\frac{P(A_{j+1})}{P(A_j)} = \exp(-k_0 t) \tag{3.32}$$

In the general form $P(ZAAY)$, the probability of finding the ZAAY sequence (Z and Y are any combinations of A and B units), can be written as

$$P(ZAAY) = \frac{P(ZAA)P(AAY)}{P(AA)} \tag{3.33}$$

It follows from this theorem that

$$P(AXA)_4 = \exp[2(k-1)(1-e^{-\tau}) - 4k\tau]\frac{2(k-1)\exp[(2k-3)\tau]-1}{2k-3} \quad (3.34)$$

$$P(A_{i+t}ZA_j) = P(A_iZA_{j+1}) = P(A_iZA_j)\exp(-k_0 t) \quad (i,j \geq 2) \quad (3.35)$$

In the particular case of the arithmetic progression $k_2 = 2k_1 - k_0$ such independence is observed on both sides from one A unit [27] or in terms of the probabilities

$$P(ZAY) = \frac{P(ZA)P(AY)}{P(A)} \quad (3.36)$$

These properties of the units sequence distribution in products of macromolecular reactions are very important, and we shall use Eqs. (3.32) to (3.36) very often.

The complete description of the units distribution includes a calculation of probabilities to find any sequences of n A and B units. Equation (3.12) allows the probabilities of the unreacted units sequences—n_A-tuplets to be calculated. Evidently that is not sufficient for the complete description because of the significant difference in the distribution of reacted and unreacted units due to the irreversibility of the reaction under consideration.

Platé and co-workers [43–46] proposed a method of calculation of the probabilities of all n_{AB}-tuplets (i.e. of sequences of both types of units) including the consideration of n_{AX}-tuplets containing A and X units, where X represents a site that can be occupied either by A or by B units.

The use of the ideas of n_{AX}-tuplets is convenient as their probabilities only decrease with time, and the probability of any sequence consisting of A and B units can be expressed in terms of probabilities of n_{AX}- and n_A-tuplets. For example:

$$P(ABAB) = P(ABA) - P(ABAA) = P(AXA) - P(AAA) - P(AXAA) + P(AAAA)$$

The simplest X-containing tuplet is AXA. $P(AXA)$ decreases with time as two fixed A units transform into B with rate constants k_0, k_1 and k_2 dependent on the nature of neighbors. As the X site can be occupied either by A or B units and on both sides of the AXA run A or B units can stay, $P(AXA)$ is a sum of $2^3 = 8$ probabilities of various 5-tuplets: (A_5), (A_4B), (BA_4), (BA_3B), (A_2BA_2), $(BABA_2)$, (A_2BAB), $(BABAB)$. Then $dP(AXA)/dt$ can be written as

$$\frac{dP(AXA)}{dt} = k_0 P(A_5) + (k_0 + k_1)[P(A_4B) + P(BA_4)]$$

$$+ 2k_1[P(BA_3B) + P(A_2BA_2)] + (k_1 + k_2)[P(A_2BAB)$$

$$+ P(BABA_2)] + 2k_2 P(BABAB) \quad (3.37)$$

One can easily show that

$$P(A_4B) = P(BA_4) = P(A_4) - P(A_5)$$
$$P(BA_3B) = P(A_3) - 2P(A_4) + P(A_5)$$
$$P(A_2BA_2) = P(A_2XA_2) - P(A_5) \tag{3.38}$$
$$P(BABA_2) = P(A_2BAB) = [P(AXA_2) - P(A_2XA_2)] - [P(A_4) - P(A_5)]$$
$$P(BABAB) = [P(AXA) - 2P(AXA_2) + P(A_2XA_2)] - [P(A_3) - 2P(A_4) + P(A_5)]$$

On introducing $\tau = k_0t$, $k = k_1/k_0$ and $k' = k_2/k_0$ and taking into account Eq. (3.34), one can obtain, after all transformations,

$$\frac{-dP(AXA)}{d\tau} = 2k'P(AXA) + 2(k'-k)[P(AXA_2) + (A_3)] - 2(k'-2k+1)P(A_4) \tag{3.39}$$

To solve Eq. (3.39) it is necessary to define $P(AXA_2)$. Using the same consideration one can obtain

$$\frac{dP(AXA_2)}{d\tau} = -(2k + k')P(AXA)_2 + (k-1)P(AXA_3)$$
$$+ (k'-k)P(A_2XA_2) + (k'-1)P(A_4) - (k'-2k+1)P(A_5) \tag{3.40}$$

To solve Eq. (3.40) it is necessary to define $P(AXA_3)$ and $P(A_2XA_2)$:

$$\frac{dP(A_2XA_2)}{d\tau} = -4kP(A_2XA_2) + (k-1)[P(A_2XA_3) + P(A_3XA_2)]$$
$$+ 2(k-1)P(A_5) \tag{3.41}$$

It follows from Eq. (3.35) that

$$P(AXA_3) = P(AXA_2)\exp(-\tau) \tag{3.42}$$

One can conclude from Eq. (3.34) that $P(A_iXA_j)$ for $i, j \geq 2$ do not depend on i and j separately, but only on the sum $i + j = m$.

We denote such probabilities as

$$P(AXA)_{i+j} \equiv P(AXA)_m \qquad (m \geq 4) \tag{3.43}$$

It follows from Eqs. (3.34) and (3.43) that

$$P(A_2XA_3) = P(A_3XA_2) = P(AXA)_5 = P(AXA)_4\exp(-\tau) \tag{3.44}$$

Then the solution of (3.41) can be written as

$$P(AXA)_4 = \exp[2(k-1)(1 - e^{-\tau}) - 4k\tau]\frac{2(k-1)\exp[(2k-3)\tau] - 1}{2k-3} \tag{3.45}$$

Using Eqs. (3.42) and (3.45) one can solve Eqs. (3.40) and (3.39). It should be pointed out that on deriving Eqs. (3.39) to (3.41) we assume that the probabilities of the mirror-image sequences are equal. This property of the chain (for the initial homopolymer consisting of A units) is intuitively obvious and can be proved by considering the equations for the probabilities of the mirror-image sequences. Thus, for example,

$$\frac{dP(AXA_2)}{d\tau} = -(2k + k')P(AXA)_2 + (k - 1)P(AXA_3)$$
$$+ (k' - k)P(A_2XA_2) + (k' - 1)P(A_4) - (k' - 2k + 1)P(A_5)$$

$$(3.46)$$

$$\frac{dP(A_2XA)}{d\tau} = -(2k + k')P(AXA)_2 + (k - 1)P(A_3XA)$$
$$+ (k' - k)P(A_2XA_2) + (k' - 1)P(A_4) - (k' - 2k + 1)P(A_5)$$

$$(3.47)$$

On substituting $P(AXA_3) = P(AXA_2)\,e^{-\tau}$ and $P(A_3XA) = P(A_2XA)e^{-\tau}$ the equivalent equations are obtained. Thus $P(AXA_2) = P(A_2XA)$ at any τ values, as this equation is valid at $\tau = 0$ when $P(AXA_2) = P(A_2XA) = 1$. In this way the probabilities of all n_{AX}-tuplets with one X unit are defined, and therefore the probabilities of all n_{AB}-tuplets containing one B unit bordered by A units can be calculated:

$$P(ABA) = P(AXA) - P(A_3) \qquad\qquad (3.48)$$

$$P(ABA_n) = [P(AXA_2) - P(A_4)]\exp[-(n-2)\tau] \qquad (n \geq 2) \qquad (3.49)$$

$$P(ABA)_m = [P(AXA)_4 - P(A_5)]\exp[-(m-4)\tau] \qquad (m \geq 4) \qquad (3.50)$$

The next step is the determination of the probabilities of sequences containing two B units. In this case it is necessary to find the probabilities $P(AXXA)$ and $P(AXAXA)$. The probabilities of the sequences in which two X units are separated by two and more A units are expressed with the aid of Eq. (3.33) by means of the probabilities of the sequences containing one X unit.

$P(AXXA)$ is a sum of $2^4 = 16$ probabilities of 6-tuplets. Generally the probability of the n-tuplet containing $(n - 2)$ X units is a sum of 2^n probabilities of $(n + 2)$-tuplets. To calculate $P(AXXA)$ it is first necessary to find $P(AXXA_2)$ and $P(AXXA)_4$. $P(AXXA)_4$ is expressed by the following equation:

$$P(AXXA)_4 = \exp[2(k-1)(1 - e^{-\tau}) - 4k\tau]$$
$$\times \left\{ \frac{2(k-1^2)\exp[2(k-2)\tau]}{(k-2)(2k-3)} + \frac{2(k-1)e^{-\tau}}{2k-3} - \frac{k}{k-2} \right\} \qquad (3.51)$$

Using Eq. (3.51) one can solve equations for $P(AXXA_2)$ and $P(AXXA)$. Similarly the probabilities $P(AXAXA)_4$, $P(AXAXA_2)$ and $P(AXAXA)$ are calculated. To calculate the probabilities of all sequences containing three B units one has first to find $P(AX_3A)$, $P(AXAX_2A)$ and $P(AXAXAXA)$. It should be pointed out that the number of types of n_{AX}-tuplets to be considered increases with an increase in the number of B units in the sequence [45, 46].

For $P(AX_mA)$ at $m \geq 2$ one can write the equation in the general form:

$$\frac{dP(AX_mA)}{d\tau} = -2k'P(AX_mA) + 2(k'-k)[P(AX_mA_2) + P(AX_{m-1}A_2)]$$

$$-2(k'-2k+1)P(AX_{m-1}A_3) \tag{3.52}$$

$$\frac{dP(AX_mA_2)}{d\tau} = -(2k+k')P(AX_mA_2) + (k-1)[P(AX_mA_3) + P(AX_{m-1}A_3]$$

$$+(k'-k)[P(AX_mA)_4 + P(AX_{m-1}A)_4]$$

$$-(k'-2k+1)P(AX_{m-1}A)_5 \tag{3.53}$$

$$\frac{dP(AX_mA)_4}{d\tau} = -4kP(AX_mA)_4 + 2(k-1)[P(AX_mA)_5 + P(AX_{m-1}A)_5] \tag{3.54}$$

All $(P(AX_mA)_m$ and $P(AX_mA)_n$ are expressed in terms of $P(AX_mA_2)$ and $P(AX_mA)_4$ following Eqs. (3.34) and (3.35).

Thus in a general form one can consecutively ascertain the probabilities of all n_{AX}-tuplets and then of all n_{AB}-tuplets, i.e. describe completely the structure of the chain. Among all the probabilities to be found only the probabilities of the $P(AX_mA)_n$ type are represented in analytical form; determination of the rest requires repeating numerical computer integration.

In the particular case of the constants ratio $k' = 2k - 1$ (the arithmetic progression), all of the probabilities are determined analytically, and the problem of the complete description of the chain structure is reduced to the calculation of the $P(AX_mA)_4$ probabilities [2, 45, 46] which can be written in the form of the recurrent expression:

$$P(AX_mA)_4 = \exp\left\{2(k-1)(1-e^{-\tau}) - 4k\tau\left[\sum_{n=1}^{m} A_n^m \exp(Bn^m\tau) + C^m\right]\right\} \tag{3.55}$$

where

$$A_n^m = \frac{2(k-1)A_n^{m-1}}{B_n^m}$$

$$B_n^m = B_n^{m-1} - 1 \qquad (n = 1, 2, \ldots, m-1)$$

$$A_m^m = -C^{m-1}2(k-1) \tag{3.56}$$

$$B_m^m = -1$$

$$C^m = 1 - \sum_{n=1}^{m} A_n^m \qquad (m \geq 2)$$

Equations (3.56) are valid for any $m \geq 1$. $P(AXA)_4$ can be derived following the same expression from $P(A_4)$, which can be written as

$$P(A_4) = \exp[2(k-1)(1-e^{-\tau}) - 4k\tau]\exp[2(k-1)\tau]$$

with

$$A_1^0 = 1$$
$$A_1^1 = 2(k-1)A_1^0/B_1^1$$
$$B_1^0 = 2(k-1)$$
$$B_1^1 = B_1^0 - 1$$
$$C^1 = 1 - A_1^1$$

Thus in this particular case of the constants ratio Eqs. (3.12) and (3.56) describe completely the sequence distribution of units.

In the general case the complete description of the chain structure includes the consecutive determination of probabilities of various combinations of A and X units on the scheme described above with increasing length n of n_{AX}-tuplets under consideration. That is an exact solution of the problem.

The values of probabilities of some sequences to a maximum length of six units for various constants ratios calculated in such a way are presented in Figs. 3.2 to 3.9. Since these calculations are too cumbersome, require much computer time and become more and more complicated with an increase in the sequence length, it is expedient to apply some approximate methods of the description of the sequence distribution of units that give results close to the exact ones.

3.3.2 MARKOV CHAINS APPROXIMATION

From the analogy of products of polymer-analogous reactions with copolymerization products one can assume that one of the possible approximations is an approximation by Markov chains of various orders. In the general case the sequence distribution in products of macromolecular reactions is not the finite Markov one due to the not-unidirectional character of the chemical acts, unlike the pure Markov process of copolymerization. However, the property (3.33) of the sequence distribution of units is characteristic for a second-order Markov chain. This fact points to the possibility of using the Markov approximations.

It should be pointed out that this property is true only for sequences separated by two A units. For other diads (AB, BA, BB) the probabilities of sequences on one side of these two units depend on the nature of units on another side of this diad. Thus the question about the resemblance of the units distribution in the chain of the product of the macromolecular reaction to the second-order Markov can be answered only by comparing the results of the second-order Markov approximation with exact ones.

One can also use the first-order Markov approximation to describe the structure of the chain of the products of macromolecular reactions [43–49]. In some cases such an approximation appears to have some grounds. As was shown above for $k' = 2k - 1$, Eq. (3.34) is valid, i.e. the sequences of unreacted units have a first-order Markov distribution. Moreover, as was shown elsewhere [47, 50, 51], the first-order Markov approximation gives results close to exact ones for some characteristics of the chain structure for the retarding effect of neighboring units $(k_0 \geq k_1 \geq k_2)$.

An interesting analogy to the structure of products of macromolecular reactions and of products of the copolymerization corresponding to the terminal model (a Markov chain of the first order) was found [43, 44] for the case $k_1 = k_2$. If we express the run number $R = 2P(AB)$ introduced by Harwood and Ritchey [52] in terms of $P(A)$ and relative reactivities r_A and r_B [2] we obtain

$$R = \frac{4P(A)[1 - P(A)]}{1 + \{1 - 4P(A)[1 - P(A)](1 - r_A r_B)\}^{1/2}} \tag{3.57}$$

Equation (3.57) permits us to calculate the dependence of R on the value $r_A r_B$ for any given copolymer composition $P(A)$. As $R = 2(AB) = 2[P(A) - P(A_2)]$ one can also calculate the dependence of R on $k = k_1/k_0 = k_2/k_0$ following Eqs. (3.12) and (3.15) at the same values of $P(A)$. It was shown [43, 44] that at equal values of k and $r_A r_B$ the corresponding values of R are also equal, and the dependences $R(k)$ and $(r_A r_B)$ coincide.

The dependence of R on k and $r_A r_B$ at $P(A) = 0.75$ is presented in Fig. 3.1. Such a coincidence of two functions is observed in the wide range of $P(A)$. The discrepancies appear only near $r_A r_B = k = 0$ at $P(A) < 0.57$, and can be explained as follows: one can easily show that at $k_1 = k_2 = 0$ the minimum value of $P(A)$ is equal to 0.57, while for the copolymerization with $r_A r_B = 0$ one can obtain a copolymer of any composition by the variation of comonomer concentrations. Though it is difficult to explain such a coincidence of $R(r_A r_B)$ and $R(k)$ dependences from stochastic considerations, this fact permits us to assume the possibility of applying the first-order Markov approximation to the description of the chain structure for the polymer-analogous reaction when $k_1 = k_2$.

Generally one can use the Markov approximations of any order. Apparently the higher the order, the closer should be the results to accurate ones. It is important to estimate which characteristics of the distribution are better and in what approximation they are described.

Platé and co-workers [45, 46] have applied the Markov approximations when calculating the structure of products of macromolecular reactions, assuming that at any fixed time the chain is an *n*th-order Markov one, i.e. the state of any unit depends only on the state of n units to the left of it (or to the right because of the mirror symmetry of the chain) and does not depend on the state of the $(n+1)$th unit. It is proposed that time transformations between

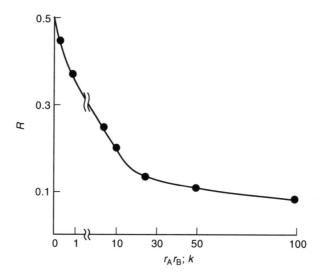

Fig. 3.1   Run number R versus $r_A r_B$ and $k = k_1/k_0 = k_2/k_0$ for $P(A) = 0.75$. (Reproduced by permission of Khimiya Moscow, Russia, from Platé, N. A., Litmanovich, A. D. and Noah, O. V., *Makromolekulyarnye Reaktsii* (*Macromolecular Reactions*), Khimiya, Moscow, 1977, p. 89)

various states occur as a macromolecular reaction with the neighboring groups effect.

Let us show how the Markov approximations of various orders for the calculation of the units distribution in a chain are derived. In a first-order Markov approximation there are two independent conditional probabilities: $P_{A/A}$ (the probability of finding A to the right of A) and $P_{A/B}$ (the probability of finding A to the right of B). All other characteristics of the chain structure are expressed in terms of these probabilities. It follows from the theory of Markov chains [5] that

$$P_{A/A} = P(A_2)/P(A)$$
$$P_{A/B} = P(BA)/P(B) \tag{3.58}$$

Differentiation of (3.58) with respect to time results in

$$\frac{dP_{A/A}}{dt} = \frac{P(A)\,dP(A_2)/dt - P(A_2)\,dP(A)/dt}{[P(A)]^2}$$

$$= \frac{dP(A_2)/dt - P_{A/A}\,dP(A)/dt}{P(A)} \tag{3.59}$$

$$\frac{dP_{A/B}}{dt} = \frac{dP(BA)/dt - P_{A/B}\,dP(B)/dt}{P(B)}$$

On substitution of

$$\frac{dP(A)}{dt} = -k_0 P(A_3) - 2k_1 P(A_2 B) - k_2 P(BAB)$$

$$\frac{dP(A_2)}{dt} = -2k_0 P(A_3) - 2k_1 P(A_2 B)$$

$$\frac{dP(B)}{dt} = -\frac{dP(A)}{dt}$$

$$\frac{dP(BA)}{dt} = \frac{dP(A)}{dt} - \frac{dP(A_2)}{dt}$$

into Eqs. (3.59), taking into account the fact that $P(A_3)$, $P(A_2 B)$ and $P(BAB)$ are expressed in terms of $P_{A/A}$ and $P_{A/B}$ [2] and introducing $\tau = k_0 t$, $k = k_1/k_0$, $k' = k_2/k_0$, one obtains

$$\frac{dP_{A/A}}{d\tau} = P_{A/A}[(k' - 2k) - P_{A/A}(k' - 2k + 1)(2 - P_{A/A})]$$

$$\frac{dP_{A/B}}{d\tau} = P_{A/B}\left\{ \frac{P_{A/A}^2}{1 - P_{A/A}} - k'(1 - P_{A/A}) \right. \tag{3.60}$$

$$\left. - P_{A/B}\left[\frac{P_{A/A}^2}{1 - P_{A/A}} + 2k P_{A/A} + k'(1 - P_{A/A}) \right] \right\}$$

The solution of the system (3.60) with initial conditions $P_{A/A} = P_{A/B} = 1$ at $\tau = 0$ gives the complete description of the chain structure in the first-order Markov approximation.

In a second-order Markov approximation there are four independent conditional probabilities: $P_{A/AA}$, $P_{A/BA}$, $P_{A/AB}$, $P_{A/BB}$. The equations for them are

$$P_{A/AA} = e^{-\tau}$$

$$\frac{dP_{A/BA}}{d\tau} = P_{A/BA}\left[\frac{e^{-\tau}}{1 - e^{-\tau}}(e^{-\tau} - P_{A/BA}) + k'(1 + P_{A/BA}) + (2k - 1)e^{-\tau} - 2k \right]$$

$$\frac{dP_{A/AB}}{d\tau} = \frac{e^{-\tau}}{1 - e^{-\tau}} P_{A/BA}(1 - P_{A/AB}) - P_{A/AB}[k' - (k' - 2k)P_{A/BA}] \tag{3.61}$$

$$\frac{dP_{A/BB}}{d\tau} = \frac{P_{A/BB}}{1 - P_{A/AB}}[2(k' - k)P_{A/BA}(P_{A/BB} - P_{A/AB})$$

$$- 2k'(P_{A/BB} - P_{A/AB}) - k'(1 - P_{A/BA})]$$

To describe the chain structure in a third-order Markov approximation one has to solve a system of eight differential equations, and so on [45, 46].

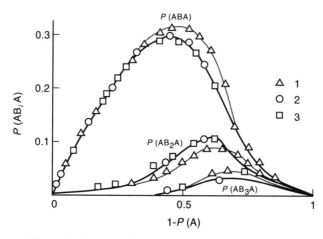

Fig. 3.2 Probabilities of blocks of i reacted units $P(AB_iA)$ versus the degree of conversion $P(B) = 1 - P(A)$ for $k_0/k_1/k_2 = 1{:}0.2{:}0.01$. Here and in Figs. 3.3 and 3.4 solid curves correspond to the exact solution; points are the results of first-order (1), second-order (2) and third-order (3) Markov approximations. (Reproduced by permission of Khimiya Moscow, Russia, from Platé, N. A., Litmanovich, A. D. and Noah, O. V., *Makromolekulyarnye Reaktsii* (*Macromolecular Reactions*), Khimiya, Moscow, 1977, p. 91)

A degree of the accuracy of Markov approximations of the first, second and third order for the calculation of the chain structure of products of macromolecular reactions can be evaluated from a comparison of results of these approximate calculations with results of the exact solution. Such a comparison is presented in Figs. 3.2–3.4. In these figures the dependences of $P(ABA)$, $P(AB_2A)$ and $P(AB_3A)$ versus the degree of conversion calculated for various ratios of rate constants $k_0/k_1/k_2$ are plotted. The initial values of the conditional probabilities are determined according to the initial condition of the exact solution ($P(A) = 1$ at $t = 0$). These particular parameters of the units distribution are chosen for an evaluation of the efficiency of the Markov approximations, as the maximum discrepancies between exact and approximate calculations should be expected, particularly for the sequences of B units.

The case of the retarding effect of reacted neighbors is presented in Fig. 3.2 ($k_0 > k_1 > k_2$). In this case the first-order Markov approximation gives results deviating from exact ones at $P(B) > 0.4$. The results of the second- and third-order Markov approximations are fairly close to the exact ones.

Figure 3.3 shows the relatively small accelerating effect of one of the reacted neighbors ($k_1/k_0 = 5$). In this case the results of the Markov approximations of the second and third orders coincide with exact results, both at the small accelerating effect of two reacted neighbors ($k_2/k_0 = 5$) and at the significant acceleration ($k_2/k_0 = 100$). The degree of deviation of results of the first-order

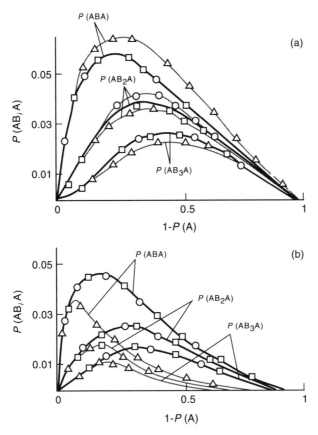

Fig. 3.3 Probabilities of blocks of i reacted units $P(AB_iA)$ versus the degree of conversion $P(B) = 1 - P(A)$ for $k_0/k_1/k_2 = 1:5:5$ (a) and 1:5:100 (b). (Reproduced by permission of Khimiya Moscow, Russia, from Platé, N. A., Litmanovich, A. D. and Noah, O. V., *Makromolekulyarnye Reaktsii* (*Macromolecular Reactions*), Khimiya, Moscow, 1977, p. 92)

Markov approximation depends on the ratio k_2/k_0 increasing with an increase in the degree of acceleration.

Figure 3.4 demonstrates the case of the strong accelerating effect of one reacted neighbor ($k_1/k_0 = 50$). In this case the accuracy of the second- and third-order Markov approximations is not sufficient. It is interesting that some regularity in the deviations of Markov approximations from the exact solution can be noted. The first-order Markov approximation gives strong deviations for all elements of the distribution: $P(ABA)$, $P(AB_2A)$, $P(AB_3A)$, presented in Fig. 3.4. The second-order Markov approximation gives strong deviations for $P(AB_2A)$ and $P(AB_3A)$, and gives results close to those of the exact solution for $P(ABA)$. The results of the

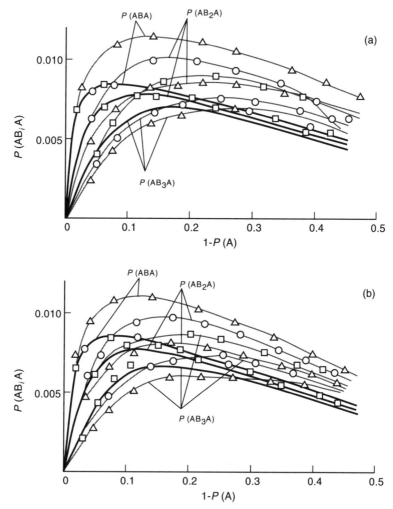

Fig. 3.4 Probabilities of blocks of i reacted units $P(AB_iA)$ versus the degree of conversion $P(B) = 1 - P(A)$ for $k_0/k_1/k_2 = 1:50:50$ (a) and $1:50:99$ (b). (Reproduced by permission of Khimiya Moscow, Russia, from Platé, N. A., Litmanovich, A. D. and Noah, O. V., *Makromolekulyarnye Reaktsii* (*Macromolecular Reactions*), Khimiya, Moscow, 1977, p. 93)

third-order Markov approximation are close to results obtained by accurate methods for $P(ABA)$ and $P(AB_2A)$; the deviations are only essential beginning with $P(AB_3A)$. Thus one can suppose that assumptions accepted at the derivation of the nth order Markov approximation allow one to obtain fairly accurate values of the probabilities $P(AX_iA)$, where $i \le n - 1$.

A description of the units distribution in the products of macromolecular reactions by Markov approximations can be quite efficient for the retarding effect of reacted neighbors and for the slight acceleration. The use of the second-order Markov approximation seems to be the most effective, as the accuracy of the first-order approximation is not sufficient, and the accuracy of results of the third-order approximation is almost the same as that of the second-order approach, but the number of equations required to be solved is twice as many.

At strong accelerating effects the Markov approximations are not very useful as they can only be applied to the calculation of the probabilities of short sequences. It should be pointed out that any approximate approaches to the calculation of the units distribution were sought only because the exact calculation of probabilities of long sequences is very complicated.

3.3.3 'B-APPROXIMATION'

One more approximate approach used to calculate the units distribution in products of macromolecular reactions was proposed by Platé and co-workers [45, 46]. This approach is based on a consideration of the kinetics of transformations of blocks of j reacted units (so-called 'B-approximation').

The shortest block of B units is ABA ($j = 1$). Such triads arise from AAA triads with the rate constant k_0 and disappear as a result of the transformation of the bordering A units with the rate constants k_1 (if this A unit is in the ABAA or AABA tetrads) and k_2 (if it is in the ABAB or BABA tetrads):

$$\frac{dP(ABA)}{dt} = k_0 P(A_3) - 2[k_1 P(ABA_2) + k_2 P(BABA)] \tag{3.62}$$

The probabilities $P(ABAA)$ and $P(ABAB)$ can be expressed as

$$P(ABA_2) = P(ABA)P_{A/ABA}$$
$$P(ABAB) = P(ABA)P_{B/ABA}$$

where $P_{A/ABA}$ and $P_{B/ABA}$ are the conditional probabilities of finding A and B to the right of ABA respectively. Then one can write

$$\frac{dP(ABA)}{dt} = k_0 P(A_3) - 2P(ABA)(k_1 P_{A/ABA} + k_2 P_{B/ABA}) \tag{3.63}$$

For $j = 2$ the tetrad ABBA disappears in the same way as ABA and arises from the ABA$_2$ and A$_2$BA tetrads with the rate constant k_1:

$$\frac{dP(AB_2A)}{dt} = 2k_1 P(ABA)P_{A/ABA} - 2P(AB_2A)(k_1 P_{A/AB_2A} + k_2 P_{B/AB_2A})$$

For the sequences AB$_j$A for $j \geq 3$ in the kinetic equation an additional term arises describing the appearance of AB$_j$A due to the recombination of shorter blocks of the B units with the rate constant k_2. This term can be written as

$$k_2 \sum_{m=1}^{j-2} P(AB_m AB_{j-m-1} A)$$

where $P(AB_m AB_{j-m-1} A)$ can be expressed in terms of conditional probabilities as

$$P(AB_m AB_{j-m-1} A) = P(AB_m A) P_{B/AB_m A} P_{B_{j-m-2} A/AB_m AB}$$

Then the equation for $P(AB_j A)$ where $j \geq 3$ can be written as

$$\frac{dP(AB_j A)}{dt} = 2k_1 P(AB_{j-1} A) P_{A/AB_{j-1} A} - 2P(AB_j A)(k_1 P_{A/AB_j A} + k_2 P_{B/AB_j A})$$

$$+ k_2 \sum_{m=1}^{j-2} P(AB_m A) P_{B/AB_m A} P_{B_{j-m-2} A/AB_m AB} \tag{3.64}$$

The transitional probabilities in Eqs. (3.63) and (3.64) can be expressed in terms of usual probabilities using the assumption made by Klepser *et al.* [53, 54] about the random distribution of blocks of two types (of reacted and unreacted units) in the chain of a product of the polymer-analogous reaction. For the random distribution of blocks one can write

$$P(ZBAY) = \frac{P(ZBA)P(BAY)}{P(BA)}$$

(a property of the independence of finding any sequences of units to the right and to the left of the BA diad). In terms of transitional probabilities this means that

$$P_{Y/ZBA} = P_{Y/BA} \tag{3.65}$$

where Z and Y are any sequences of A and B units.

The expression (3.65) is not an exact one [44] and can be accepted only as an assumption. The results of checking this assumption by Monte Carlo calculations [53, 54] permit us to say that it does not lead to the essential error, but a final conclusion about the accuracy of the approximation based on this assumption can be made only after a comparison of the results of approximate and exact solutions. It follows from Eq. (3.65) that

$$P_{A/ABA} = P_{A/AB_2 A} = P_{A/AB_m A} = P_{A/BA} = P(BA_2)/P(BA)$$

$$P_{B/ABA} = P_{B/AB_2 A} = P_{B/AB_m A} = P_{B/BA} = P(BAB)/P(BA) \tag{3.66}$$

$$P_{B_{j-m-2} A/AB_m AB} = P_{B_{j-m-2} A/AB} = P(AB_{j-m-1} A)/P(AB)$$

After substitution of Eqs. (3.66) into Eqs. (3.63) to (3.64) and introducing $\tau = k_0 t$, $k = k_1/k_0$ and $k' = k_2/k_0$, one can write the system for $dP(AB_jA)/d\tau$ in a general way:

$$\frac{dP(AB_jA)}{d\tau} = \delta_{j,1}P(A_3) + 2P(AB_{j-1}A)(A - \delta_{j,1})\,k\,\frac{P(A_2B)}{P(AB)}$$

$$- 2P(AB_jA)\frac{kP(A_2B) + k'P(BAB)}{P(AB)} \tag{3.67}$$

$$+ \frac{k'P(BAB)}{[P(AB)]^2}\sum_{m=1}^{j-2} P(AB_mA)P(AB_{j-m-1}A)$$

where

$$P(AB) = P(A) - P(A_2)$$

$$P(A_2B) = P(A_2) - P(A_3)$$

$$P(BAB) = P(A) - 2P(A_2) + P(A_3)$$

$$\delta_{j,1} = \begin{cases} 1 & \text{for } j = 1 \\ 0 & \text{for } j \neq 1 \end{cases}$$

and $P(A_j)$ are determined from Eqs. (3.12) and (3.16).

Now we will consider what other elements should be included into the complete description of the units sequence distribution in the 'B-approximation'. The probabilities of 'pure' n_A-tuplets $P(A_n)$ are determined from Eqs. (3.12). The probabilities $P(A_nB)$ and $P(BA_nB)$ are expressed in terms of $P(A_n)$:

$$P(A_nB) = P(A_n) - P(A_{n+1})$$

$$P(BA_nB) = P(A_n) - 2P(A_{n+1}) + P(A_{n+2})$$

The probabilities $P(AB_jA)$ of blocks of reacted B units surrounded by A units are determined by the solution of the system (3.67). The probabilities $P(B_jA)$ and $P(B_j)$ can be expressed through $P(AB_jA)$, $P(A)$ and $P(A_2)$:

$$P(B_jA) = P(B_{j-1}A) - P(AB_{j-1}A) = P(B_{j-2}A) - P(AB_{j-2}A) - P(AB_{j-1}A) = \cdots$$

$$= P(BA) - \sum_{m=1}^{j-1} P(AB_mA) = P(A) - P(A_2) - \sum_{m=1}^{j-1} P(AB_mA)$$

$$P(B_j) = P(B_{j-1}) - P(AB_{j-1}) = P(B_{j-2}) - P(AB_{j-2}) - P(AB_{j-1}) = \cdots$$

$$= 1 - jP(A) + (j-1)P(A_2) + \sum_{m=1}^{j-2} (j-m-1)P(AB_mA)$$

Now the probabilities of sequences with alternating blocks of A and B units are to be determined. Let us consider any sequence $AB_jA_nB_iA_mB$. The probability of such a sequence can be written as

$$P(AB_jA_nB_iA_mB) = P(AB_jA)P_{A_{n-1}B/AB_jA}P_{B_{i-1}A/AB_jA_nB}P_{A_{m-1}B/AB_jA_nB_iA} \quad (3.68)$$

Taking the assumption [53, 54] about the random distribution of blocks (3.65) one obtains

$$P(AB_jA_nB_iA_mB) = P(AB_jA)P_{A_{n-1}B/BA}P_{B_{i-1}A/AB}P_{A_{m-1}B/BA}$$
$$= \frac{P(AB_jA)P(BA_nB)P(AB_iA)P(BA_mB)}{[P(BA)]^3} \quad (3.69)$$

Thus the probability of the sequence of N blocks BA_jB and AB_iA is equal to the product of probabilities of these blocks divided by $[P(BA)]^{N-1}$ (for Eqs. (3.68) and (3.69), $N = 4$). One can therefore calculate all of the parameters of the units distribution with the aid of 'B-approximation'.

In Figs. 3.5 to 3.8 the dependences of probabilities $P(ABA)$, $P(AB_2A)$ and $P(AB_3A)$ on the degree of conversion $1 - P(A)$ calculated in the 'B-approximation' and using the exact equations are compared. It can be seen from these figures that results of the approximate approach coincide with exact results in a wide range of constants ratios. However, for nonmonotonous constants ratios ($k_0 > k_1, k_1 < k_2$ (Fig. 3.8a) and $k_0 < k_1$, $k_1 > k_2$ (Fig. 3.8b)) this approximation gives some

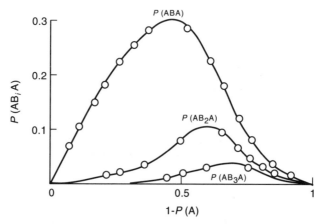

Fig. 3.5 Probabilities of blocks of i reacted units $P(AB_iA)$ versus the degree of conversion $P(B) = 1 - P(A)$ for $k_0/k_1/k_2 = 1{:}0.2{:}0.01$. Here and in Figs. 3.6 to 3.8 curves correspond to the exact solution; points are the results of the 'B-approximation'. (Reproduced by permission of Khimiya Moscow, Russia, from Platé, N. A., Litmanovich, A. D. and Noah, O. V., *Makromolekulyarnye Reaktsii* (*Macromolecular Reactions*), Khimiya, Moscow, 1977, p. 97)

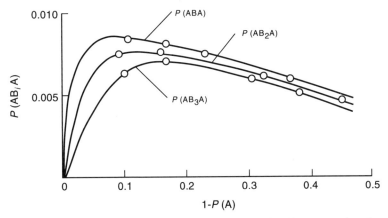

Fig. 3.6 Probabilities of blocks of i reacted units $P(AB_iA)$ versus the degree of conversion $P(B) = 1 - P(A)$ for $k_0/k_1/k_2 = 1:50:50$. (Reproduced by permission of Khimiya Moscow, Russia, from Platé, N. A., Litmanovich, A. D. and Noah, O. V., *Makromolekulyarnye Reaktsii* (*Macromolecular Reactions*), Khimiya, Moscow, 1977, p. 97)

deviations from the exact solution (for the second case the deviations are larger than for the first one). Evidently such constants ratios are of the formal type, but are not of practical interest. The slight deviations are observed at $k_2 \gg k_1$ (Fig. 3.7b and c). For all other constants ratios the results of the 'B-approximation' are very close to those obtained by the exact method for all elements of the distribution under consideration.

It should be pointed out that the calculation by the 'B-approximation' is rather simple, and the numerical solution of Eqs. (3.67) for the characteristics of the 'B-approximation' is not too difficult to calculate, even for long sequences. The fairly good agreement of results of the 'B-approximation and of the exact solution allows us to conclude that assumption (3.65) does not lead to considerable numerical errors, and equations of the type of Eq. (3.69) in combination with Eqs. (3.12), (3.15) and (3.67) can be recommended for a general description of the sequence distribution of units in products of macromolecular reactions.

One more point that should be mentioned here is the possibility of applying all the statistical approaches described above to the cases when the initial polymer at $t = 0$ is not a homopolymer but a copolymer containing both A and B units. At first we shall consider how the equations of McQuarrie *et al.* [26] giving the sequence length distribution of the unreacted units have to be solved in this case.

This problem was first described by Platé *et al.* [46]. Later Serdyuk [41] proposed an approach that was valid for any number of neighbors affecting the rate of the reaction and for any initial distribution of units. Gonsalez and

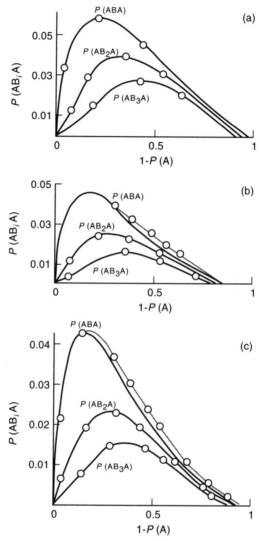

Fig. 3.7 Probabilities of blocks of i reacted units $P(AB_iA)$ versus the degree of conversion $P(B) = 1 - P(A)$ for $k_0/k_1/k_2 = 1:5:5$ (a), 1:5:100 (b) and 1:5:1000 (c). (Reproduced by permission of Khimiya Moscow, Russia, from Platé, N. A., Litmanovich, A. D. and Noah, O. V., *Makromolekulyarnye Reaktsii* (*Macromolecular Reactions*), Khimiya, Moscow, 1977, p. 98)

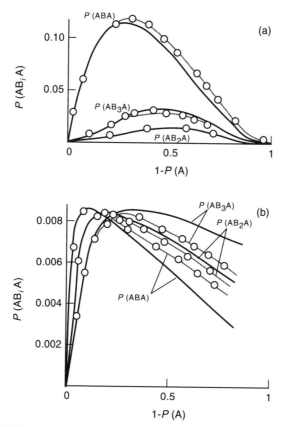

Fig. 3.8 Probabilities of blocks of i reacted units $P(AB_iA)$ versus the degree of conversion $P(B) = 1 - P(A)$ for $k_0/k_1/k_2 = 1:0.2:100$ (a) and $1:50:0$ (b). (Reproduced by permission of Khimiya Moscow, Russia, from Platé, N. A., Litmanovich, A. D. and Noah, O. V., *Makromolekulyarnye Reaktsii* (*Macromolecular Reactions*), Khimiya, Moscow, 1977, p. 99)

co-workers [55–58] also considered the kinetics of the reactions with the neighboring groups effect for an initial copolymer.

The solution of Eq. (3.9) for initial conditions $P(A_j)_{t=0} = 1$ was given by McQuarrie *et al.* in the form:

$$P(A_j) = \exp(-jk_0t)\Psi(t) \qquad (3.12)$$

This validity of Eq. (3.12) was proved by Mityushin [27]. One can suppose that for some initial units distribution that does not correspond to the condition

$P(A_j)_{t=0} = 1$ the solution of Eqs. (3.9) is presented in the form:

$$P(A_j) = \exp(-jk_0 t)\Psi(t)P(A_j)^0 \qquad (j \geq 2) \tag{3.70}$$

where $P(A_j)_0$ is the value of $P(A_j)$ at $t = 0$.

Apparently such a presentation is valid if Eq. (3.32) is valid for the initial distribution of units, i.e. if the last is the AA Markov one. Such a property characterizes the products of macromolecular reactions corresponding to the model under consideration and the products of copolymerization with random, first- and second-order Markov distributions of units.

After the substitution of Eq. (3.70) into Eqs. (3.9) one obtains, after all transformations [46],

$$\frac{d\Psi}{dt} = 2(k_0 - 1)\Psi(t)\left[1 - \frac{P(A_{j+1})^0}{P(A_j)^0}\exp(-k_0 t)\right] \qquad (j \geq 2) \tag{3.71}$$

For the products of macromolecular reactions and products of copolymerization corresponding to random, terminal and penultimate models,

$$\frac{P(A_{j+1})^0}{P(A_j)^0} = M \qquad (j \geq 2)$$

For the random initial distribution $M = P(A)^0$, for the first-order Markov one $M = P_{A/A}$, for the second-order Markov one $M = P_{A/AA}$ and for the products of polymer-analogous reactions $M = \exp(-k_0 t)$.

The solution of Eq. (3.71) is

$$\Psi(t) = \exp\left\{\frac{2(k_1 - k_0)}{k_0}M[1 - \exp(-k_0 t)] - 2(k_1 - k_0)t\right\} \qquad \text{for } j \geq 2 \tag{3.72}$$

$$P(A_j) = P(A_j)^0 \exp(-jk_0 t)\exp\left\{\frac{2(k_1 - k_0)}{k_0}M[1 - \exp(-k_0 t)] - 2(k_1 - k_0)t\right\}$$

Substitution of Eqs. (3.72) into Eq. (3.15) gives

$$\frac{dP(A)}{dt} = -k_2 P(A) + P(A_2)^0[2(k_2 - k_1) - (k_2 - 2k_1 + k_0)M\exp(-k_0 t)]$$

$$\times \exp\left\{\frac{2(k_1 - k_0)}{k_0}M[1 - \exp(-k_0 t)] - 2(k_1 - k_0)t\right\} \tag{3.73}$$

The solution of Eq. (3.73) with the initial condition $P(A)_{t=0} = P(A)^0$ allows the kinetics of the reaction for this case to be described.

Apparently all equations of the exact solution can be solved with any known initial conditions if Eq. (3.32) is valid for the initial distribution and the probabilities of mirror-image sequences are equal (i.e. valid for the random, first- and second-order Markov distributions). One can assume that the approximate methods of the calculation of the sequence distribution are good enough for the initial distributions corresponding to these conditions, as well as for the condition $P(A)_{t=0} = 1$. It looks as though it would be possible to use them under any initial conditions, but in this case some additional source of errors appear.

One more approach to the calculation of the units sequence distribution in products of macromolecular reactions was proposed by Brun and Kuchanov [59, 60]. This approach is similar to that described above, but the so-called 'base sequences' are chosen as the main parameters. Such a base sequence $U_{ij}(\mathbf{n})$ is a sequence of the $A^i B^{n_1} AB^{n_2} \cdots AB^{n_m} A^j$ type consisting of m blocks of B units separated by isolated A units. The probability of any sequence of A and B units can be expressed through the probabilities of these base sequences and $P(A_j)$ using the property (3.32). The equations for the probabilities of base sequences $P_{ij}(n)$ are derived using the same consideration of their change in time as the one described above. The solution of these equations is written in the form of recurrent expressions analogous to (3.55). The results can be obtained by numerical integration. Thus we can say that the problem of a complete description of the units distribution in the chain of products of polymer-analogous reactions corresponding to the model described above is solved.

A possible complication of the model is the consideration of reversible reactions. The same thermodynamic approach which was used to derive the equilibrium conditions [36] was proposed by Berlin *et al.* [61] to calculate the sequence distribution in products of reversible polymer-analogous reactions. These authors assumed that the free energy of polymer chains is a sum of free energies of all units and of the energy related to the interchange of m reacted B units and n unreacted A units. Taking into account the fact that each unit of the chain can have one of six different energy states with different values of the free energy:

AAA	BAA	BAB	ABA	BBA	BBB
F_1	F_2	F_3	F_4	F_5	F_6

one can obtain the following expression for F.

$$F = F_3 n_1 + F_4 m_1 + \sum_{i=2}^{\infty} [2F_2 + (i-2)F_1] n_i + \sum_{i=2}^{\infty} [2F_5 + (i-2)F_6] m_i$$
$$- RT \ln \frac{(\sum_{i=1}^{\infty} n_i)! + (\sum_{i=1}^{\infty} m_i)!}{\prod_{i=1}^{\infty} (n_i)! \prod_{i=1}^{\infty} (m_i)!} \tag{3.74}$$

where m_i and n_i are the concentrations of blocks of reacted and unreacted units of the length i.

After the minimization of F using the Lagrange method with the condition of material balance

$$\sum_{i=1}^{\infty} i(n_i + m_i) = n_0$$

and the condition being valid for the infinite chains:

$$\sum_{i=1}^{\infty} n_i = \sum_{i=1}^{\infty} m_i = u$$

one can derive equations for the sequence distribution in the chain:

$$
\begin{aligned}
n_1/n_0 &= uba(\varphi_3\varphi_4) \\
n_i/n_0 &= u(\varphi_3\varphi_4)(\varphi_2^2/\varphi_3)ba^i \\
m_i/n_0 &= uab^{-1} \\
m_i/n_0 &= u(\varphi_5^2/\varphi_4)b^{-1}\varphi_6^{i-2}a^i
\end{aligned}
\tag{3.75}
$$

where $\varphi_i = \exp(-F_i/kT)$, a and b are the Lagrange parameters and n_0 is the initial concentration of monomer units.

The following equations determine a, b and u:

$$\left(\frac{\varphi_2^2/\varphi_3 a^2}{1-a} + a\right)\varphi_3\varphi_4 = \left(\frac{\varphi_5^2/\varphi_4 a^2}{1-a\varphi_6} + a\right)^{-1}$$

$$b = \left(\frac{\varphi_5^2/\varphi_4 a}{1-a\varphi_6} + 1\right)a$$

$$
\frac{n_0}{u} = 2 + \varphi_3\varphi_4 a \left[\left(\frac{\varphi_5^2/\varphi_4 a}{1-a\varphi_6} + 1\right)\frac{\varphi_2^2/\varphi_3 a^2}{(1-a)^2}\right.
\tag{3.76}
$$

$$\left. + \left(\frac{\varphi_2^2/\varphi_3}{1-a} a + 1\right)\frac{\varphi_5^2/\varphi_4 a}{1-a\varphi_6}\right]$$

As φ_i are expressed through the equilibrium constants [36], the parameters of the units distribution can be calculated in terms of these equilibrium constants or F values. On the other hand, the kinetic and thermodynamic parameters of the reaction can be found from the experimentally determined parameters of the distribution of units.

Krieger and Klesper [37–39] described the sequence units distribution for the steady state in a reversible polymer-analogous reaction using Monte Carlo simulation. They showed that for the case of equal equilibrium constants $K_1 = K_2 = K_3$ the units distribution is a Bernullian one, while for $K_1^*K_3 = K_2^2$ this distribution is a first-order Markov one.

3.4 COMPOSITION HETEROGENEITY

If the calculation of sequence distribution in the products of polymer-analogous reactions is a more complicated problem than the kinetic description, the calculation of the composition heterogeneity is even more difficult. As there is no exact analytical solution of this problem we will discuss here only the approximate approaches. In such a case a problem arises in evaluating the accuracy of approximations. Let us look at how this problem can be solved in the absence of an exact solution.

3.4.1 MONTE CARLO SIMULATION

A very convenient approach used to solve complicated problems of the statistical theory of polymers is the Monte Carlo method [62–64]. In recent years this approach has been widely used for the calculation of macromolecule conformations, dynamics of a polymer chain, cooperative interactions in macromolecular systems and so on. This method can also be applied to kinetic calculations [65]. A number of authors show the possibility of applying the Monte Carlo method to the simulation of macromolecular reactions [37–39, 53, 54, 62–70].

The point of the matter in using this method to solve a particular problem is to build some model stochastic process with parameters corresponding to the values that should be found. Observation of the model process and calculation of its characteristics allow the unknown parameters to be evaluated approximately. In other words the Monte Carlo approach uses the relation between stochastic characteristics and analytical functions to replace the calculation of complicated analytical expressions with the experimental determination of corresponding probabilities or expected values. It should be pointed out that the nature of the model process itself does not affect the Monte Carlo results; it is necessary only for this process to be a stochastic one.

The Monte Carlo method allows many factors defining the parameters of the system to be included in a model. One can say that the results of Monte Carlo simulation have an analogy with experimental results, and this method is considered as the mathematical experiment. Therefore if a model is rather correct, the Monte Carlo calculation is equivalent to the exact analytical solution, and its results can be used as a criterion of the accuracy of other approximate analytical approaches.

Now we describe how a model of the polymer-analogous reaction with the neighboring groups effect should be constructed for results obtained to be comparable to those of the exact solution [66, 67]. The problem is to simulate the polymer-analogous reaction with gradual substitution of all A units in a homopolymer chain by B units and with the calculation of the parameters of the composition distribution on intermediate stages.

A polymer chain can be represented in the computer storage as a sequence of cells, each cell corresponding to one unit. The end effects are neglected by representing chains in the form of cycles. It is evident that for long chains this assumption is valid. The procedure of the simulation of the reaction includes three stages: a selection of some unit of a chain, checking a probability of the reaction for the unreacted units and the substitution of the unreacted unit by the reacted one, if the reaction occurs [66, 67]. The characteristics of the simulated process are calculated after many repetitions of the procedure from the beginning to the end, i.e. after the simulation of the reaction in many chains.

To calculate the composition heterogeneity after each set of N trials (N is an analog of time) the fraction of reacted units y is calculated for each chain. The y values are stored, and when the procedure is completed the functions of the composition distribution and dispersions are calculated. The dispersion of the composition distribution is calculated as

$$D_n = \frac{n^2 \sum_{i=1}^{m} (y_i - \bar{y})^2}{m}$$

where y_i is the degree of conversion for the ith chain, $\bar{y}$ is the average value of y for all chains, n is the chain length and m is the number of chains. The results of such a mathematical experiment depend on the length n of the model chains and on their number m. What should the n and m values be for the results of the Monte Carlo simulation to be good enough to describe the composition heterogeneity in the case of high-molecular (practically infinite) chains?

One approach to answer this question is that of computer simulation with increasing values of n and m until no further change in the composition distribution is observed. However, such a procedure requires too much computer time.

One can use another approach [66]. It is reasonable to assume that the composition heterogeneity may be strictly calculated using values of n and m that are suitable to describe the sequence distribution of A and B units in the chain of infinite length. Therefore in the procedure described above the parameters of the sequence length distribution are calculated after each N trials. The average values of these parameters can be considered as approximate values of $P(A)$, $P(AB)$ and so on.

A comparison of these values with results obtained by exact analytical solution permits values of n and m corresponding to the simulation of practically infinite chains to be found. Both these approaches give the same results, but the second one is more suitable for the optimization of the simulation procedure.

In Figs. 3.9 and 3.10 the choice of optimum values of the number of chains m and chain length n by a comparison of results of Monte Carlo and analytical calculations is demonstrated. In Fig. 3.9 the probabilities of blocks of one, two and three reacted units and the run number $R = 2P(AB)$ as functions of the conversion are shown. Cases are presented of the absence of the neighboring groups effect ($k_0/k_1/k_2 = 1:1:1$), of the slight retardation ($k_0/k_1/k_2 = 1:03:0.3$) and

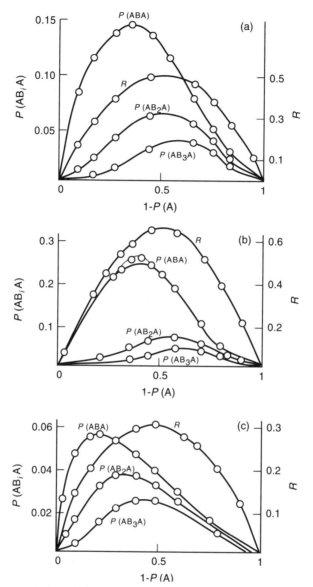

Fig. 3.9 Characteristics of the sequence length distribution versus the degree of conversion $P(B)$ for $k_0/k_1/k_2 = 1{:}1{:}1$ (a), $1{:}0.3{:}0.3$ (b) and $1{:}5{:}5$ (c). Curves correspond to the exact solution; points are the results of the Monte Carlo calculation. (Reproduced by permission of Khimiya Moscow, Russia, from Platé, N. A., Litmanovich, A. D. and Noah, O. V., *Makromolekulyarnye Reaktsii* (*Macromolecular Reactions*), Khimiya, Moscow, 1977, p. 103)

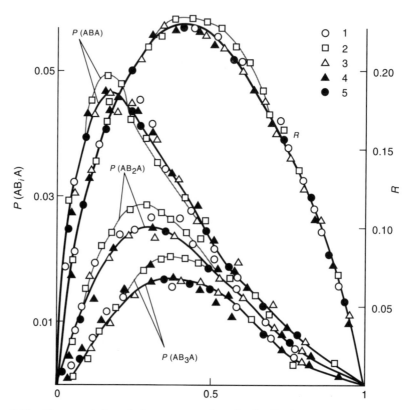

Fig. 3.10 Characteristics of the sequence length distribution versus the degree of conversion $P(B)$ for $k_0/k_1/k_2 = 1:5:100$. Curves correspond to the exact solution; points are the results of the Monte Carlo calculation for 50 (1), 100 (2–4), 200 (5) chains with a length of 50 (2), 100 (3, 5) and 200 units (1, 4). (Reproduced by permission of Khimiya Moscow, Russia, from Platé, N. A., Litmanovich, A. D. and Noah, O. V., *Makromolekulyarnye Reaktsii* (*Macromolecular Reactions*), Khimiya, Moscow, 1977, p. 104)

of the slight acceleration ($k_0/k_1/k_2 = 1:5:5$). The results of the analytical calculation are compared with results of the Monte Carlo simulation with $n = 50$–200 and $m = 50$–200. A good agreement of all results indicates with sufficient accuracy that the mathematical experiment with 50 model chains consisted of 50 units. For these rate constants ratios the results of such a procedure ($n = 50$, $m = 50$) can be considered as an equivalent of the exact solution. For stronger acceleration effects ($k_0/k_1/k_2 = 1:5:100$ in Fig. 3.10) the lower limit is $n = 100$, $m = 100$. For greater acceleration effects the minimal chain length should be longer. The results of the Monte Carlo calculation of the composition heterogeneity are presented in Figs. 3.11 to 3.15.

An analogous simulation procedure was used by Smidsrod *et al.* [69] for the study of the oxidation of amylose by the periodate ion. By varying the simulation parameters up to the best agreement with experimental kinetic data these authors calculated the rate constants of the amylose oxidation and the number-average length of reacted sequences.

3.4.2 MARKOV CHAIN APPROXIMATION

One of the analytical methods of calculation of the composition heterogeneity was proposed by Frensdorf and Ekiner [49]. These authors used a Markov approach to the calculation of statistics of substitution assuming that the probability of the reaction in the $(n+1)$th unit does not depend on the state of the $(n-1)$th unit. This assumption corresponds to the first-order Markov approximation. Introducing two independent parameters—the probability of the substitution in the sequence of more than one unit distance from other substituents (f) and the parameter of the interaction with one nearest neighbour (μ)—one can express in terms of these the quantities $\phi(i, j, k)$, which are the conditional probabilities of finding j substituents in the nth unit, if in the $(n-1)$th and $(n+1)$th units there are respectively i and k substituents $(i, j, k = 0$ or $1)$. For the model under consideration there are eight such probabilities (Frensdorf and Ekiner take into account a possibility of double substitutions in one unit):

$$
\begin{aligned}
&\phi(0, 0, 0) = 1 - f && \phi(0, 1, 1) = \phi(1, 1, 0) = \mu f \\
&\phi(0, 1, 0) = f && \phi(1, 0, 1) = 1 - \mu^2 f && (3.77) \\
&\phi(1, 0, 0 = \phi(0, 0, 1) = 1 - \mu f && \phi(1, 1, 1) = \mu^2 f
\end{aligned}
$$

These expressions contain a rather strong assumption about the multiple interaction with nearest neighbors. It is reasonable to introduce an independent parameter v (instead of μ^2 in Eqs. (3.77)) corresponding to the presence of substituents in both neighboring units $(v/\mu = k_2/k_1)$.

The Markov transitional probabilities can be further expressed in terms of $\phi(i, j, k)$:

$$
\Gamma_{ij}\Pi_i = \sum_i \phi(i, j, k) \sum_{j'} P\{k/i, j'\}\Gamma_{ij'}\Pi_i \qquad (3.78)
$$

where Π_i is the probability of finding i substituents in the $(n-1)$th unit, Π_i is the conditional probability of finding j substituents in the nth unit, if there are i substituents in the $(n-1)$th unit, and $P\{k/ij'\}$ is the conditional probability of finding k substituents in the $(n+1)$th unit, if there are i and j' substituents in the $(n-1)$th and nth units respectively.

An assumption about the independence of finding any substituents in the

$(n+1)$th unit on the state of the $(n-1)$th unit (an approximation by the first-order Markov chain), $P\{k/i, j'\} = \Gamma_{j'k}$ permits Eq. (3.78) to be simplified:

$$\Gamma_{ij} = \sum_k \phi(i, j, k) \sum_{j'} \Gamma_{ij'} \Gamma_{j'k} \tag{3.79}$$

The expression (3.79) is nothing else but a system of four equations. The solution of this system gives the Markov transitional probabilities as functions of f, μ and v. As the polymer chain is considered in this case as a first-order Markov one, the composition heterogeneity can be calculated with the aid of corresponding equations [5].

After replacing the statistical parameters f, μ and v by more convenient kinetic constants k_0, k_1 and k_2 one can use Eqs. (3.60) for the Markov transitional probabilities $\Gamma_{00} = P_{A/A}$, $\Gamma_{01} = P_{B/A}$, $\Gamma_{10} = P_{A/B}$, $\Gamma_{11} = P_{B/B}$:

$$\frac{dP_{A/A}}{d\tau} = P_{A/A}[(k' - 2k) - P_{A/A}(k' - 2k + 1)(2 - P_{A/A})]$$

$$\frac{dP_{A/B}}{d\tau} = P_{A/B}\left\{\frac{P_{A/A}^2}{1 - P_{A/A}} - k'(1 - P_{A/A}) \right. \tag{3.60}$$

$$\left. - P_{A/B}\left[\frac{P_{A/A}^2}{1 - P_{A/A}} + 2kP_{A/A} + k'(1 - P_{A/A})\right]\right\}$$

The values of $P_{A/A}$ and $P_{A/B}$ obtained by solution of the system (3.60) with initial conditions $P_{A/A} = P_{A/B} = 1$ at $\tau = 0$ can be substituted into the equation for the dispersion of the compositional distribution [5]:

$$\lim_{n \to \infty} \frac{D_n}{n} = \frac{(1 - P_{A/A})P_{A/B}(1 - P_{A/B} + P_{A/A})}{(1 - P_{A/A} + P_{A/B})^3} \tag{3.80}$$

It is known from the general theory of Markov chains [5] that the composition distribution for the first-order Markov chain is close to the Gaussian one. The functions of the composition distribution can be derived for any conversion using the dispersion calculated following (3.80). The values of this function at the point y ($y = 1 - P(A)$ is a degree of conversion) are determined as

$$w(y) = \frac{1}{\sqrt{D_n}} \varphi\left(\frac{y - \bar{y}}{D_n}\right)$$

where $\varphi(x) = (1/\sqrt{D_n})\exp(-x^2/2)$ is a probability density of the Gaussian distribution.

Let us evaluate the accuracy of such a first-order Markov approximation by comparing its results with those of the Monte Carlo calculation. It is of interest to elucidate a possibility of applying this approximation to calculate the composition heterogeneity for such ratios of the rate constants, when the first-order Markov approximation does not describe in the correct way the kinetics of the

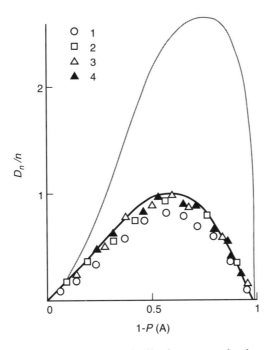

Fig. 3.11 Dispersion of the composition distribution versus the degree of conversion $P(B)$ for $k_0/k_1/k_2 = 1:5:100$. The thin curve corresponds to the first-order Markov approximation and the solid curve to the modified first-order Markov approximation; points are the results of the Monte Carlo calculation for 100 (1,2,4) and 200 (3) chains with a length of 50 (1), 100 (2, 3) and 200 units (4). (Reproduced by permission of Khimiya Moscow, Russia, from Platé, N. A., Litmanovich, A. D. and Noah, O. V., *Makromolekulyarnye Reaktsii* (*Macromolecular Reactions*), Khimiya, Moscow, 1977, p. 105)

reaction [48, 50, 51] and the sequence distribution, i.e. for the accelerating neighboring groups effect.

Results of the calculation of D_n/n for the case $k_0/k_1/k_2 = 1:5:100$ are presented in Fig. 3.11. As mentioned above for this constants ratio the model chain of 100 units can be considered as an infinite one in the mathematical experiment. Therefore the dispersion calculated for such a model chain can be compared with the dispersion calculated in the first-order Markov approximation. It can be seen from the figure that the results of the approximation deviate essentially from the data obtained by the Monte Carlo simulation (the maximum value of D_n/n is almost 2.5 times larger than the Monte Carlo result).

In Fig. 3.12 the composition distribution functions for the same constants ratio (1:5:100) are compared. Again the same deviations are observed. One can conclude that the first-order Markov approximation cannot be used for the calculation of the composition heterogeneity. However, since this method is very

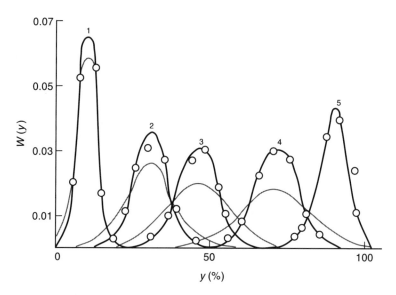

Fig. 3.12 Composition distribution functions $W(y)$ at $\bar{y} = 10$ (1), 30 (2), 46 (3), 71 (4), 89% (5) for $k_0/k_1/k_2 = 1:5:100$. Thin curves correspond to the first-order Markov approximation and solid curves to the modified first-order Markov approximation; points are the results of the Monte Carlo calculation. (Reproduced by permission of Khimiya Moscow, Russia, from Platé, N. A., Litmanovich, A. D. and Noah, O. V., *Makromolekulyarnye Reaktsii* (*Macromolecular Reactions*), Khimiya, Moscow, 1977, p. 105)

simple it seems reasonable to modify it to a form that gives accurate results at any $k_0/k_1/k_2$ ratios.

Platé, Litmanovich and Noah proposed a modified form of the first-order Markov approximation [66, 67]. In this approach the assumption about a Gaussian distribution form is maintained and the dispersion is calculated according to Eq. (3.80). However, the Markov transitional probabilities $P_{A/A}$ and $P_{A/B}$ are proposed to be calculated following the expressions:

$$P_{A/A} = P(AA)/P(A) \qquad P_{A/B} = P(BA)/P(B)$$

with $P(A)$, $P(B) = 1 - P(A)$, $P(AA)$ and $P(BA) = P(A) - P(AA)$ obtained from the solution of the exact Eqs. (3.9). In other words, the modified first-order Markov approximation assumes that in the time moment under consideration, the chain is a first-order Markov one, but all its prehistory is described by the exact equations.

In Figs. 3.11 and 3.12 the results of the calculation of D_n/n and of the composition distribution functions obtained by such a modified approximation are presented. The figures show that this method gives results very close to those of the mathematical experiment. Thus the modified first-order Markov approxi-

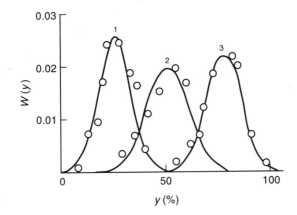

Fig. 3.13 Composition distribution functions $W(y)$ at $\bar{y} = 25$ (1), 50 (2), 77% (3) for $k_0/k_1/k_2 = 1:50:99$. Curves correspond to the modified first-order Markov approximation and points to the Monte Carlo calculation. (Reproduced by permission of Khimiya Moscow, Russia, from Platé, N. A., Litmanovich, A. D. and Noah, O. V., *Makromolekulyarnye Reaktsii* (*Macromolecular Reactions*), Khimiya, Moscow, 1977, p. 106)

mation is a rather simple method allowing the calculation of the dispersion of the composition distribution for a wide range of constants ratios.

One can see from Figs. 3.12 and 3.13 that the Gaussian distribution is a fairly good approximation for the composition distribution functions if the degree of conversion is not too close to 0 or 1. Therefore it is sufficient to consider the composition heterogeneity for different constants ratios only from the viewpoint of the dispersion of the composition distribution determined in the first-order Markov approximation from Eq. (3.80). As this equation includes the limit at the chain length $n \to \infty$, it is of interest to analyze the relation between the chain length and the accuracy of the first-order Markov approximation.

One can see from the comparison with results of the computer simulation (Figs. 3.11 and 3.14) that the minimum chain length that can be considered as an infinite one and permits application of the first-order Markov approximation depends on the constants ratio $k_0/k_1/k_2$ increasing with an increase in the acceleration effect. For $k_0/k_1/k_2 = 1:5:100$ the minimum chain length of the 'infinite' chain is equal to 100 units; for $k_0/k_1/k_2 = 1:50:99$ it is equal to 200 units.

It should be pointed out that for the same values of the chain length the parameters of the sequence distribution calculated by the Monte Carlo method coincide with results of the solution of analytical equations derived for infinite chains. For n less than some critical value the possibility of the existence of long blocks of reacted units is excluded. The propagation of such long blocks typical for the strong acceleration effect is artificially depressed, resulting in a change of

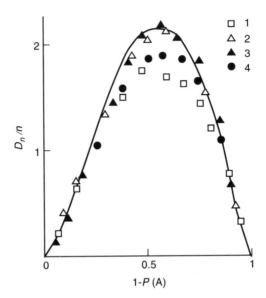

Fig. 3.14 Dispersion of the composition distribution versus the degree of conversion $P(B)$ for $k_0/k_1/k_2 = 1:50:99$. The curve corresponds to the modified first-order Markov approximation; points are the results of the Monte Carlo calculation for 100 (1, 3, 4) and 200 (2) chains with a length of 50 (1) and 100 (4) and 200 units (2, 3). (Reproduced by permission of Khimiya Moscow, Russia, from Platé, N. A., Litmanovich, A. D. and Noah, O. V., *Makromolekulyarnye Reaktsii* (*Macromolecular Reactions*), Khimiya, Moscow, 1977, p. 107)

parameters of the sequence distribution and in a decrease in the composition heterogeneity. Therefore D_n/n for the short chains is less than for the chains of the limiting length.

Thus for the short chains (20–30 units) in the case of a strong acceleration effect the modified first-order Markov approximation is not suitable (as well as the analytical approaches to the calculation of parameters of the sequence distribution described above) and the Monte Carlo procedure is the only accurate method of calculation. However, the most practically important polymers have sufficiently high molecular mass, which permits all the theoretical descriptions described above to be applied.

Summarizing the results of the calculation of the composition heterogeneity by both methods (Monte Carlo and modified first-order Markov approximation) one can note some regularities of the chemical reaction itself [50, 51, 66, 67]. As an illustration we consider Fig. 3.15, where the dependences of the dispersion of the composition distribution on the degree of conversion for the wide range of the constants ratios are shown. It can be seen from the figure that the copolymers obtained by polymer-analogous reactions are heterogeneous, even in the absence of the neighboring groups effect, because of the statistical nature of the process

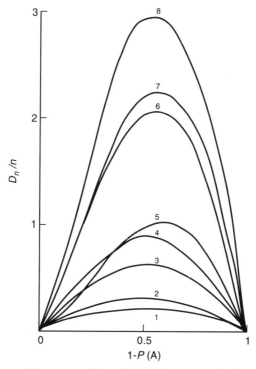

Fig. 3.15 Dispersion of the composition distribution versus the degree of conversion $P(B)$ for $k_0/k_1/k_2 = 1{:}0.3{:}0.3$ (1), $1{:}1{:}1$ (2), $1{:}5{:}5$ (3), $1{:}10{:}10$ (4), $1{:}5{:}100$ (5), $1{:}50{:}50$ (6), $1{:}50{:}99$ (7) and $1{:}100{:}100$ (8). (Reproduced by permission of Khimiya Moscow, Russia, from Platé, N. A., Litmanovich, A. D. and Noah, O. V., *Makromolekulyarnye Reaktsii* (*Macromole-cular Reactions*), Khimiya, Moscow, 1977, p. 108)

which is exhibited in the existence of different numbers of substituted units in different macromolecules at some definite average degree of conversion. This results in the appearance of the more or less wide composition distribution. The composition heterogeneity depends essentially on the neighboring groups effect. It can be seen from Fig. 3.15 that for the accelerating effect of reacted neighbors ($k_0 \leq k_1 \leq k_2$) the heterogeneity is more than in the absence of the neighboring groups effect. For the retardation effect ($k_0 \geq k_1 \geq k_2$) the heterogeneity is lower, but still exists and has maximal values at the degree of conversion close to 0.5.

It should also be pointed out that an increase in the accelerating effect is accompanied by a general increase in the compositional heterogeneity and by a slight change in the shape of the curve of the dependence of the dispersion of the composition distribution versus the degree of conversion. The maximum on the curve is displaced a little toward higher conversions, and the curve becomes the nonsymmetrical one. For the content of reacted groups more than 50% the

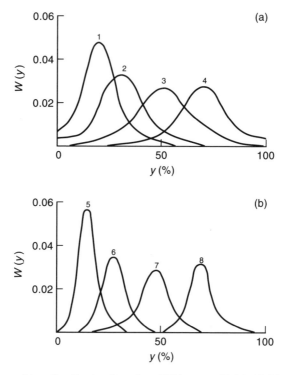

Fig. 3.16   Composition distribution functions $W(y)$ at $\bar{y} = 20$ (1), 30 (2), 50 (3), 70 (4), 19 (5), 30 (6), 51 (7), 72% (8) for the polymolecular sample with $\bar{n} = 50$ (a) and for the monomolecular polymer of the chain length 50 units (b) for $k_0/k_1/k_2 = 1{:}5{:}100$. (Reproduced by permission of Khimiya Moscow, Russia, from Platé, N. A., Litmanovich, A. D. and Noah, O. V., *Makromolekulyarnye Reaktsii* (*Macromolecular Reactions*), Khimiya, Moscow, 1977, p. 111)

heterogeneity is higher than for the same content of unreacted groups (Fig. 3.15). One can note from the same figure that for the accelerating effect of the neighboring reacted units the composition heterogeneity depends more on the k_1/k_0 ratio than on the k_2/k_0 ratio.

All these results were obtained for practically infinite chains (the modified first-order Markov approximation) or for sets of chains of the same length, i.e. for the monodisperse polymer samples (Monte Carlo). The question arises about the possibility of applying these results to real polymers containing macromolecules of different lengths.

In Fig. 3.16 the composition distribution function for a polydisperse sample with the number-average degree of polymerization $n = 50$ and for the corresponding monodisperse polymer are compared ($k_0/k_1/k_2 = 1{:}5{:}100$) [50, 51].

The most probable length distribution for the polydisperse sample was taken to be

$$w(n) = [n/(\bar{n})^2] \exp(-n/\bar{n})$$

It can be seen from the figure that the polydispersity results in an increase in the compositional heterogeneity.

One can see from Fig. 3.15 that the dependences of D_n/n on the degree of conversion are rather standard curves with the maxima near $P(A) = 0.5$. This fact allows the results of the calculations of the composition heterogeneity to be summarized in the form of diagrams [70], presented in Fig. 3.17. The curves correspond to the constant values of D_n/n at the 50% degree of conversion, and on the axes the values of $k = k_1/k_0$ and $k' = k_2/k_0$ are plotted. The nonequivalency of the influence of the constants ratios on the composition heterogeneity mentioned above can also be seen in Fig. 3.17. Here it shows in the different slopes of the curves toward the axes. The diagrams of Fig. 3.17 can be directly used for the calculation of dispersions and composition distribution functions for the products of particular macromolecular reactions with the neighboring groups effect if the ratios if the kinetic constants corresponding to k and k' values in the range from 0 to 100 are known.

It was shown above how to calculate the parameters of the sequence distribution of units and of the composition heterogeneity when the values of kinetic constants $k_0, k_1/k_2$ are known. However, the experimental determination of these constants in a rather difficult task. Therefore it is very useful to find the mutual relationship between the parameters of the sequence distribution and of the composition heterogeneity, which allows us to calculate, for example, the dispersion of the composition distribution from the experimental data on the triad content and vice versa. In the frames of the modified first-order Markov approximation described above this relation between the limit value $D_\infty = \lim_{n \to \infty} D_n/n$ and the run number $R = 2P(AB)$ can easily be found.

One can show that the substitution of $P_{A/A} = P(AA)/P(A)$ and $P_{A/B} = P(BA)/P(B)$ into Eq. (3.80) gives, after all simplifications,

$$D_\infty = \frac{P(A)[1 - P(A)]}{P(AB)} \{2P(A)[1 - P(A)] - P(AB)\}$$

Introducing $X = P(A)[1 - P(A)]$ one obtains

$$D_\infty = \frac{X[2X - P(AB)]}{P(AB)} = \frac{X(4X - R)}{R} \tag{3.81}$$

$$R = \frac{4X^2}{D_\infty + X} \tag{3.82}$$

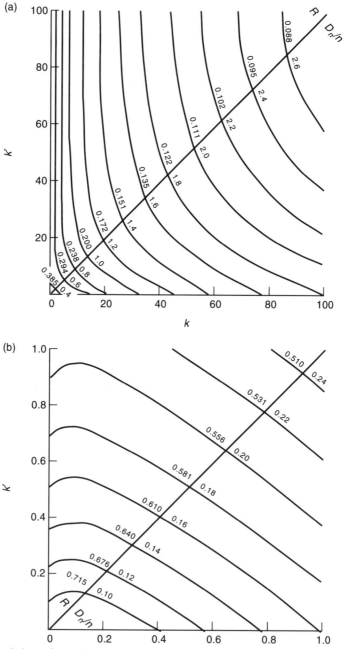

Fig. 3.17 D_n/n and $R = 2P(AB)$ dependences on $k_0/k_1/k_2$ for the accelerating effect $k_1 \geq k_0$, $k_2 \geq k_0$ (a) and for the retarding effect $k_1 \leq k_0$, $k_2 \leq k_0$ (b) at $P(A) = 0.5$ ($k = k_1/k_0$, $k' = k_2/k_0$). (Reproduced by permission of Khimiya Moscow, Russia, from Platé, N. A., Litmanovich, A. D. and Noah, O. V., *Makromolekulyarnye Reaktsii* (*Macromolecular Reactions*), Khimiya, Moscow, 1977, p. 112)

If the degree of conversion is equal to 0.5, $X = 0.25$ and

$$D_\infty = \frac{1-R}{4R} \qquad R = \frac{1}{4D_\infty + 1}$$

i.e. the diagrams of Fig. 3.17 also summarize the results of the calculation of the run number R.

3.4.3 OTHER APPROACHES

One more approach used to calculate the composition heterogeneity for products of polymer-analogous reactions was proposed by Kuchanov and Brun [59, 60, 71]. These authors introduced for the description of the statistical properties of polymer chains of finite length N the N-dimensional stochastic vector $\mathbf{n}$, the components of which are equal to the number of j-clusters in the chain. Introducing the distribution function of probabilities of the vector $\mathbf{n}$ $f(\mathbf{n}, t)$ and the generating function of this distribution

$$g(s, t) = \sum_n f(\mathbf{n}, t) \prod_{i=1}^N s_i^{n_i}$$

(where s_i are the components of the auxiliary vector $\mathbf{s}$), and summing f over definite values $\mathbf{n}$, one can obtain some important characteristics of the chain structure. Therefore the summation of f with the condition $\sum_i n_i = m$ results in the fraction of the composition distribution of unreacted units $f^k(m, t)$. The condition $\sum_i n_i = n$ results in the distribution functions of molecules toward the total number of clusters n, $f(n, t)$. The summation over all values of components $\mathbf{n}$ except the jth one results in the function of the distribution toward the number of j-clusters.

The first moments are the most interesting of all these functions and correspond to the average values of the composition and of the structure parameters, while the second moments determine the width of the corresponding distributions:

$$B_j(t) = \sum_i n_j f(\mathbf{n}, t)$$

$$N_{i,j}(t) = \sum_i n_j(n_j - \delta_{i,j}) f(\mathbf{n}, t) \tag{3.83}$$

$$\delta_{i,j} = \begin{cases} 1 & \text{if } i = j \\ 0 & \text{if } i \neq j \end{cases}$$

With known moments (3.83) one can easily calculate other characteristics of the chain structure. For a solution of Eq. (3.83) one can use the equation for the generating function $g(s, t)$, which is derived from the densities of the probabilities of all clusters.

As $N \to \infty$ the authors obtained the asymptotic expressions for the composition heterogeneity which is described, for example, by the Gaussian law:

$$
\sqrt{\frac{N}{2\pi D_m}} \int_{\xi_1}^{\xi_2} \exp\left(-\frac{N(\xi - \bar{\xi})^2}{2D_m} \right) d\xi = \frac{1}{2} \left\{ \mathrm{erf}\left(\sqrt{\frac{N}{2D_m}} (\xi_2 - \bar{\xi}) \right) \right.
$$

$$
\left. - \mathrm{erf}\left(\sqrt{\frac{N}{2D_m}} (\xi_1 - \bar{\xi}) \right) \right\}
$$

where ξ_1 and ξ_2 are the deviations from the average composition $\bar{\xi}$ and D_m are calculated as functions of f.

REFERENCES

1. Moore, J. A. (Ed.), *Reactions on Polymers*, Reidel, Dordrecht, 1973.
2. Platé, N. A., Litmanovich, A. D. and Noah, O. V., *Makromolekulyarnye Reaktsii* (*Macromolecular Reactions*), Khimiya, Moscow, 1977.
3. Platé, N. A. and Noah, O. V., *Adv. Polymer Sci.*, 1979, **31**, 133–73.
4. Soutif, J.-C. and Brosse, J.-C., *Reactive Polymers*, 1990, **13**, 1–26.
5. Kemeny, J. G. and Snell, J. L., *Finite Markov Chains*, Springer, Berlin, 1976.
6. Flory, P. J., *J. Am. Chem. Soc.*, 1939, **61**, 1518.
7. Cohen, E. R. and Reiss, H., *J. Chem. Phys.*, 1963, **38**, 680.
8. Gordon, M. and Hillier, J. H., *J. Chem. Phys.*, 1963, **38**, 1376.
9. Lee, D. F., Scanlan, J. and Watson, W. F., *Proc. Roy. Soc.*, 1963, **A-273**, 345.
10. McQuiston, R. B. and Lichtman, D., *J. Math. Phys.*, 1968, **9**, 1680.
11. Lewis, C. W., *J. Polymer Sci. A-2*, 1972, **10**, 377.
12. Barron, T. H. K. and Boucher, E. A., *Trans. Faraday Soc.*, 1969, **65**, 3301; 1970, **66**, 2320.
13. Boucher, E. A., *J. Chem. Soc. Faraday Trans. 1*, 1972, **68**, 2295.
14. Boucher, E. A., *J. Chem. Phys.*, 1973, **59**, 3848.
15. Boucher, E. A., *Chem. Phys. Letters*, 1972, **17**, 221; *J. Chem. Soc. Faraday Trans. 2*, 1973, **69**, 1839.
16. Boucher, E. A., *Makromol. Chem.*, 1973, **173**, 253.
17. Barron, T. H. K., Bawden, R. J. and Boucher, E. A., *J. Chem. Soc. Faraday Trans. 2*, 1974, **70**, 651.
18. Boucher, E.A., *J. Chem. Soc. Faraday Trans. 2*, 1976, **72**, 1697.
19. Gonzalez, J. J. and Hemmer, P. C., *J. Polymer Sci. Polymer Letters*, 1976, **14**, 645; *J. Polymer Sci. Polymer Phys.*, 1977, **15**, 321.
20. Fuoss, R. M., Watanabe, M. and Coleman, B. D., *J. Polymer Sci.*, 1960, **48**, 5–15.
21. Keller, J. B., *J. Chem. Phys.*, 1962, **37**, 2584–90.
22. Alfrey, T. Ir. and Lloyd, W. G., *J. Chem. Phys.*, 1963, **38**, 318–21.
23. Arends, C. B., *J. Chem. Phys.*, 1963, **38**, 322–4.
24. Keller, J. B., *J. Chem. Phys.*, 1963, **38**, 325–6.
25. Lazare, L., *J. Chem. Phys.*, 1963, **39**, 727–31.
26. McQuarrie, D. A., McTague, J. P. and Reiss, H., *Biopolymers*, 1965, **3**, 657–63.
27. Mityushin, L. G., *Problemy peredachi informatsii*, 1973, **9**, 81–5.
28. Ueda, M., *Polymer J.*, 1980, **12**, 841–2.

29. Noah, O. V., PhD Thesis, Moscow, 1972.
30. Silberberg, A. and Simha, R., *Biopolymers*, 1968, **6**, 479–90.
31. Rabinowitz, P., Silberberg, A., Simha, R. and Loftus, E., in *Stochastic Processes in Chemical Physics*, Ed. K. E. Shuler, Interscience, New York, 1969.
32. Simha, R. and Lacombe, R. H., *J. Chem. Phys.*, 1971, **55**, 2936–9.
33. Silberberg, A. and Simha, R., *Macromolecules*, 1972, **5**, 332–4.
34. Lacombe, R. H. and Simha, R., *J. Chem. Phys.*, 1973, **58**, 1043–53.
35. Lacombe, R. H. and Simha, R., *J. Chem. Phys.*, 1974, **61**, 1899–911.
36. Vainshtein, E. F., Berlin, A. A. and Entelis, S. G., *Vysokomol. Soed B*, 1975, **17**, 835–7.
37. Krieger, D. and Klesper, E., *Makromol. Chem. Rap. Com.*, 1985, **6**, 693–8.
38. Krieger, D. and Klesper, E., *Makromol. Chem.*, 1987, **188**, 155–70.
39. Krieger, D. and Klesper, E., *Makromol. Chem.*, 1988, **189**, 1819–34.
40. Krishnaswami, P. and Vadav, D. P., *J. Appl. Polymer. Sci.*, 1976, **20**, 1175–85.
41. Serdyuk, O. V., *Vysokomol. Soed B*, 1978, **20**, 380–2.
42. Dobrushin, R. L., *Problemy peredachi informatsii*, 1971, **7**, 57–65.
43. Platé, N. A. and Litmanovich, A. D., XXIII International Congress on *Pure and Applied Chemistry*, Boston, 1971, Vol. 8, pp. 123–50.
44. Platé, N. A. and Litmanovich, A. D., *Vysokomol. Soed A*, 1972, **14**, 2503–17.
45. Noah, O. B., Toom, A. L., Vasil'yev, N. B., Litmanovich, A. D. and Platé, N. A., *Vysokomol. Soed A*, 1973, **15**, 877–88.
46. Plate, N. A., Litmanovich, A. D., Noah, O. V., Toom, A. L. and Vasil'yev, N. B., *J. Polymer Sci. Polymer Chem.*, 1974, **12**, 2165–85.
47. Litmanovich, A. D., *Dokl. AN SSSR*, 1965, **165**, 354–6.
48. Krentsel' L. B. and Litmanovich, A. D., *Vysokomol. Soed B*, 1967, **9**, 175–9.
49. Frensdorf, H. K. and Ekiner, O., *J. Polymer Sci. A*, 1967, **2**, 1157–75.
50. Platé, N. A., Litmanovich, A. D., Noah, O. V. and Golyakov, V. I., *Vysokomol. Soed A*, 1969, **11**, 2204–10.
51. Litmanovich, A. D., Platé, N. A., Noah, O. V. and Golyakov, V. I., *Europ. Polymer J. Suppl.*, 1969, 517–21.
52. Harwood, H. J. and Ritchey, W. M., *J. Polymer Sci. B*, 1964, **2**, 601–8.
53. Klesper, E., Gronski, W. and Barth, V., *Makromol. Chem.*, 1971, **150**, 223–49.
54. Klesper, E., Johnsen, A. and Gronski, W., *Makromol. Chem.*, 1972, **160**, 167–81.
55. Gonzalez, J. J. and Hemmer, P. C., *J. Polymer Sci. Polymer Letters*, 1976, **14**, 645.
56. Gonzalez, J. J. and Hemmer, P. C., *J. Chem. Phys.*, 1977, **67**, 2496–508, 2509–26.
57. Hemmer, P. C. and Gonzalez, J. J., *J. Polymer Sci. Polymer Phys.*, 1977, **15**, 321–33.
58. Gonzalez, J. J. and Kehr, K. W., *Macromolecules*, 1978, **11**, 996–1000.
59. Brun, E. B. and Kuchanov, S. I., *Zh. P. Kh.*, 1977, **50**, 1065–9.
60. Kuchanov, S. I., *Metody Kineticheskikh Raschetov v Khimii Polimerov* (*Methods of Kinetic Calculations in Polymer Chemistry*), Khimiya, Moscow, 1978, Chap. 10.
61. Berlin, Al. Al., Vainshtein, E. F. and Entelis, S. G., *Vysokomol. Soed B*, 1978, **20**, 275–7.
62. Lowery, G. G. (Ed.), *Markov Chains and Monte Carlo Calculations in Polymer Science*, Marcel Dekker, New York, 1970.
63. Dashevskii, V. G., *Itogi Nauki i Techniki Org. Khimiya*, 1975, **1**, 5–98.
64. Klesper, E. and Johnsen, A. O., *Comput. Chem. Instrum.*, 1977, **6**, 1–55.
65. Cohen, G. D., *Indust. Engineer Chem.*, 1965, **F4**, 471–80.
66. Noah, O. V., Toom, A. L., Vasil'yev, N. B., Litmanovich, A. D. and Platé, N. A., *Vysokomol. Soed A*, 1974, **16**, 412–18.

67. Noah, O. V., Litmanovich, A. D. and Platé, N. A., *J. Polymer Sci. Polymer Phys.*, 1974, **12**, 1711–25.
68. Goldshtein, B. N., Goryunov, A. N., Gotlib, Yu. Ya., Elyashevich, A. M., Zubova, T. P., Koltsov, A. I., Nemirovskii, V. D. and Skorokhodov, S. S., *J. Polymer Sci. A*, 1971, **9**, 769–77.
69. Smidsrod, O., Larsen, B. and Painter, T. J., *Acta Chem. Scand.*, 1970, **24**, 3201–12.
70. Platé, N. A. and Noah, O. V., *Vysokomol. Soed B*, 1977, **19**, 483–4.
71. Kuchanov, S. I. and Brun, Ye. B., *Dokl AN SSSR*, 1976, **227**, 662–5.

CHAPTER 4

Neighboring Groups Effect: Experimental Approaches

4.1 PECULIARITIES OF EXPERIMENTAL INVESTIGATION OF THE KINETICS AND THE MECHANISM OF MACROMOLECULAR REACTIONS

The peculiarities of macromolecular reactions described in Chapter 2, which are caused by the polymeric nature of reagents, greatly complicate not only the theoretical description of the chemical transformations of polymers but also the experimental study of their kinetics and mechanism. Various aspects of this problem are analyzed in references [1] to [4].

Gaylord [1] stresses that the kinetics of polymer reactions is determined by both the reactivity of the functional groups of a macromolecule and their accessibility to the low-molecular reagent. From this point of view he regards the influence exerted upon the course of macromolecular reactions by such factors as crystallinity and orientation of chains in polymeric reagents, solubility and compatibility of polymers in solutions. Indeed, as established by studying poly(ethylene terephtalate) hydrolysis [5], the chlorination [6] and oxidation [7] of polyethylene, the reactions proceed in amorphous areas more readily than in crystalline ones. The orientation of crystalline and amorphous polymers hinders access of reagents to the functional groups of macromolecules (Gaylord notes that in polymer melts a similar effect is caused by the entanglement of chains.) The degree of penetration of reagents into a polymer sample and, therefore, the rate and degree of conversion depend also on the proximity of the reaction temperature to the polymer glass temperature.

When macromolecular reactions are conducted in solution, the medium, although in other respects suitable for the reaction to be performed in, may be a poor solvent for the polymer. In this case (to say nothing of a possible heterogeneity of the system) the macromolecules form compact coils and the functional groups can be as inaccessible to the reagent as in melts or a solid polymer. Of importance here is the variation of polymer solubility in the course of the reaction. Thus, when polyethylene is brominated in the presence of appropriate solvents and the degree of substitution reached is that at which the crystallinity of a polymer disappears completely, the initially heterogeneous system becomes homogeneous [8]. Conversely, during the methanolysis of poly(vinyl acetate) the solubility of the polymer decreases with the conversion and a homogeneous system turns into a heterogeneous one [9].

When estimating the possible influence of the polymer's incompatibility on the course of macromolecular reactions one should bear in mind the fact that even polymers of similar structure, e.g. polystyrene and poly(tert-butyl styrene) or copolymers differing in composition by as little as 5%, can be incompatible in solution and that incompatibility can manifest itself not only in the separation of phases but also in the compaction of polymer coils. The initial polymer and the product of its transformation may therefore become incompatible at a certain degree of conversion, and the phase separation or compaction of the coils occurs. In the latter case the reaction is localized on the surface of compact coils and/or becomes diffusion controlled.

The phenomena examined by Gaylord [1] are thus seen to amount to intermolecular, mainly interchain, interactions varying during a reaction and affecting the accessibility of the functional groups of macromolecules to low-molecular reagents.

Harwood [2] attributes the difficulties of experimental quantitative study of the reactivity of macromolecules to a number of effects, distinguishing the reactions of polymers from those of their low-molecular analogs:

(a) The steric environment of the functional groups of a macromolecule. The main chain of vinyl polymer close to the functional group creates considerable steric hindrances. This results in a low rate and the incompleteness of macromolecular reactions.
(b) The effect of neighboring units.
(c) The polymer–solvent interaction. In dilute solutions the reacting macromolecules constitute quasi-isolated coils separated by the solvent. The composition of the solvent is different in the coils and in solution. However, it is the solvent composition in the coils that affects such factors of determining the reaction kinetics as the formation of hydrogen bonds, and the electrostatic, hydrophobic and nonbonding interactions.
(d) The variation of the reactivity of the functional groups of a macromolecule with conversion, caused by the changes in the structure and the conformation of a polymer and its interaction with the solvent which take place in the course of a reaction.

Harwood [2] thus analyzes the role of the factors that characterize the nearest environment of the functional groups of a macromolecule and change by degree the chemical transformation of a polymer. He considers it necessary in kinetic studies to use such methods that would permit effects (a) and (b) to be separated from effects (c) and (d). The contribution of effect (c) can then be estimated in conditions when the structure and conformation of a polymer and its interaction with the solvent do not vary significantly in the course of an experiment. For this purpose, it is recommended that the initial stages of polymer-analogous transformations or the catalytic activity of polymers be studied. Among such kinds of methods Harwood also places the method of double hydrolysis of labeled poly(methyl methacrylate) [10], as well as the method of studying the isotopic hydrogen exchange in poly-*N*-vinylacetamide used by Hvidt and Corett [11]. It should be noted, however, that a study of only the initial stages of macromolecular reactions does not make it possible to determine the kinetic parameters with the required accuracy (as will be shown in the next sections) and the other above-mentioned methods are applicable in particular cases only.

Works [1] and [2] are mainly devoted to stating the difficulties arising in experimental investigations of the chemical transformations of polymers, whereas Platé and Litmanovich [3, 4] made an attempt at a constructive approach to solve the problem of a quantitative study of the kinetics and mechanism of macromolecular reactions. Their concept is presented below.

Theoretical and experimental investigations of macromolecular reactions, with all the inherent peculiarities taken into account in their totality, indeed comprise an extremely complicated poblem. It is therefore expedient to elaborate the theoretical and experimental methods of investigating these processes stage by stage, examining the effect of every factor separately. For example, attempts have been made to describe quantitatively the interaction of remote groups in a macromolecule [12–14] or the electrostatic effect [15, 16]. In our opinion it is the effect of neighboring units that primarily deserves a detailed investigation.

The mutual influence of atoms and groups of atoms in a molecule on its reactivity is one of the most general problems of chemistry and manifests itself in both radical and ionic reactions. Therefore, a possibility of the effect of neighboring units appearing has to be reckoned with when one investigates a broad range of macromolecular reactions. Moreover, in many cases configurational, conformational and electrostatic effects are superimposed, as it were, on the action of nearest neighbors or are its component parts. Thus, in the hydrolysis of polymethacrylates [17, 18] and polyacrylates [19] the acceleration of the reaction with a degree of conversion was observed to be much greater for isotactic polymers than for syndiotactic or atactic samples; evidently (though this may not be the only manifestation of the configurational effect), it is the isotactic configuration that is most favorable for the anchimeric assistance of the nearest neighboring carboxylic groups which is a cause of the acceleration.

Such an effect of chain microstructure was also observed in the hydrolysis of copolymers of small amounts of *p*-nitrophenyl or *p*-methoxyphenyl acrylates with methacrylic acid [20]. The rate of the hydrolysis of copolymers of *p*-nitrophenyl or *p*-methoxyphenyl methacrylates with acrylic acid does not, however, depend on the microstructure of macromolecules [20], although it would appear that anchimeric assistance in the acrylic acid–methacrylic ester and methacrylic acid–acrylic ester diads must be accompanied by the same steric tensions.

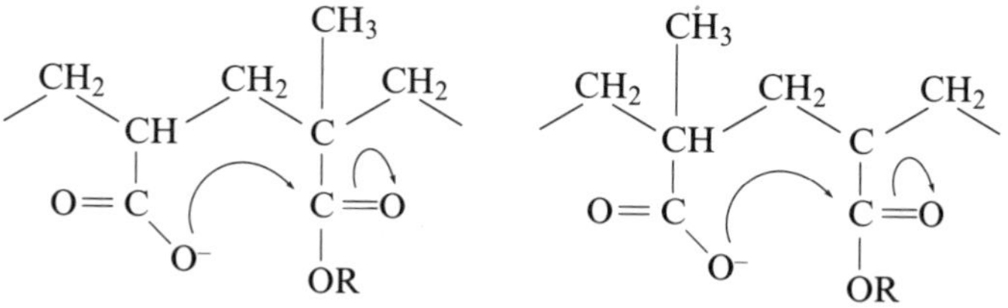

Scheme 4.1

Gaetjens and Morawetz [20] explain this interesting result as being due to the relatively greater rigidity of the poly(methacrylic acid) chain which in syndiotactic triads hinders the realization of local conformations favorable to the reaction and facilitates their realization in isotactic triads; i.e. in this case the conformational characteristics of the chain directly affect the interaction of neighboring units.

The superposition of electrostatic interactions on the effect of neighboring units can be exemplified by the hydrolysis of polymethacrylamide. In a number of works [21, 22] the hydrolysis of amide groups in macromolecules was shown to be accelerated by the neighboring carboxylic groups. As established by Arcus [23], however, the alkaline hydrolysis of polymethacrylamide is not brought to a completion, and Pinner [24] later showed that the portion of amide groups incapable of hydrolysis in methacrylamide–methacrylic acid copolymers increases with the decreasing overall content of methacrylamide in the sample. These results agree with the assumption that amide groups located between two methacrylic acid units lose their reactivity as a result of their electrostatic repulsion of the OH$^-$ ions catalyzing the hydrolysis.

The above facts prove the impossibility of interpreting a wide range of phenomena caused by the configurational, conformational and electrostatic effects in isolation from a quantitative investigation of the effect of neighboring units. It is also obvious that one has to study the behavior of quasi-isolated macromolecules before estimating the contribution of interchain interaction to the kinetics of macromolecular reactions. Therefore, quantitative investigation of the effect of neighboring units appears to be the key stage in solving theoretical and practical problems in the field of chemical transformations of polymers.

This chapter is devoted to an experimental investigation of the effect of neighboring units in polymer-analogous transformations. The methods developed for this purpose can also be applied to study the kinetics and the mechanism of a number of other macromolecular reactions proceeding with the length of the macromolecular chain remaining unchanged. In such investigations the reactivity of the functional group of a macromolecule is assumed to depend only on the state of the nearest neighboring units—whether they have reacted or not. In this case, as shown in Chapter 3, the kinetics of a process, as well as the distribution of units in the chain and the composition heterogeneity of the reaction products, are the functions of three individual rate constants, k_0, k_1 and k_2, that characterize the reactivity of the initial units having 0, 1 and 2 reacted neighboring units respectively. A question naturally arises: can this model of polymer-analogous transformation be realized in the conditions of kinetic experiment?

It appears comparatively easy to exclude, or at least reduce to a minimum, the effect of interchain interactions by conducting kinetic investigations in dilute solutions of polymers. Moreover, the interchain interactions may also not affect the kinetics in solutions of moderate concentrations and, in some cases, in heterogeneous systems either. It is only important that the rate of diffusion of the low-molecular reagent into the polymer associates should exceed substantially the rate of the chemical reaction, i.e. that the reaction should proceed in the kinetic region.

More complicated is the situation with the other effects inherent in the quasi-isolated macromolecules themselves. Here, however, it is necessary first of all to formulate the problem precisely.

When one studies the kinetics and the mechanism of chemical transformations of polymers the primary task is to find the parameters characterizing quantitatively the reactivity of functional groups in macromolecules. The parameters found can then be used for various purposes and, in particular, to compare the reactivity of polymers with that of the coresponding low-molecular compounds. This comparison is undoubtedly of fundamental importance because it stimulates the search for concrete physical models based on which one could interpret quantitatively the relationship between the long-chain nature of polymers (i.e. the principal feature that distinguishes a polymer from its low-molecular analog) and their reactivity. In many works devoted to the analysis of the peculiarities of macromolecular reactions, it is the comparison between the kinetics of chemical transformations of polymers and of their analogs on which the main attention is focused. However, the complex course of macromolecular reactions is not infrequently regarded as an insurmountable obstacle, which makes it impossible to reveal the relatively simple regularities of reactivity variations occurring with an increase in a conversion degree (as can be done in the case of low-molecular compounds).

Harwood in the previously cited work [2] claims that a quantitative description of the kinetics of polymer-analogous transformations with the neighbor effect is impossible if one does not take into account the polymer–solvent interaction and its variation, as well as the change taking place in the conformation of a polymer with conversion. This assertion appears to have resulted from the confusion of two closely interrelated, but actually different, problems of a quantitative investigation of the neighbor effect: the formal kinetic description of a process and the ascertainment of the very mechanism of the influence exerted by neighboring units upon the reactivity of functional groups in macromolecules.

The equations presented in Chapter 3 describing the kinetics of a reaction with the neighbor effect do not contain any assumption on the mechanism of this effect. When they were derived it was only assumed that, in accordance with the model of a process, the reactivity of functional groups depends solely on the nature of the nearest neighbors (whether they reacted or not). These are essentially conventional equations of formal kinetics, and the possibility of using them to describe adequately the kinetics of a concrete reaction is determined by the nature of changes undergone by the macromolecule in the conditions of experiment.

The transformation of a functional group can be accompanied by substantial changes in the chemical and physical properties, primarily in the portion of polymer chain with which this group is directly connected. In this portion, among other things, the conformational characteristics of the chain and the polymer–solvent interaction are changed. If ionogenic groups are formed (e.g. by the hydrolysis of polyacrylates or polymethacrylates), this gives rise to electrostatic interactions of these groups with the low-molecular components of the reaction mixture. The whole point concerns how far an effect of this kind of change extends along the macromolecular chain.

If, as a result of the transformation of each functional group, only the reactivity of its nearest neighbors changes, it follows that under experimental conditions the reaction model of the neighbor effect is actually realized and its kinetics may be described with corresponding equations given in Chapter 3. The ascertainment of such a possibility is the primary task of kinetic study and makes it possible to assume that the variation of the polymer–solvent interaction in the course of the reaction and the conformational and electrostatic effects manifest themselves in this case exclusively as a component part of the neighbor effect. On this basis one can now begin to solve the next problem: interpretation of the reaction mechanism which includes, in particular, an estimation of the contribution of each of the above-mentioned factors to the resulting change in reactivity of the functional groups of a polymer during its chemical transformation.

When, however, the transformation of a functional group affects the reactivity of not only the nearest but also remote parts along the chain units, the reaction model taken is not realized. To describe quantitatively the kinetics of such a process it is indeed necessary to develop a theory that would take into account, along with the neighbor effect (and when the polymer reactivity changes with conversion this factor must be taken into consideration), the 'long-distance' role of conformational and other effects.

It can also happen that during the chemical transformation of a polymer the long-distance action will influence the reaction kinetics only within a narrow interval of conversion, beyond which the 'pure' effect of neighboring units is realized. This situation was observed in experiments when investigating the kinetics of isotactic poly(diphenylmethyl methacrylate) hydrolysis [25] and poly(4-vinyl pyridine) quaternization [26] (see Sections 4.4 and 4.6). In the region where the 'pure' neighbor effect manifests itself, one can obtain quantitative information on polymer reactivity using the theory based on a model of a macromolecular reaction, which only takes into account the effect of the nearest neighbors. The analysis of discrepancies between the predictions of the theory and experiment, observed outside this region, may permit a preliminary assumption to be made on the contribution of other effects to reaction kinetics. It is the result of such an analysis that should stimulate both the staging of experiments proving directly the actual existence of the effects assumed to be taking place (e.g. the conformational changes or the cooperative nature of electrostatic interaction between the polyelectrolyte macromolecules and the components of a reaction mixture) and the development of a theory describing the influence of these factors upon polymer reactivity.

It would therefore not be correct to state *a priori* that the kinetics of any macromolecular reaction could never be described with a model of the neighbor effect. On the contrary, we definitely have grounds to believe that in appropriate experimental conditions this model can be realized. Needless to say, in each concrete case one has to prove the fact that the model of a neighbor effect is actually realized. There exists a sufficiently clear-cut criterion for this; viz. if the regularities of a macromolecular reaction are determined in given conditions by the neighbor effect only, it must then be possible to describe (using equations presented in Chapter 3) its kinetics, as well as the units distribution and the composition heterogeneity of the products obtained, using one and the same set of individual rate constants k_0, k_1, k_2. When choosing the conditions of a kinetic experiment one must, naturally, take into account the requirements of the process model.

We have already noted that in order to avoid the effects of interchain interactions it is desirable to conduct the reaction in dilute solutions. Usually 1–2% solutions of a polymer ($\sim 10^{-1}$ base mol/l) are taken, but it also possible to work with much lower concentrations if one can apply a sufficiently sensitive method of determining the degree of conversion of a polymeric reagent [27, 28].

Special attention should be given to the choice of solvent. If both the initial polymer and the product of its transformation are readily dissolved in the chosen solvent, it can then be anticipated that not only the homogeneity of the system will be preserved but also any significant changes in the size of macromolecular coil will be prevented during the reaction; i.e. the part of conformational effects that could not be described within the framework of the chosen model of the process will be suppressed.

Some experimental methods of studying units distributions and composition heterogeneity of the reaction products are described in Chapter 8.

As far as the kinetic measurements proper are concerned, as a rule they are quite conventional. The degree of conversion is found by determining, with the help of appropriate analytical methods, either the polymer composition [17–19, 29] or the decay of the low-molecular reagent [30], or else the accumulation of low-molecular reaction product [27–29]. The methods to be preferred are those not involving the time-consuming isolation of polymer.

Tutorsky *et al.* [30] studied the kinetics of polydiene epoxydation by Prilezhaev's reaction, determining the unconsumed peracid in aliquots. Morawetz and Zimmering [27] studied the hydrolysis of copolymers of acrylic or methacrylic acids with a small (~ 1 mol%) amount of *p*-nitrophenyl methacrylate, determining the formed *p*-nitrophenol with the help of UV spectrometry. A similar technique (phenol determination) was applied by Yun [31] when she investigated the hydrolysis of poly(phenyl methacrylate). A high sensitivity of the analytical method made it possible to study the reaction kinetics in very dilute polymer solutions ($\sim 10^{-4}$–10^{-3} base mol/l).

Usmanov [32] investigated the chlorination of partially chlorinated polyethylene samples in CCl_4 (CH_2 groups concentration $\sim 10^{-2}$ mol/l) by determining photocolorimetrically the concentration of dissolved chlorine. In this case a continuous recording of the kinetic curve was obtained.

4.2   EVALUATION OF INDIVIDUAL RATE CONSTANTS FROM UNITS DISTRIBUTION DATA

To determine the individual rate constants it is in principle possible to use data both on the kinetics and on the distribution of units or the composition heterogeneity. It should be noted, however, that the dispersion of errors in experimental methods of studying the composition heterogeneity is comparable to the dispersion of the composition distribution proper in the products of macromolecular reactions, especially for the retarding effect of the neighbor when heterogeneity is relatively low (see Chapter 3). That is why the data on composition heterogeneity are most often only used for qualitative or semiquantitative comparison with the results of investigating the kinetics and the units distribution (see Section 4.5). Here we consider estimations of the rate constants based on units distribution data; the kinetic methods will be discussed in the next section.

To determine the rate constants from the data on the units distribution it is necessary to have information on the content of various triads in the polymer depending on the degree of conversion. Since the units distribution in the products of a macromolecular reaction is caused not by the absolute values of the individual rate constants but by their ratios, the data on the triads content are most often used to find the relative constants k'_0, k'_1 and k'_2, where $k'_0 = k_0/(k_0 + k_1 + k_2)$, etc. Note that

$$k'_0 + k'_1 + k'_2 = 1 \qquad (4.1)$$

Barth and Klesper [33] developed a method to estimate the relative constants using differential equations describing the variations in the fractions of different diads with conversions:

$$\frac{dP(AA)}{dP(A)} = \frac{2k_0'P(AAA) + k_1'P(AAB^+)}{N} \tag{4.2}$$

$$\frac{dP(AB^+)}{dP(A)} = -\frac{2k_0'P(AAA) - 2k_2'P(BAB)}{N} \tag{4.3}$$

$$\frac{dP(BB)}{dP(A)} = -\frac{k_1'P(AAB^+) + 2k_2'P(BAB)}{N} \tag{4.4}$$

Here

$$N = -k_0'P(AAA) - k_1'P(AAB^+) - k_2'P(BAB)$$

$$P(AB^+) = P(AB) + P(BA)$$

$$P(AAB^+) = P(AAB) + P(BAA)$$

Since only one of the equations (4.2) to (4.4) is independent, any one of them can be used to find the relative constants. If, using Eq. (4.1), we substitute k_2' by $1 - k_0' - k_1'$ and denote

$$S = \frac{dP(AA)}{dP(A)}$$

$$c_1 = \frac{SP(BAB)}{SP(AAA) - SP(BAB) - 2P(AAA)}$$

$$c_2 = \frac{SP(BAB) - SP(AAB^+) + P(AAB^+)}{SP(AAA) - SP(BAB) - 2P(AAA)}$$

it will then be easy to obtain, from Eq. (4.2),

$$k_0' = c_1 + c_2 k_1' \tag{4.5}$$

At a fixed $P(A)$ value c_1 and c_2 are constants and, according to Eq. (4.5), the dependence between k_0' and k_1' is linear.

Triad fractions are found from experiment, most often from NMR data. Diad fractions are easily calculated using well-known statistical relations (see Chapter 3), e.g.

$$P(AA) = P(AAA) + P(AAB^+)/2$$

By graphic differentiation of the dependence of $P(AA)$ on $P(A)$ the values of S and, therefore, c_1 and c_2 for a fixed $P(A)$ value are found. A straight line is then plotted in $k_0' - k_1'$ coordinates according to Eq. (4.5). It is the intersection point of these straight lines plotted for different $P(A)$ values that gives the required values of k'_0 and k'_1 and, therefore, of k'_2 also. By a similar method the absolute values of k_0, k_1 and k_2 constants can be found [34].

Since this interesting method includes graphic differentiation it is desirable that a sufficiently large number of experimental points should be used to plot the $P(AA)-P(A)$ curve. It is also to be noted that if the curves (4.5) do not intersect at one point (and this is a rather typical situation [33, 34]), it is necessary to estimate the adequacy of alternative sets of constants k'_0, k'_1 and k'_2.

Such kinds of estimation can be performed by comparing the calculated and the experimental values of the triad fractions in a polymer. In general, comparison of the calculated and the experimental values of units distribution parameters forms the basis of any methods used to find the most probable values of rate constants.

A simple and obvious method of finding the relative rate constants was proposed by Harwood *et al.* [35]. Any set of k'_0, k'_1 and k'_2 constants has an unambiguously corresponding point in an equilateral triangle (see Fig. 4.1).

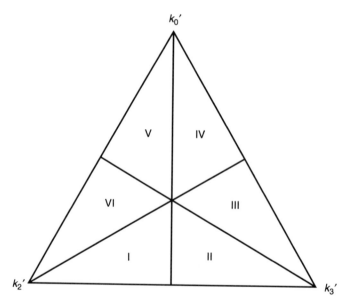

Fig. 4.1 Application of triangular coordinates to depict the set of individual relative rate constants of a macromolecular reaction. Six areas of the triangle characterize the neighbor effect: I, $k'_0 < k'_1 < k'_2$; II, $k'_1 > k'_2 > k'_0$; III, $k'_1 > k'_0 > k'_2$; IV, $k'_0 > k'_1 > k'_2$; V, $k'_0 > k'_2 > k'_1$; VI, $k'_2 > k'_0 > k'_1$. (Reproduced by permission of John Wiley & Sons Ltd from *J. Polymer Sci. Lett.*, 1978, **16**, 109)

Altitudes dissect the triangle into six areas with a characteristic ratio of constants, as shown in the figure, and their intersection point corresponds to the absence of the neighbor effect, $k'_0 = k'_1 = k'_2$. For the chosen sets of constants the triad distribution in a polymer is calculated for different degrees of conversion using any of the methods described in Chapter 3 (in particular, Harwood *et al.* [35] recommend the Monte Carlo simulation improved by them [36]). The sum of the absolute deviations of the calculated values from the experimental ones ('error sums') are written out in the triangle at the points corresponding to the tested sets of constants. This procedure is repeated several times, with the values of constants at every subsequent stage being varied within relatively narrow limits in the vicinity of the sets of k'_0, k'_1 and k'_2 values to which minimum error sums corresponded at the preceding stage. The set of constants to which the absolute minimum of error sums corresponds is taken as the most probable (it may be better to minimize not the sum of absolute deviations but that of quadratic deviations of the calculated data from the experiment).

Harwood *et al.* [35] also proposed a variant of their method which includes graphical plotting. First, k'_0, k'_1 and k'_2 are varied at a fixed k'_2 value, and through the points corresponding to the minimum error sums, at different fixed k'_2 values, a straight line is drawn. Then a second straight line is plotted in a similar way, with k'_0 and k'_2 being varied at different fixed k'_1 values. The intersection point of the straight lines corresponds to the most probable set of k'_0, k'_1 and k'_2 (see Fig. 4.2). Unfortunately, the authors [35] do not present any mathematical proof that the points corresponding to the minimum error sums must actually lie on one straight line.

Bauer [37] used Keller's equations (see Chapter 3) to find the ratio of rate constants from the data on triad distribution. These equations can be written out in the following way:

$$\frac{dP(AAA)}{dt} = -k_0 P(AAA) - \frac{[2k_0 P(AAA) + k_1 P(AAB^+)]P(AAA)}{P(AAA) + P(AAB^+)/2} \quad (4.6)$$

$$\frac{dP(AAB^+)}{dt} = -k_1 P(AAB^+)$$

$$+ \frac{[2k_0 P(AAA) + k_1 P(AAB^+)][P(AAA) - P(AAB^+)/2]}{P(AAA) + P(AAB^+)/2} \quad (4.7)$$

$$\frac{dP(BAB)}{dt} = -k_2 P(BAB) + \frac{[2k_0 P(AAA) + k_1 P(AAB^+)]P(AAB^+)/2}{P(AAA) + P(AAB^+)/2} \quad (4.8)$$

Assuming $k_0 = 1$ and varying k_1 Bauer simultaneously solves Eqs. (4.6) and (4.7) by the numerical predictor–corrector method for different k_1 values until the calculated values of $P(AAA)$ and $P(AAB^+)$ are sufficiently close to those found in experiment. Using these values of k_1, k_2 is then varied until, by solving Eq. (4.8)

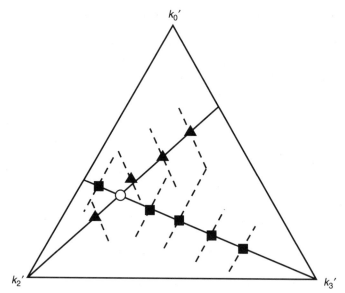

Fig. 4.2 Illustration of the graphical method of determining k'_0, k'_1 and k'_2. The squares (■) and triangles (▲) represent the minimum of error sums on the dashed straight lines corresponding to constant values of k'_1 and k'_2 respectively. The intersection point of solid lines (○) corresponds to the most probable set of k'_0, k'_1 and k'_2. (Reproduced by permission of John Wiley & Sons Ltd from *J. Polymer Sci. Lett.*, 1978, **16**, 109)

using the same method, a $P(BAB)$ value close to that found in experiment is obtained. Standard deviations in calculating k_1 and k_2 by this method are found from the formula

$$\sigma_k^2 = \sigma_{AAA}^2 \left[\frac{\delta k}{\delta P(AAA)} \right]^2 + \sigma_{(AAB^+)}^2 \left[\frac{\delta k}{\delta P(AAB^+)} \right]^2 + \sigma_{BAB}^2 \left[\frac{\delta k}{\delta P(BAB)} \right]^2 \quad (4.9)$$

where σ_{AAA}, $\sigma_{(AAB^+)}$ and σ_{BAB} are standard deviations in the experimental determination of triad distribution. The partial derivatives are calculated by the numerical method, varying (within narrow limits) the triad fractions.

Bauer's method makes it possible to estimate the ratios of constants from the triad distribution of a single sample. When making calculations for a series of samples corresponding to different degrees of conversion the ratios of constants found in this way will, naturally, vary as a result of the errors in experimental determination of the triad distribution. However, in reference [37] the question of the criterion for choosing the most probable ratio of constants is not considered.

Merle [38] suggested that the relative rate constants $k = k_1/k_0$ and $k' = k_2/k_0$ could be calculated using the improved graphical method allowing a linear regression. The following equation is derived in reference [38]:

$$\frac{(1 - S)P(AAA)}{(1 + S)P(BAB)} = \left[\frac{SP(AAB^+)}{(1 + S)P(BAB)} \right] k + k' \tag{4.10}$$

where

$$S = \frac{dG}{dP(A)}$$

$$2G = 1 - P(AB^+) = 1 - P(AAB^+) - 2P(BAB) = 1 - P(ABB^+) - 2P(ABA) \tag{4.11}$$

According to Eq. (4.10) k and k' are determined graphically as the slope and intersection with the Y axis respectively. G is calculated from Eqs. (4.11) using one of the possible relations depending on the character of NMR spectra: whether A-centered or B-centered triads are better resolved.

The proper method for determining the individual rate constants has been developed based on the approach used by Stroganov, Taran and Platé [39] for interpretation of poorly resolved NMR spectra. The method [40] consists in searching for a set of k_0, k_1 and k_2 constants (which we shall denote as $\mathbf{k}$) and minimizes the sum of square deviations (SSD) of the values of triad fractions calculated for different values of time, $T_m(t)$, from the experimental values, $T_e(t)$. By means of the least-squares method it is possible to obtain unambiguous estimations of rate constants and their dispersions.

The efficiency of the method is tested in a simulation to estimate the influence of experimental errors on the magnitude of confidence intervals of the estimation of constants. This simulation is performed on a computer in accordance with Fig. 4.3.

For a prescribed set of $\mathbf{k}^*$ the dependence of triad distribution in reaction products on time $T_m(t|\mathbf{k}^*)$ is calculated from exact equations (see Chapter 3). The calculations are performed by numerical methods with a high degree of accuracy. Normally distributed random numbers with a prescribed dispersion S^2 and a zero mathematical expectation are then added to the $T_m(t|\mathbf{k}^*)$ values (the normal distribution of random errors is characteristic of many experimental methods, that is why it is used in simulations, but one can certainly introduce random numbers with a different type of distribution as well). The $T_m(t|\mathbf{k}^*)$ functions modified in this way serve as an analog of a given experimental dependence, $T_e(t)$, and are inserted in block 3 of the search for the minimum SSD. An arbitrary initial set of constants k^0 is also inserted in this block, and the iteration process of searching for $\mathbf{k}$ which minimizes SSD begins. At every stage of the iteration process $T_m(t|\mathbf{k}^*)$ functions are calculated in block 3 and SSD in block 2. After the most probable set $\mathbf{k}$ has been found, in block 4 the statistical processing is performed by calculating the dispersion of the constants estimations.

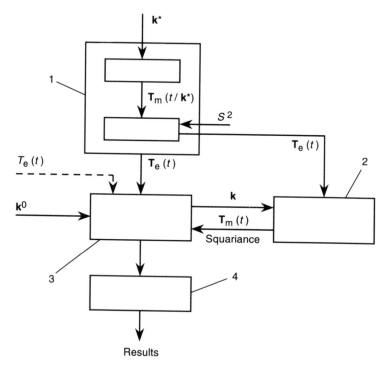

Fig. 4.3 Diagram of the algorithm to estimate the rate constants: 1, block of modeling the dependence of triad distribution for a prescribed set of k^* constants; 2, block of calculating SSD for the current sets of k constants; 3, block of SSD minimization; 4, block of statistical processing. (Reproduced by permission of Nauka, Moscow, Russia, from *Vysokomol. Soed. A*, 1983, **25**, 882)

The same program is used to calculate k from experimental data. In this case the experimental values of $T_e(t)$ are directly inserted in block 3 (dashed line on Fig. 4.3). The results of a typical simulation are shown in Table 4.1. The prescribed sets k^* correspond here to the decelerating and accelerating effects of the neighbor. The set of experimental $T_e(t)$ values was simulated by calculating the fractions of all six triads for six polymeric samples with a different degree of conversion in the range of $1 \geq P(A) \geq 0.4$, with the errors of experimental determination of triad fractions being taken into account. The values of relative errors, taken equal to 2.5 and 7.5%, are characteristic for conventional methods of processing of the experimental determination of triad distribution.

It can be seen from the table that standard deviations for the constants increase with an increase in the experimental error. The k_2 is determined in the simulation with the lowest accuracy. The results of simulations indicate the necessity of using a sufficiently large set of samples of different degrees of conversion to determine

Table 4.1. Model calculation of individual rate constants from the data on triad distribution in polymeric samples

$k_0^*:k_1^*:k_2^*$	Error of 'experiment' (%)	Prescribed k_i^* and calculated $\hat{k}_i$ rate constants and standard deviations σ_{k_i}								
		k_0^*	$\hat{k}_0$	σ_{k_0}	k_1^*	$\hat{k}_1$	σ_{k_1}	k_2^*	$\hat{k}_2$	σ_{k_2}
		$10^3\,\min^{-1}$			$10^4\,\min^{-1}$			$10^5\,\min^{-1}$		
1:0.1:0.01	2.5	1.00	1.01	0.06	1.00	0.94	0.16	1.00	1.18	0.26
1:0.1:0.01	7.5	1.00	1.07	0.20	1.00	1.05	0.58	1.00	0.81	0.78
		$10^5\,\min^{-5}$			$10^4\,\min^{-1}$			$10^3\,\min^{-1}$		
1:10:100	2.5	1.00	1.04	0.16	1.00	0.99	0.18	1.00	0.63	0.95
1:10:100	7.5	1.00	0.89	0.37	1.00	1.13	0.53	1.00	0.40	1.12

the constants, as well as the necessity to improve the accuracy of the experimental determination of triad distribution. The latter can be attained both by raising the instrument resolution and by improving the methods of interpreting the primary experimental data.

The estimation of individual rate constants from the units distribution data appears to be very promising, especially in connection with the rapid progress of NMR spectrometry. However, until it becomes possible to determine reliably the triad distribution in the product of any macromolecular reaction, the problem of elaborating kinetic methods for the rate constant determination preserves its urgency (in addition, not only the rate constants ratios but also their absolute values are needed). This task, closely associated with the problem of choosing the model compounds, is discussed in the next section.

4.3 POSSIBILITIES OF USING LOW-MOLECULAR AND POLYMERIC MODELS TO DETERMINE REACTION RATE CONSTANTS

Many of the effects peculiar to reactions of polymers can also manifest themselves in the reactions of low-molecular compounds, and the data obtained by investigating the latter are widely used to interpret the mechanism of the macromolecular reactions.

Thus, Bender and co-workers [41–43] observed the anchimeric assistance of neighboring carboxylic groups when they investigated the hydrolysis of phtalic acid esters and amides. The intramolecular interaction of functional groups separated by a flexible chain was studied by Ruzicka *et al.* [44] on thermal cyclization of dicarboxylic acids as early as 1926. Later Sisido [45] showed that the data on the lactonization of ω-hydroxyacids [46] can be quantitatively interpreted using the Morawetz and Goodman method [13] developed to describe the intramolecular interactions of remote (along the chain) functional groups of polymer. The effect of electrostatic interaction on the ionization equilibrium of dicarboxylic acids [47] is a generally known fact. Intermolecular association is regarded as one of the causes resulting in the variation of the reactivity of methacrylic acid in the course of copolymerization [48].

In the light of these facts it is natural to assume that, to determine the rate constants of macromolecular reactions, one could use low-molecular compounds modeling a corresponding part of the macromolecular chain. Of course, when one investigates the neighbor effect, monofunctional analogs are known in advance to be inadequate, so only polyfunctional models can be used.

Thus, to interpret the accelerating effect of reacted neighboring units in the hydrolysis of polyacrylates and polymethacrylates, the data of the hydrolysis of monoesters of dicarboxylic acids have been resorted to repeatedly. The accelerating effect observed on the models proved, however, to be largely dependent on the monoester structure; e.g. the ratio of the rate constants for the hydrolysis of monophenyl esters of succinic and glutaric acids is 160:1 [49], whereas for the

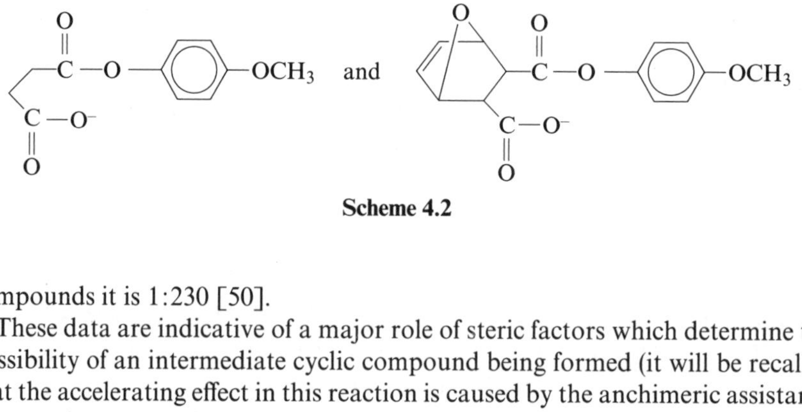

Scheme 4.2

compounds it is 1:230 [50].

These data are indicative of a major role of steric factors which determine the possibility of an intermediate cyclic compound being formed (it will be recalled that the accelerating effect in this reaction is caused by the anchimeric assistance of reacted groups). It is therefore not surprising that in a comparative study of the kinetics of the reactions of bifunctional models and those of the corresponding polymers (in whose long chain there can be substantially differing steric tensions, i.e. different from those in the low-molecular analog potentials of rotation about the bonds and combinations of possible conformations) the observed effects of neighboring groups do not, as a rule, coincide in magnitude. Thus, Sakurada *et al.* [51] observed a smaller accelerating effect of OH groups in the hydrolysis of ethyleneglycoldiacetate as compared with poly(vinyl acetate) hydrolysis. In the hydrolysis of glutaric acid mono-*p*-nitroanilide the carboxylic group does not affect the reactivity of the nitroanilide group, whereas in the hydrolysis of copolymers of acrylic acid and its *p*-nitroanilide the accelerating effect is actually observed [52].

Therefore, it is not only the sequence of groups but also their mutual spatial arrangement that has to be modeled. A number of compounds have been synthesized that modeled diads and triads in macromolecules of differing stereochemical configuration for poly(vinyl chloride) [53, 54], poly(vinyl alcohol) [55], poly(vinyl acetate) [56] and poly(methyl methacrylate) [57]. Models of this type have been applied to investigate the effect of neighboring units.

Shibatani and Fujii [58] compared the rates of hydrolysis for the formals of pentane-2,4-diols and heptane-2,4,6-triols:

$$
\begin{array}{ccc}
CH_3\!\diagdown{}\diagup CH_2\diagdown{}\diagup CH_3 & & CH_3\!\diagdown{}\diagup CH_2\diagdown{}\diagup CH_2\diagdown{}\diagup CH_3 \\
CH\quad CH & \text{and} & CH\quad CH\quad CH \\
| \qquad | & & | \qquad | \qquad | \\
O\diagdown{}\diagup O & & O\diagdown{}\diagup O\quad O \\
CH_2 & & CH_2\quad H
\end{array}
$$

Scheme 4.3

It turned out that the OH group in the formals of triols does not accelerate the reaction, whereas in the hydrolysis of poly(vinyl alcohol) formals acceleration was observed. The authors [58] believe the influence of the OH group in the latter case to be insignificant and explain the acceleration by a certain 'polymeric effect'. The idea is not excluded, however, that the acceleration is nevertheless actualized by the OH group. Simply, the conformation and the steric tension of the chain in the polymer are favorable for this effect to manifest itself, while in the model they are not.

Skorokhodov and co-workers [59, 60] synthesized tricarbonates of mannitol, dulcitol and sorbitol, modeling iso-, syndio- and hetero-triads of poly(vinylene carbonate) respectively. The kinetic curve, calculated at the constants ratios $k_0 : k_1 : k_2 = 1 : 0.2 : 0.02$, adequately described the experimental data on the aminolysis of sorbitol tricarbonate (for the other trimers, the constants ratios obtained were close to those found for sorbitol). The kinetics of poly(vinylene carbonate) aminolysis, however, was not described by these constants ratios. In analyzing the causes of this discrepancy Skorokhodov [59] considers it necessary, in particular, to take into account the steric tensions in the polymer chain.

It can be seen that the analogs reproducing the stereochemical configuration of triads can neither be regarded in the general case as suitable models to determine the rate constants of macromolecular reaction, since the contribution of steric factors to the reactivity of the polymer and the model can be markedly different due to the dissimilar steric tension in the chain of the macromolecule and its analog. It is probably possible to control the steric tension of model compounds by imparting, for example, a cyclic structure to them or by inserting bulky substituents. However, the principles of selecting adequate models have not been studied yet, and until this problem has been solved one cannot estimate the suitability of a model *a priori*. This fact, as well as the difficulties of synthesizing model compounds, sharply restrict at present the possibilities of determining the individual rate constants with the help of low-molecular analogs.

This conclusion appears to be quite justified, since the problem of modeling the chemical properties of a polymer is closely associated with the peculiarities of macromolecular reactions. It is precisely the experimentally established differences in the reactivity of polymers and their analogs that make it difficult (though not hopeless) to select adequate low-molecular models. It is therefore expedient to develop the methods of determining the rate constants of macromolecular reactions, based on investigating the kinetics of chemical transformations of the polymeric reagents proper.

At first sight the problem seems to be extremely simple. Kinetics of the reaction, within the framework of the process model in question, is described by (see Chapter 3)

$$P(A) = e^{-k_2 t} \left(2(k_2 - k_1) e^{-2(k_0 - k_1)/k_0} \int e^{(k_2 - 2k_1)t} \exp\left\{ [2(k_0 - k_1)/k_0] e^{-k_0 t} \right\} dt \right.$$

$$+ (2k_1 - k_0 - k_2) e^{-2(k_0 - k_1)/k_0} \int e^{-(2k_1 + k_0 - k_2)t}$$

$$\left. \cdot \exp\left\{ [2(k_0 - k_1)/k_0] e^{-k_0 t} \right\} dt + C \right) \tag{4.12}$$

where $P(A)$ is the fraction of unreacted units and C is the integration constant. The rate constants k_0, k_1 and k_2 are the parameters of this equation. One would think it is sufficient simply to select the values of k_0, k_1 and k_2 at which the curve calculated from Eq. (4.12) describes the kinetics of homopolymer A transformation to a high degree of conversion for the problem of determining the rate constants to be solved. However, due to a complicated form of the three-parameter $P(A)$ function, several sets of k_0, k_1 and k_2 are practically always found to describe, within the experimental errors, the kinetics of homopolymer A transformation. The problem has been studied by Platé *et al.* [61, 62]; their final results are shown in Fig. 4.4.

It is seen that even for very high accuracy of experimental kinetic measurements, 1–2%, there are quite narrow regions of the relative rate constants ratios for which the only set of individual constants may be unambiguously found. These regions belong to the retarding effect or nonmonotonous ratios, but not to the accelerating neighbor effect. Actually this is a problem of finding a global minimum among several local minima in a classical solution of so-called reverse tasks.

In this respect the use of polymeric models, i.e. polymer samples with a known distribution of unreacted A and reacted B units in a macromolecule chain, appears to be very promising for determining the rate constants. For such

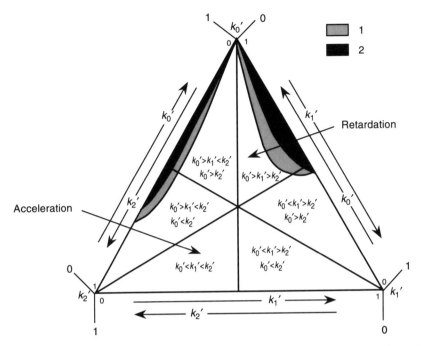

Fig. 4.4 Ambiguity of the individual rate constants evaluation. Shadings show the unambiquity regions of the constants estimation for experimental accuracy equal to 1% (1) and 2% (2) respectively. (Reproduced by permission of Nauka, Moscow, Russia, from *Vysokomol. Soed. A*, 1983, **25**, 2243)

samples the values of N_0, N_1 and N_2 are known, i.e. the fractions of A units having 0, 1 and 2 B neighbors and reacting with rate constants k_0, k_1 and k_2 respectively. The initial reaction rate for this type of polymeric model can be described by

$$\left[-\frac{dP(A)}{dt} \right]_0 = k_0 N_0 + k_1 N_1 + k_2 N_2 \qquad (4.13)$$

Having found the initial rates for three (or more) polymeric models of different compositions one can solve a set of equations of the type of Eq. (4.13) and find the rate constants. The kinetic curves calculated from these values with the help of Eq. (4.12) can be compared with experimental kinetic data on the above-mentioned models transformation to a high degree of conversion (see the next sections). The proposed method places rather rigid requirements on the set of constants found: a description of the initial rate and the entire kinetic curve of reaction not for one but for a series of polymeric models of different composition. What are the methods that can be used to prepare polymeric models?

Certainly, if the triad distribution of investigated polymer can be determined experimentally, the AB copolymer obtained by any method will serve as a suitable polymer model. (For such an object, k_0, k_1 and k_2 can be determined from the data on the units distribution, and it would seem needless to investigate the reaction kinetics in this case. However, as emphasized in the preceding section, in investigating the polymer reactivity one has to compare the units distribution or the composition heterogeneity with kinetic data. It is therefore extremely important to be convinced that with the help of the set of constants found from the data of the units distribution it is possible to describe the reaction kinetics for a series of polymeric models of different composition, and vice versa.) At present, however, for most of the polymers it remains necessary to elaborate such methods of obtaining polymeric models that would make it possible to find the values N_0, N_1 and N_2 by calculation.

For this purpose one can use, for example, the copolymerization of A and B monomers. If the reactivity ratios r_A and r_B are known, the values needed for the copolymer of $P(A)$ composition obtained at the monomer ratio $A/B = x$ are calculated from [63]

$$N_0 = \frac{(r_A x)^2 P(A)}{(1 + r_A x)^2}$$

$$N_1 = \frac{2 r_A x P(A)}{(1 + r_A x)^2} \tag{4.14}$$

$$N_2 = \frac{P(A)}{(1 + r_A x)^2}$$

It is to be noted, however, that, as a rule, stereoregular copolymers cannot be obtained by copolymerization; therefore, such models can only be applied, within the framework of the problem in question, to those reactions where the neighbor effect does not depend on the stereochemical configuration of the chain.

Smets and De Loecker [64] used radical poly(methyl methacrylate-co- methacrylic acid) to study the kinetics of poly(methyl methacrylate) hydrolysis. However, they were unable to determine the set of k_0, k_1 and k_2 constants, which is due, along with other causes, to the above-mentioned restrictions on the applicability of copolymerization products as polymeric models (it will be recalled that stereochemical effects make a substantial contribution to the kinetics of this reaction; see Chapter 2). To exclude stereochemical effects it is necessary to use stereoregular polymeric models that can be obtained from stereoregular polymers as a result of polymer-analogous transformations proceeding without any changes in the stereochemical configuration of the chain.

Klesper *et al.* [65] studied the interaction of syndiotactic poly(methacrylic acid) with diazomethane. Using the NMR method they established that the copolymers formed had a random (Bernoullian) distribution of units. Similar results have been obtained [66] in investigating the esterification of syndiotactic poly(methacrylic acid) with methanol, ethanol or 2,2,2-trifluoroethanol in concentrated sulfuric acid. For the products of such reactions with composition $P(A)$ (A is an ester unit) the values of N_0, N_1 and N_2 can easily be calculated from obvious relationships:

$$N_0 = [P(A)]^3$$
$$N_1 = 2[P(A)]^2[1 - P(A)]$$
$$N_2 = P(A)[1 - P(A)]^2 \qquad (4.15)$$

Such samples can be used as polymeric models when investigating the effect of neighboring units in the hydrolysis of the corresponding esters of syndiotactic poly(methacrylic acid).

In this case stereoregular polymeric models suitable for investigating the hydrolysis could be synthesized in an indirect way from the reactions of diazomethylation and esterification proceeding without the neighbor effect (i.e. $k_0 = k_1 = k_2$), which is indicated by the nature of the units distribution in the chain of copolymers formed. For most polymers this very favorable situation can not always be realized. There is reason to believe, however, that stereoregular polymeric models with a random distribution of units can be obtained with the help of the very reaction that is being studied. At first sight the formulation of the problem itself appears to be paradoxical: it is suggested that a reaction proceeding with the neighbor effect $(k_0 \neq k_1 \neq k_2)$ should be used to synthesize models that are only formed in the absence of the neighbor effect $(k_0 = k_1 = k_2)$. However, the point is that the character of the influence of neighboring units upon the reactivity of the functional groups depends substantially on the conditions of experiment.

Arcus and Hall [67], for example, found that the decelerating effect in the quaternization of poly-4-vinylpyridine with n-butylbromide, observed when the reaction is conducted in sulpholane, weakens when dimethylformamide is added to the system, and in pure dimethylformamide the reaction proceeds without any retardation.

The chlorination of polyethylene with molecular chlorine usually proceeds with retardation [29, 68, 69]. However, if a mixture of chlorine and SO_2 is used as a chlorinating agent, the reaction product, as it follows from NMR data [70], has a random distribution of CH_2 and $CHCl$ units. The random distribution of units was also established by the NMR method in the products of hydrolysis, under certain conditions, of syndiotactic and isotactic poly(metyl methacrylate) [71, 72].

These are the facts, based on which it can be hoped that it should be possible to carry out various polymer-analogous transformations in conditions when the neighbor effect for some reason or an other does not manifest itself, and the products formed can be used as polymeric models with a random distribution of units. Using this method of obtaining the polymeric model one has to be convinced that in the conditions of experiment the neighbor effect really does not exist (it is to be recalled that we are speaking of polymers of which the units distribution has, as yet, not been determined experimentally). For this purpose one can recommend a simple kinetic criterion. When $k_0 = k_1 = k_2$, the reaction rate obeys the conventional kinetic relations. In particular, for a reaction of the first order with respect to the polymer and with an excess of the low-molecular reagent the reaction kinetics is described by a linear semilogarithmic anamorphosis (for details see the next section).

Let us now consider in greater detail the procedure of calculating the units distribution in polymeric models. Equations (4.14) and (4.15) are recommended to calculate the values of N_0, N_1 and N_2 in macromolecules of fixed composition $P(A)$. However, both the samples obtained by copolymerization and by polymer-analogous transformation are heterogeneous in composition. It is, therefore, necessary to estimate to what extent the composition heterogeneity of polymeric models should be taken into account when one calculates their units distribution.

Let us first consider polymeric models with a random units distribution. As shown in Chapter 3, the composition heterogeneity of the products of polymer-analogous transformations is described with a good approximation by the normal distribution function

$$\phi(y) = \frac{1}{(2\pi)^{1/2}\sigma} \exp\left[-\frac{(y-\mu)^2}{2\sigma^2} \right] \tag{4.16}$$

where y is the number of B units in the macromolecule with a degree of polymerization n, μ is the mathematical expectation (mean value) of y, σ is the standard deviation and σ^2 is the distribution dispersion. Calculate now the mean value of N_0 or, which is the same, the $P(AAA)$ probability of the AAA triad for a sample with such a composition heterogeneity. According to Eqs. (4.15),

$$P(AAA) = [P(A)]^3 = \left[\frac{(n-y)}{n} \right]^3$$

The mean $P(AAA)$ value

$$\overline{P(AAA)} = \int_{-\infty}^{+\infty} \left[\frac{(n-y)}{n} \right]^3 \phi(y)\,dy$$

After integration and not complicated transformation we get

$$\overline{P(AAA)} = (1 - \bar{\beta})^3 + \left(\frac{3\sigma^2}{n^2} \right)(1 - \bar{\beta}) \tag{4.17}$$

where $\bar{\beta}$ is the mean value of the mole fraction of B units in the sample. For the value σ^2/n (see Chapter 3),

$$\lim_{n \to \infty} \frac{\sigma^2}{n} = \frac{(1 - P_{aa})P_{ba}(1 - P_{ba} + P_{aa})}{(1 - P_{aa} + P_{ba})^3} \tag{4.18}$$

where $P_{aa} = P(AA)/P(A)$ and $P_{ba} = [P(A) - P(AA)]/[1 - P(A)]$ are the conditional probabilities of finding unit A in the chain to the right of A and to the right of B respectively. In the case of $k_0 = k_1 = k_2$, $P_{aa} = P_{ba} = P(A)$ and

$$\frac{\sigma^2}{n} = P(A)[1 - P(A)] = \bar{\beta}(1 - \bar{\beta}) \tag{4.19}$$

Inserting Eq. (4.19) into Eq. (4.17) we finally get

$$\overline{P(AAA)} = (1 - \bar{\beta})^3 + \frac{3\bar{\beta}(1 - \bar{\beta})^2}{n} \tag{4.20}$$

At the same time, calculation of $\overline{P(AAA)}$ without taking into account the composition heterogeneity yields according to Eqs. (4.15) the following relation:

$$\overline{P(AAA)} = (1 - \bar{\beta})^3 \tag{4.21}$$

By comparing Eqs. (4/21) and (4.20) we find the correction for composition heterogeneity

$$\delta H = \frac{3\bar{\beta}(1 - \bar{\beta})^2}{n} \tag{4.22}$$

Table 4.2 gives the expressions for the probabilities of all six triads calculated from Eqs. (4.15) and also those with composition heterogeneity taken into account, X being the place that can be occupied by A or B. The value of the correction is in all cases inversely proportional to the degree of polymerization.

Table 4.2. Probabilities of triads in polymeric models with random distribution of units

Triad XXX	From Eqs. (4.15)	With correction for heterogeneity
AAA	$(1 - \bar{\beta})^3$	$(1 - \bar{\beta})^3 + 3\bar{\beta}(1 - \bar{\beta})^2/n$
AAB$^+$	$2\bar{\beta}(1 - \bar{\beta})^2$	$2\bar{\beta}(1 - \bar{\beta})^2 - 2\bar{\beta}(1 - \bar{\beta})(2 - 3\bar{\beta})/n$
BAB	$\bar{\beta}^2(1 - \bar{\beta})$	$\bar{\beta}^2(1 - \bar{\beta}) + \bar{\beta}(1 - \bar{\beta})(1 - 3\bar{\beta})/n$
BBB	$\bar{\beta}^3$	$\bar{\beta}^3 + 3\bar{\beta}^2(1 - \bar{\beta})/n$
BBA$^+$	$2\bar{\beta}^2(1 - \bar{\beta})$	$2\bar{\beta}^2(1 - \bar{\beta}) + 2\bar{\beta}(1 - \bar{\beta})(1 - 3\bar{\beta})/n$
ABA	$\bar{\beta}(1 - \bar{\beta})^2$	$\bar{\beta}(1 - \bar{\beta})^2 - \bar{\beta}(1 - \bar{\beta})(2 - 3\bar{\beta})/n$

Table 4.3. Probabilities of triads from Eqs. (4.15) and corrections for composition heterogeneity for models with degree of polymerization $n = 100$

Triad XXX	$\bar{\beta} = 0.2$		$\bar{\beta} = 0.5$		$\bar{\beta} = 0.8$	
	$P(XXX)$	$\delta H \times 10^2$	$P(XXX)$	$\delta H \times 10^2$	$P(XXX)$	$\delta H \times 10^2$
AAA	0.512	0.384	0.125	0.375	0.008	0.096
AAB$^+$	0.256	-0.450	0.250	-0.250	0.064	0.128
BAB	0.032	0.064	0.125	-0.125	0.128	-0.225
BBB	0.008	0.096	0.125	0.375	0.512	0.384
BBA$^+$	0.064	0.128	0.250	-0.250	0.256	-0.450
ABA	0.128	-0.225	0.125	-0.125	0.032	0.064

Table 4.3 gives the $P(XXX)$ values for models with a relatively low degree of polymerization, $n = 100$, at mean values of composition $\bar{\beta} = 0.2$, 0.5 and 0.8, as well as the correction for composition heterogeneity δH calculated from the data of Table 4.2. As seen from the table, the correction for heterogeneity is in most cases not more than 3% of the $P(XXX)$ value calculated from Eqs. (4.15) and is, therefore, within the margin of error of contemporary methods for experimental determination of units distribution in copolymers. Only for very small values of $P(XXX) = 0.008$ does the correction equal 12%, but $P(XXX)$ values of such small magnitude are determined experimentally with an even greater error. For samples with a higher molecular weight ($n = 10^3–10^4$) the δH values do not exceed fractions of a percent of the $P(XXX)$ values. The values of N_0, N_1 and N_2 in polymeric models with a random units distribution can thus be calculated from Eqs. (4.15) without heterogeneity corrections.

Examine now the polymeric models synthesized by copolymerization, assuming that the copolymerization can be described as the first-order Markovian process. What is the role of composition heterogeneity here?

If copolymerization is carried out at a strictly constant ratio of A and B comonomers—copolymerization in a continuous-flow stirred reactor [73, 74], azeotropic copolymerization, low conversions (at not excessively great differences in relative reactivities r_A and r_B)—the 'instantaneous' composition heterogeneity of the copolymers obtained will then be described with a good approximation by Eq. (4.16). The probabilities of triads in copolymers can be expressed as a function of composition $P(A)$ and the run number $R = 2P(AB)$ [63]; e.g.

$$P(BAB) = R^2/4P(A)$$

The value of σ^2/n can be calculated from Eq. (4.18) using the relationships

$$P_{aa} = \frac{P(A) - R/2}{P(A)}$$

$$P_{ba} = \frac{1 - P(A) - R/2}{1 - P(A)}$$

Parameter R is either determined experimentally [75] or calculated from [63]

$$R = \frac{2}{r_A x + 2 + r_B/x} \tag{4.23}$$

where $x = A/B$ is the comonomers ratio in reaction mixture. R and the triads probabilities can also be presented as functions of the fraction of B units in macromolecules β:

$$R = \frac{4\beta^2(1 - \beta)}{1 + [1 - 4(1 - r_A r_B)\beta(1 - \beta)]^{1/2}} \tag{4.24}$$

$$P(BAB) = \frac{4\beta^2(1 - \beta)}{\{1 + [1 - 4(1 - r_A r_B)\beta(1 - \beta)]^{1/2}\}^2} \tag{4.25}$$

etc.

However, if the copolymerization is carried out, as is most often the case, in a batch reactor to relatively high conversion, the composition heterogeneity of copolymers obtained is determined by the change of the comonomers ratio during reaction. 'Conversional' composition heterogeneity has been examined in references [76] to [83]. Meyer and Lowry [79] found the analytical function of the copolymer differential composition distribution

$$\frac{1}{M_0}\frac{dM}{dF_A} = \frac{M}{M_0}\left(\frac{\alpha}{f_A} - \frac{\beta}{1 - f_A} - \frac{\gamma}{f_A - \delta}\right)\frac{[(r_A + r_B - 2)f_A^2 + 2(1 - r_B)f_A + r_B]^2}{(r_A + r_B - 2r_A r_B)f_A^2 + 2r_B(r_A - 1)f_A + r_B} \tag{4.26}$$

where M_0 and M are the initial and the current total monomer concentrations; f_A and F_A are the 'instantaneous' mole fractions of monomer A in the monomer mixture and in the copolymer; $\alpha = r_B/(1 - r_B)$, $\beta = r_A/(1 - r_A)$, $\gamma = (1 - r_A r_B)/(1 - r_A)(1 - r_B)$ and $\delta = (1 - r_B)/(2 - r_A - r_B)$. By solving Eq. (4.26) simultaneously with the well-known equation of copolymer composition written in the following form:

$$F_A = \frac{(r_A - 1)f_A^2 + f_A}{(r_A + r_B - 2)f_A^2 + 2(1 - r_B)f_A + r_B} \tag{4.27}$$

one can find the $M_0^{-1}(dM/dF_A) - F_A$ relationship, i.e. the differential function of composition distribution, which then should be used to calculate the mean values of triad probabilities in polymeric models of the type in question.

Calculating the mean values for the parameters of the units distribution in copolymerization products with the composition heterogeneity of samples being taken into account is a very cumbersome process [84]. Based on the very few published results [85, 86], it is only possible to make some preliminary recommendations with respect to the calculation of N_0, N_1 and N_2 values for polymeric models that are heterogeneous in composition.

For the samples obtained at a constant ratio of monomers, these values should most probably be calculated, disregarding the 'instantaneous' heterogeneity, from

$$N_0 = \frac{[P(A) - R/2]^2}{P(A)}$$

$$N_1 = \frac{R[P(A) - R/2]}{P(A)}$$

$$N_2 = \frac{R^2}{4P(A)} \tag{4.28}$$

In copolymerization to relatively high conversion the composition heterogeneity must be taken into account. Moreover, when there is a strong difference between r_A and r_B the heterogeneity affects the values of units distribution parameters, even at low conversions.

Johnston and Harwood [85] studied the intramolecular cyclization in poly(vinyl chloride-co-methyl methacrylate) which takes place at the boundary of sequences:

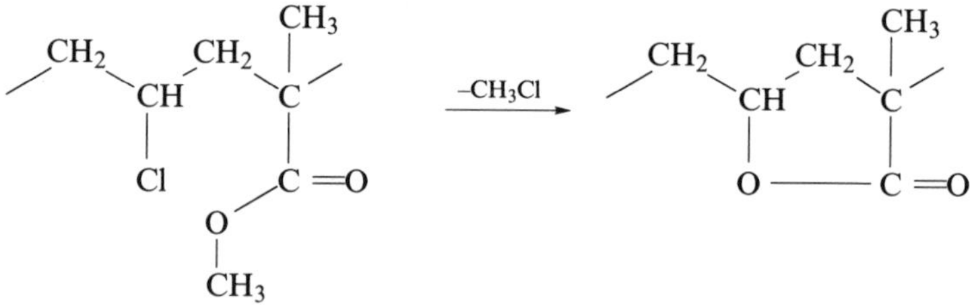

Scheme 4.4

The degree of completion of this reaction naturally depends on the units distribution in the chain. When the degree of reaction completion was calculated it turned out that for copolymer obtained at 9–13% conversion the correction for heterogeneity amounts to about 10% of the value being measured (with relative reactivities $r_A = 0.044$ and $r_B = 11.2$ for vinyl chloride and methyl methacrylate respectively). The program for the calculation of units distribution parameters in copolymers that are heterogeneous in composition is given in reference [84].

Polymers made up of A-centered triads of a single type can also be used as models; e.g. homopolymer A consists of N_0 type of units, and from the initial rate of its transformation one can, in a number of cases, find the k_0 constant. Strictly alternating copolymer AB can serve as a model to determine the k_2 constant. If polymeric models were available for each of the three types of triads it would be possible to determine the complete set of constants. As yet such examples are unknown. Even with an incomplete set of models, however, one can find individual rate constants with the help of the kinetic equation.

Arends [87] analyzed the experimental data of Fuoss *et al.* [88] on the kinetics of poly(4-vinyl pyridine) quaternization with n-butyl bromide having made two assumptions: (a) $k_0 = k_1$; (b) since the reaction proceeds with retardation, at high degrees of conversion ($> 80\%$) all the unreacted units are of the N_2 type. From the initial and the final parts of the kinetic curve Arends then found the constants $k_0 = k_1$ and k_2 respectively. Substituting the found values into Eq. (4.12), he calculated the kinetic curve of the process that had good agreement with experiment.

Arends' assumption of the equality of k_0 and k_1 appears to be unnecessary, since having determined k_0 and k_2, calculated the kinetic curves from Eq. (4.12) for different k_1 values and compared the calculations with experiment, one can find the most probable value of the latter constant. It is to be noted, however, that Arends' method has a substantial limitation, since k_2 can only be estimated in this way when there is a retarding neighbor effect, and even in that case the hypothesis taking all unreacted units to be of the N_2 type at 80% conversion is not correct.

Experimental investigations of the kinetics and mechanism of macromolecular reactions in which the above-mentioned methods for determining the individual rate constants have been used are examined in the next sections.

4.4 HYDROLYSIS OF POLYMETHACRYLATES

The hydrolysis of polymethacrylates is of particular interest for studies concerned with the effect of neighboring units. To begin with, the pattern of influence exerted by carboxylic groups on the rate of hydrolysis of neighboring ester units is markedly dependent on experimental conditions, so that both significant accelerating and retarding effects could be observed [71, 89]. Next, the rate of hydrolysis of polymethacrylates is essentially affected by chain configuration [90] (cf. also Chapter 2), and quantitative studies of the hydrolysis kinetics of samples having divers microstructures make possible an estimation of the contribution of configurational effects to the process specificity. It should also be noted that hydrolysis of polymethacrylates proceeds without any changes in the stereochemical configuration of the chain [91, 92], thereby rendering the esters of poly(methacrylic acid) particularly suited for studying both the neighbor effect and the configurational effect (a change in the chain configuration during the hydrolysis, as in the case of isotactic polyacrylonitrile [93], would make any quantitative interpretation of experimental data extremely difficult).

However, quantitative investigations of the hydrolysis of polymethacrylates involve major experimental problems. During the reaction, a relatively weakly polar polymer undergoes gradual transformation into a polyelectrolyte, while changes in the composition of macromolecules may essentially alter their interaction with solvent. In this case there also arise difficulties associated with the preparation of polymer models which are required for determining the individual rate constants. For example, it is hardly feasible to find the constant k_0 from the initial rate of hydrolysis of the poly(methyl methacrylate) (PMMA). Indeed, in aqueous alkalis, in which the retarding effect is expected [71], PMMA is insoluble, whereas in organic solvents, in the case of the accelerating effect [89], the influence of reacted units manifests itself at the earliest stages of hydrolysis and renders the determination of k_0 difficult.

Evidently stereoregular polymeric models with a random units distribution would be the most suitable for determining the rate constants of the reaction. Klesper and co-workers [71, 72] succeeded in obtaining such models by hydrolysis of syndiotactic and isotactic PMMA, the units distribution being directly determined by NMR. Litmanovich, Platé and co-workers [25, 94] employed the kinetic criterion in searching for the condition of obtaining stereoregular models having a random units distribution.

In this treatment of the problem, stereoregular polymethacrylate undergoes hydrolysis under the conditions selected so that the rate constants are equal to one another, $k_0 = k_1 = k_2$. As mentioned in the preceding section, in the presence of a hydrolyzing agent excess, the kinetics would be described by the usual equation of the first-order reaction and the products would have a random units distribution.

It would be highly desirable to verify the existence of such a relationship between the kinetic behavior of the process and the pattern of units distribution in the case of PMMA hydrolysis, for which it is possible to compare the kinetic data with the results of a direct determination of the units distribution by the NMR method. Agreement between the kinetic data and NMR findings would justify the employment of a simple kinetic criterion (semilogarithmic anamorphosis linearity) for searching the conditions of obtaining stereoregular polymeric models with random units distribution also in the investigation of the hydrolysis of other polymethacrylates, for which the pattern of units distribution has so far escaped experimental determination. Therefore Litmanovich, Platé and co-workers [25, 94] studied the kinetics of hydrolysis of syndiotactic and isotactic PMMA under the conditions in which, according to the NMR data [71, 72], there are formed copolymers with a random units distribution.

Syndiotactic (85% of syndio-triads, molecular weight 3.5×10^5) PMMA obtained using dipiperidyl magnesium as catalyst [95] was hydrolyzed in concentrated H_2SO_4 at 50 °C; the kinetic data are presented in Fig. 4.5. Solvolysis of isotactic (100% of iso-triads, molecular weight 2.5×10^6) PMMA prepared in the presence of a dipyrryl magnesium catalyst [95] was carried out in a dioxane–methanol–KOH system at 85 °C; the results are shown in Fig. 4.6. It can be seen in the figures that in both instances semilogarithmic anamorphoses are linear,

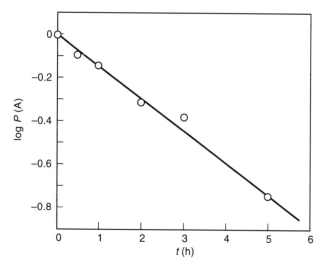

Fig. 4.5 Kinetics of hydrolysis of syndiotactic poly(methyl methacrylate) in 96% H_2SO_4 at 50 °C, PMMA concentration 5 g/100 ml. (Reproduced by permission of Nauka, Moscow, Russia, from *Vysokomol. Soed. A*, 1972, **14**, 2503)

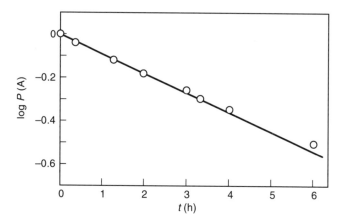

Fig. 4.6 Kinetics of solvolysis of isotactic poly(methyl methacrylate) in a mixture of dioxane and 20% KOH solution in methanol (2:1 by volume) at 85 °C, PMMA concentration 2.1 g/100 ml. (Reproduced by permission of Nauka, Moscow, Russia, from *Vysokomol. Soed. A*, 1975, **17**, 1112)

this being consistent with the assumption with regard to the equality of constants $k_0 = k_1 = k_2$ made on the basis of NMR data on the random units distribution in the reaction products formed under the aforementioned conditions.

This technique of preparing polymer models for studying the kinetics of hydrolysis of stereoregular poly(phenyl methacrylate) (PPMA) and poly(diphenylmethyl methacrylate) (PDPMMA) was employed in references [94], [96] and [97].

Highly isotactic PPMA (90% iso-triads, molecular weight 1.7×10^4) obtained using butyl lithium as catalyst [98] was subjected to hydrolysis in a dioxane–water–KOH system at 65 °C. Syndiotactic (*ca.* 87% of syndio-triads, molecular weight 2.3×10^4) and isotactic (94% of iso-triads, molecular weight 8×10^3) samples of PDPMMA synthesized using butyl lithium in tetrahydrofurane and toluene respectively [98] were hydrolyzed in a H_2SO_4–water–tetrahydrofurane system at 66 °C. Thus for all stereoregular samples, appropriate hydrolysis conditions were selected to obtain polymeric models having a random units distribution.

The following considerations provide insight into the fact that the rate constants of polymethacrylates hydrolysis are independent of conversion during preparation of the polymeric models. In the case of acid hydrolysis of PDPMMA and syndio-PMMA, the reaction appears to proceed under the action of an external agent, viz. hydroxonium ions having a sufficiently high concentration that the effect of undissociated neighboring carboxylic groups is negligibly small. It will be noted that the rate of hydrolysis is independent of the microstructure of PDPMMA and, accordingly, the ester groups in the iso- and syndio-triads of the polymer chain, under the specified conditions, are equally available to the attack of hydroxonium ions.

When iso-PPMA and iso-PMMA are subjected to alkaline hydrolysis, the COO^- groups are capable of exerting both retarding (electrostatic repulsion of external agent, i.e. OH^- ions) and accelerating (anchimeric assistance) effects on the hydrolysis of neighboring ester groups. Under the conditions used, these factors appear to compensate each other and the neighbor effect fails to manifest itself.

The influence of reaction conditions on the magnitude of the neighbor effect is, of course, not only typical for the hydrolysis of polymethacrylates (see Section 4.3). It is therefore reasonable to expect that the method of preparing polymeric models with the random units distribution described above may find application in the investigation of the kinetics and mechanism of divers macromolecular reactions.

Such models lend themselves to employment for determining the individual rate constants by divers techniques. Thus, Klesper *et al.* [71, 99] estimated the rate constants ratios using the data on units distribution in the products of hydrolysis of syndiotactic methyl methacrylate–methacrylic acid polymeric models. Klesper also demonstrated [72, 100–102] the feasibility of determining the triad composition of methyl methacrylate–methacrylic acid syndiotactic, isotactic and atactic copolymers by the NMR technique.

Syndiotactic models were hydrolyzed by Klesper *et al.* [71] at 145 °C in 0.2 M and 0.05 M aqueous alkali. The triad fractions in the products found by NMR are presented in Table 4.4, which are compared with the values calculated for

Table 4.4. Parameters of units distribution of hydrolyzed syndiotactic methyl methacrylate (A)–methacrylic acid (B) polymeric models (145 °C) [71]

Sample number[a]	Hydrolysis conditions		P(A)	Triad probabilities and their deviations from random values (in brackets)						Run number[b]		Distribution pattern
	KOH (mol/l)	time (h)		P(BBB)	P(BAB)	P(ABB+)	P(AAB+)	P(ABA)	P(AAA)	R	R_{rand}	
2	—	0	0.63	0.04 (−0.01)	0.10 (0.02)	0.15 (−0.02)	0.29 (0.00)	0.15 (0.00)	0.26 (0.01)	47.1	46.6	Random
8	0.05	10	0.53	0.21 (0.11)	0.05 (−0.07)	0.17 (−0.07)	0.26 (−0.01)	0.08 (−0.05)	0.24 (0.09)	33.8	49.8	
9	0.05	24	0.41	0.34 (0.14)	0.03 (−0.11)	0.16 (−0.13)	0.22 (0.02)	0.05 (−0.05)	0.20 (0.13)	27.0	48.4	Tendency to block formation
10	0.05	60	0.29	0.52 (0.16)	0.04 (−0.11)	0.14 (−0.15)	0.15 (0.03)	0.03 (−0.03)	0.12 (0.09)	21.4	41.2	
11	0.05	96	0.15	0.68 (0.07)	0.05 (−0.06)	0.11 (−0.10)	0.09 (−0.05)	0.02 (0.00)	0.05 (0.09)	16.4	25.4	
1	—	0	0.83	0.03 (0.01)		0.04 (0.00)	0.20 (−0.03)	0.11 (0.00)	0.61 (0.03)	25.8	27.6	Random
14	0.183	85	0.39	0.14 (−0.09)	0.25 (0.11)	0.28 (−0.02)	0.14 (−0.05)	0.17 (−0.08)	0.03 (−0.03)	62.8	47.6	Tendency to alternation

[a] Samples 8 to 11 were prepared by hydrolyzing model 2, and sample 14 as a result of model 1 hydrolysis.
[b] R is the number of AB^+ diads per 100 units in the chain; R_{rand} denotes the R value for random units distribution.
(Reproduced by permission of Hüthig & Wepf Verlag, Basel, Switzerland, from *Die Makromol. Chem.*, 1970, **139**, 1).

corresponding samples with a random units distribution. The comparison reveals that in the products of the models hydrolysis at a KOH concentration of 0.2 M, units alternation is prevalent, while at 0.05 M KOH concentration there is a tendency to the block-type distribution of units. Hence, the neighboring units exert a retarding effect in the first case and an accelerating effect in the second case.

To estimate the retarding effect quantitatively, Klesper *et al.* [99] used the Monte Carlo simulation to calculate the content of triads in the products of hydrolysis for various rate constants ratios. The experimental data fitted satisfactorily the $k_0:k_1:k_2$ ratios 1:0.5:0.25 and 1:0.33:0.11 (in reference [99] preference is given to the first ratio).

To determine the individual rate constants from kinetic data, the following method has been suggested [4, 25]. The first step comprises determining the initial rate of reaction and calculating the N_0, N_1 and N_2 values for each of the three (or more) polymeric models of a given series, followed by writing a system of equations of the type

$$-\frac{dA^{(0)}}{dt} = k_0 A_0^{(0)} + k_1 A_1^{(0)} + k_2 A_2^{(0)} \tag{4.29}$$

where $A_0^{(0)}$, $A_1^{(0)}$ and $A_2^{(0)}$ stand for concentration of ester units of the type N_0, N_1 and N_2 respectively, in mol/l at $t = 0$, and $A^{(0)} = A_0^{(0)} + A_1^{(0)} + A_2^{(0)}$. Solving the system of equations (4.29) yields approximate values of the constants k_0, k_1 and k_2. Next, recourse is made to a computer to search the unique set of constants k_0, k_1 and k_2 that would provide the description of reaction kinetics for all models over the entire range of conversion degrees. In so far as the starting samples for kinetic experiments are polymeric models with the random units distribution, their hydrolysis kinetics is described as follows (cf. Chapter 3):

$$\frac{dP(A)}{dt} = -k_2 P(A) + [P(A)^0]^2 [2(k_2 - k_1) - (k_2 - 2k_1 + k_0)P(A)^0 e^{-k_0 t}]$$
$$\times \exp\left\{\left[\frac{2(k_1 - k_0)}{k_0}\right]P(A)^0(1 - e^{-k_0 t}) - 2k_1 t\right\} \tag{4.30}$$

Here $P(A)^0$ denotes the composition of the starting model.

It should be emphasized that the initial rate for each polymeric model calculated on the basis of experimental data generally varies within certain limits depending on the kinetic experimental errors. Thus the magnitude of the largest individual constant (k_0 for retardation and k_2 for acceleration) found by solving Eq. (4.29) would be the most accurate, whereas the magnitude of the smallest constant, especially for a great neighbor effect, may vary within a very broad range. Therefore in references [4] and [25] searching for the refined values of constants involved fixing the value of the largest constant and varying the other two constants.

For retardation, the value of k_0 is fixed and Eq. (4.30) is transformed as follows:

$$\frac{dP(A)}{d\tau} = -k'P(A) + [P(A)^0]^2[2(k' - k) - (k' - 2k + 1)P(A)^0 e^{-\tau}]$$

$$\times \exp[2(k - 1)P(A)^0(1 - e^{-\tau}) - 2k\tau] \qquad (4.31)$$

where $k = k_1/k_0$, $k' = k_2/k_0$ and $\tau = k_0 t$. Usually for retardation $k \geq k'$ and the magnitudes of k and k' (varied from 0 to 1) that minimize the sum of square deviations of calculated meanings of $P(A)$ from experimental points are found.

For acceleration, the value of k_2 is fixed and the transformation of Eq. (4.30) yields

$$\frac{dP(A)}{d\tau} = -P(A) + [P(A)^0]^2[2(1 - k) - (1 - 2k + k')P(A)^0 e^{-k\tau}]$$

$$\times \exp\left\{\left[\frac{2(k - k')}{k'}\right]P(A)^0(1 - e^{-k\tau}) - 2k\tau\right\} \qquad (4.32)$$

where $k = k_1/k_2$, $k' = k_0/k_2$ and $\tau = k_2 t$. Usually $k \geq k'$ and the procedure used to refine the constants is as described above.

It is worth noting that the suggested method involving polymeric models is especially valuable for a great acceleration by two reacted neighbors. Indeed, in this case, the instantaneous content of BAB triads during the hydrolysis of homopolymer is very small due to high rate of their decay. Thus the kinetics may be described by varying the k_2/k_1 ratio in a broad range (say from 2 to 10) and estimating with the least accuracy. In the meantime, using polymeric models with a random distribution it is possible to prepare models with a high content of BAB triads and, from their hydrolysis kinetic data, to estimate k_2 with the most accuracy.

For the largest constants found from the initial rates of the model reaction the errors estimated in reference [4] according to reference [103] are about $\pm 15\%$. The errors for the other constants were estimated in reference [25] in the following manner. Using Eqs. (4.31) or (4.32), for various values of k and k' (in the vicinity of their most probable magnitudes) the kinetic curves were calculated and their deviations from experimental data served as a criterion for estimation of the possible errors.

Let us consider now the results of determining the constants by the method in question. We shall operate with the conditional constants having the $\min^{-1}$ dimension, since their values include the hydrolyzing agent concentration. However, this fact does not distort the conclusions about the influence of various factors on the reaction kinetics and units distribution which are based not on the absolute values of the rate constants but on their ratios.

The results of hydrolysis of the syndiotactic methyl methacrylate (MMA)–methacrylic acid (MAA) models in 0.2 M aqueous alkali at 145 °C [25] are

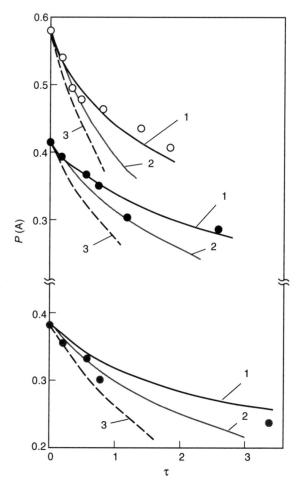

Fig. 4.7 Kinetics of hydrolysis of syndiotactic MMA–MAA copolymers at 145 °C, 250 mg of polymer per 12 ml of 0.2 M aqueous KOH. Dots, experimental data; curves, calculated for ratios $k_0:k_1:k_2$ equal to 1:0.2:0.05 (1), 1:0.33:0.11 (2) and 1:0.5:0.25 (3). (Reproduced by permission of Nauka, Moscow, Russia, from *Vysokomol. Soed. A*, 1975, **17**, 1112)

presented in Fig. 4.7. The constant $k_0 = 5.8 \times 10^{-4} \, \text{min}^{-1}$ and the most probable ratio for $k_0:k_1:k_2$ is 1:0.2:0.05. Hence there is a retarding neighbor effect.

Compare these results with Klesper *et al.* data [89]. Figure 4.8 presents their NMR data for the content of all six triads in the products of syndiotactic MMA–MAA copolymer hydrolysis under the same conditions (points) and also calculations for various rate constants ratios (curves). It is seen that the data on the content of AAA, BBB, AAB$^+$ and ABB$^+$ triads can not be used for precise

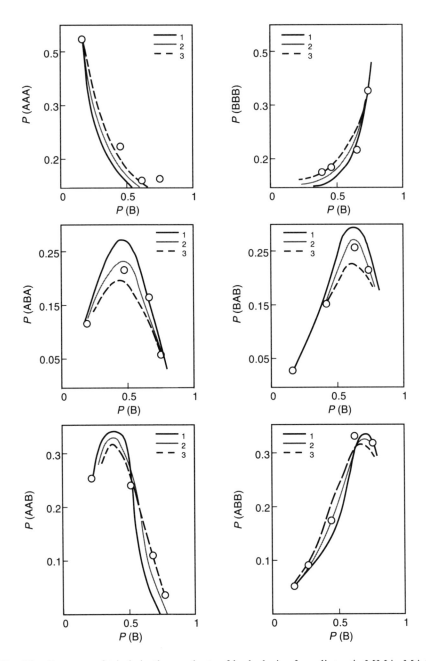

Fig. 4.8 Contents of triads in the products of hydrolysis of syndiotactic MMA–MAA copolymers at 145 °C, 250 mg of polymer per 12 ml of 0.2 M aqueous KOH. Dots, NMR data [89]; curves, calculated for ratios $k_0:k_1:k_2$ equal to $1:0.2:0.05$ (1), $1:0.33:0.11$ (2) and $1:0.5:0.25$ (3). (Reproduced by permission of khimiya, Moscow, Russia, from Platé, N.A., Litmanovich, A.D. and Noah, O.V., *Makromolekulyarnye Reaktsii (Macromolecular Reactions)*, Khimiya, Moscow, 1977, p. 194)

determination of the constants ratios since the three curves are close to one another. The content of ABA and BAB triads provides more accurate information on constants ratios. Unfortunately, only one experimental point for each of these triads is available in the area where the curves diverge. Judging by these two points, the data are best described by curve 2 ($k_0:k_1:k_2 = 1:0.33:0.11$), though curve 3 ($k_0:k_1:k_2 = 1:0.5:0.25$) is also in satisfactory agreement with the experimental data.

Kinetic curves for these constants ratios are plotted in Fig. 4.7. It is evident that the ratio $k_0:k_1:k_2 = 1:0.5:0.25$ is inconsistent with kinetic experiment, while the ratio $k_0:k_1:k_2 = 1:0.33:0.11$ describes, within the experimental errors, these data, although the agreement is somewhat inferior to that of the ratio $1:0.2:0.05$.

In view of the errors of determining the constants by various methods, it is reasonable to assume the ratios found from kinetics ($1:0.2:0.05$) and units distribution ($1:0.33:0.11$) to be sufficiently close to each other, this agreement being indicative of the dominant role of the effect of neighboring units in the reaction under the conditions studied. For the hydrolysis of isotactic MMA–MAA copolymers in $0.2\,M$ aqueous alkali, $k_0 = 90 \times 10^{-4}\,min^{-1}$ and the most probable ratio is $k_0:k_1:k_2 = 1:0.4:0.4$ [25], so here there is also a retarding effect.

Hydrolysis retardation of both syndiotactic and isotactic copolymers appears to be caused by the electrostatic repulsion of OH^- ions by carboxylate anions. Since the reaction kinetics, and for syndiotactic copolymers the units distribution as well, lends itself to description by equations accounting for the effect of the nearest neighbor exclusively, it is reasonable to assume that under the conditions examined the electrostatic interaction of OH^- ions is effective with the COO^- groups closest to the ester group being attacked, but not with the charged macromolecular coil as a whole.

The stereochemical configuration of MMA–MAA copolymers affects the kinetic parameters of their hydrolysis. To begin with, k_0 for the isotactic sample is by an order of magnitude greater than that for the syndiotactic sample, so the isotactic configuration of AAA triads is sterically more favorable for the reaction than the syndiotactic one.

Next, the retarding effect in isotactic models ($1:0.4:0.4$) is less pronounced than in syndiotactic models ($1:0.2:0.05$). As pointed out earlier, the COO^- groups, apart from the retarding effect (repulsion of OH^- ions), may accelerate the reaction by means of anchimeric assistance, thereby counterbalancing to some extent the overall retarding action. Evidently the isotactic configuration is more favorable for anchimeric assistance than the syndiotactic one.

For the hydrolysis of syndiotactic MMA–MAA copolymers in $0.05\,M$ aqueous alkali, $k_0 = 1.9 \times 10^{-4}\,min^{-1}$ and $k_0:k_1:k_2 = 1:0.7:0.7$ [25], so a weak retardation is observed. Meanwhile, Klesper *et al.* [71,89] found from NMR spectra that under these conditions the products of syndiotactic MMA–MAA copolymer hydrolysis display a tendency to the block distribution of units (see Table 4.4).

This tendency is typical for the accelerating neighbor effect. Hence the kinetic data are inconsistent with the data on the units distribution. It follows that, under the conditions used, the specificity of the reaction is not governed by the neighbor effect alone. System microheterogeneity [94] is a likely cause of discrepancy between the kinetic and NMR data. If the macromolecules in the surface layer of microgel particles undergo hydrolysis to a greater degree of conversion than the macromolecules inside the particles, the relatively long sequences of MAA units would be formed in the chain of the former macromolecules, while the latter macromolecules would retain the short sequences formed at low conversion. For a sample with a heterogeneous composition the NMR spectrum would naturally reveal a tendency to the block-type units distribution, though under the conditions in question the COO^- groups presumably retard the hydrolysis of neighboring ester units.

The accelerating effects have been observed for the hydrolysis of stereoregular esters of poly(methacrylic acid) in the pyridine–water medium.

Figure 4.9 illustrates the results of hydrolysis of MMA–MAA isotactic models. Here $k_2 = 530 \times 10^{-4}\,min^{-1}$ and $k_0:k_1:k_2 = 1:8:100$. The fact that the reactivity

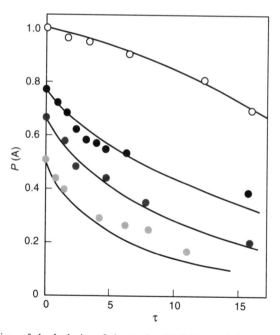

Fig. 4.9 Kinetics of hydrolysis of isotactic PMMA and isotactic MMA–MAA copolymers at 145 °C in a pyridine–water system (95:5 by volume), 250 mg of polymer per 12 ml of a solution. Dots, experimental data; curves, calculated for ratio $k_0:k_1:k_2 = 1:8:100$. (Reproduced by permission of Nauka, Moscow, Russia, from *Vysokomol. Soed. A*, 1975, **17**, 1112)

of ester units under the effect of a single carboxylic group increases by an order of magnitude ($k_1:k_0 = 8$) could be naturally explained in terms of anchimeric assistance according to the nucleophilic mechanism generally accepted for hydrolysis [49].

In the case of the two reacted neighbors, the probability of a nucleophilic attack of carboxylate anions would apparently increase by a factor 2, this increase corresponding to the ratio $k_2:k_1 \approx 2$. However, k_2 is actually of an order of magnitude greater than k_1 ($k_2:k_1 = 12.5$). The large ratio is likely to be caused by the simultaneous interaction of an ester group with two reacted neighbors, viz. an undissociated and a dissociated carboxylic group, and the interaction proceeds according to the mechanism of bifunctional electrophilic–nucleophilic catalysis:

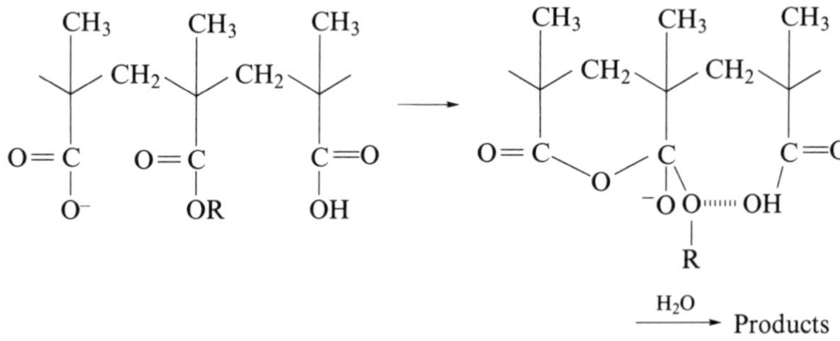

Scheme 4.5

Indeed, as early as in 1958 Morawetz and Oreskes [104] showed the hydrolysis of aspirin (I) to proceed at pH 7–9 at a rate an order of magnitude slower than the hydrolysis of its derivative (II):

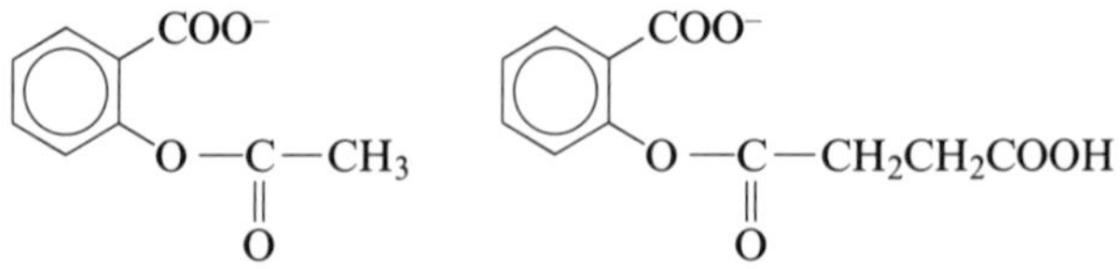

Scheme 4.6

and proposed the following explanation of this phenomenon: I undergoes hydrolysis according to the nucleophilic anchimeric assistance mechanism, while the hydrolysis of II proceeds by the mechanism of intramolecular electrophilic–nucleophilic catalysis:

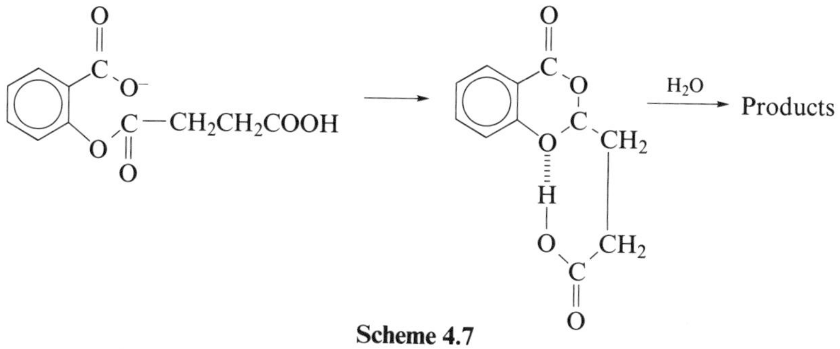

CH_2CH_2COOH → $\xrightarrow{H_2O}$ Products

Scheme 4.7

Hydrogen bond formation between the ester and carboxylic groups facilitates the nucleophilic attack of carboxylate anion on the carbonyl carbon atom and additionally stabilizes the transient state, thereby providing an increased reaction rate as compared to the nucleophilic anchimeric assistance rendered by one neighboring ionized group only.

It is noteworthy that the rate constants found from the data on the hydrolysis of MMA–MAA polymeric models adequately describe the hydrolysis kinetics of the homopolymer, isotactic PMMA (cf. Fig. 4.9). In view of this situation, it is reasonable to assume that, under these conditions, despite the alteration of the composition of macromolecules during the hydrolysis (from 100 to 20 mol% MMA), the conformational factors do not affect the reaction kinetics; in other words, the accessibility of ester groups to the attack of an external agent and their interaction with the neighboring units are independent of the degree of conversion.

For the hydrolysis of isotactic phenyl methacrylate (PMA)–MAA copolymers, $k_2 = 6000 \times 10^{-4}$ min^{-1} and $k_0:k_1:k_2 = 1:40:1000$ (k_0 is so small compared to k_2 that only its order of magnitude can be assessed). These ratios correspond to the same mechanism of influence of one and two reacted neighbors which has just been discussed for the hydrolysis of MMA–MAA models under these conditions.

Figure 4.10 presents the results of hydrolysis of seven isotactic diphenylmethyl methacrylate (DPMMA)–MAA copolymers of various composition. Processing the data demonstrated the inadequacy of a single set of constants to describe the hydrolysis kinetics of all seven models. The models may be separated into two groups.

The hydrolysis kinetics for four models having the composition $P(A) \leq 0.55$ (Fig. 4.10b) is described by the constant $k_2 = 33 \times 10^{-4}$ min^{-1} and $k_0:k_1:k_2 = 1:20:100$. The acceleration by one reacted neighbor can be explained by nucleophilic anchimeric assistance. The constant k_2 is 5 times as great as k_1 (and not by an order of magnitude as in the case of the MMA–MAA and PMA–MAA copolymers). Presumably due to the bulky substituent in the ester group of

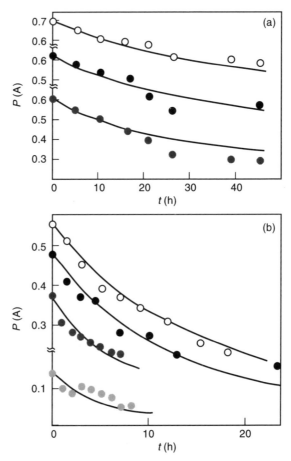

Fig. 4.10 Kinetics of hydrolysis of isotactic DPMMA–MAA copolymers at 145 °C in a pyridine–water system (95:5 by volume), 250 mg of polymer per 12 ml of a solution. Curves, calculated for the ratios: (a) $k_2:k_1:k_0 = 1:0.1:0.01$ at $k_2 = 20 \times 10^{-4}$ min^{-1}, (b) $k_2:k_1:k_0 = 1:0.2:0.01$ at $k_2 = 33 \times 10^{-4}$ min^{-1}. (Reproduced by permission of Nauka, Moscow, Russia, from *Vysokomol. Soed. A*, 1975, **17**, 1112)

DPMMA, the effect of bifunctional electrophilic–nucleophilic catalysis for the DPMMA–MAA models is somewhat hindered.

For the models having the composition $P(A) \geq 0.6$ (Fig. 4.10a), the constant $k_2 = 20 \times 10^{-4}$ min^{-1} and $k_0:k_1:k_2 = 1:10:100$, the copolymer of composition $P(A) = 0.68$ being hydrolyzed slightly slower and the model of composition $P(A) = 0.60$ slightly faster than would be expected from calculation of the aforementioned constants ratios.

A comparison of respective constants and kinetic curves (for the sake of clarity, the curves in Fig. 4.10 are plotted in the $P(A)$–t coordinates instead of the $P(A)$–τ

coordinates, as in Figs. 4.7 and 4.9) demonstrates that on transition from the models of composition $P(A) \geq 0.6$ to the models of composition $P(A) \leq 0.55$ the reactivity of DPMMA groups is markedly increased. This fact enabled Litmanovich *et al.* [25] to assume that in the range of model composition $P(A)$ from 0.60 to 0.55, provided the units distribution is random, some conformational factors might become apparent.

Figure 4.11 shows the dependence of reduced viscosity on the models composition for 1% solutions of isotactic DPMMA–MAA copolymers in a pyridine–water mixture (95:5 by volume) at 60 °C. In the 0.6–0.5 range of $P(A)$ values the viscosity increases sharply (at 28 and 40 °C the pattern of η_{sp}/C dependence on the model composition remained as shown in Fig. 4.11 and can be expected to be preserved at the temperature of kinetic experiments). A corresponding growth of coil size is apparently caused by the electrostatic repulsion of hydrolyzed groups in the copolymer. Such a coil expansion is likely to render the ester groups more accessible to an external agent attack and to facilitate the anchimeric assistance of neighboring units.

Hence, viscometry data are not at variance with the assumption concerning a contribution of conformational effects in the hydrolysis of DPMMA–MAA copolymers, the contribution of these effects being vanished where the content of randomly distributed MAA units in the model exceeds 45 mol%.

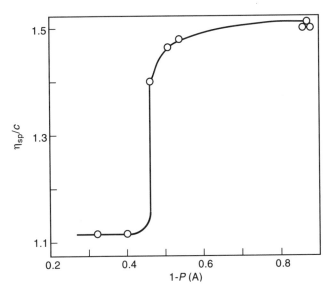

Fig. 4.11   Dependence of reduced viscosity of 1% solutions of isotactic DPMMA–MAA copolymers in a pyridine–water system (95:5 by volume) at 60 °C on their composition. (Reproduced by permission of Nauka, Moscow, Russia, from *Vysokomol. Soed. A*, 1975, **17**, 1112)

A comparison of constants k_2 for isotactic MMA–MAA, PMA–MAA and DPMMA–MAA models shows the ester group in the BAB triads for PMA–MAA copolymers to be an order of magnitude more reactive than in the case of MMA–MAA models (the constants $k_2 \times 10^{-4}\,\text{min}^{-1}$ are 6000 and 530 respectively). This difference can be explained as follows. First, the positive inductive effect of the methyl group causes electron density enhancement at the C carbonyl atom and, therefore, the nucleophilic attack of the neighboring COO^- group is hindered. Second, the $C_6H_5O^-$ anion undergoes stabilization as a result of conjugating the electron pair of the O atom with the π-electrons of the phenyl nucleus, so that the probability of O-acyl cleavage is increased. Both factors are responsible for raising the reactivity of PMA units as compared to MMA units.

Conversely, k_2 for the isotactic DPMMA–MAA copolymers is by an order of magnitude smaller than that for MMA-MAA models, this being presumably the result of steric hindrances due to a larger volume of the substituent in the ester group of DPMMA.

The influence exerted by the stereochemical configuration of a chain on the reaction kinetics can be assessed from the results of examining the hydrolysis of syndiotactic MMA–MAA [105] and DPMMA–MAA [25] polymeric models under the aforementioned conditions (pyridine–water mixture, 95:5 by volume, 145 °C). Platé et al. [105] measured the rate constants of syndiotactic PMMA hydrolysis both by the kinetic method and by determining the triad composition from NMR spectra. Polymeric models having a random units distribution were obtained by hydrolyzing syndiotactic PMMA containing 93% of syndio-diads in concentrated sulfuric acid [106]. The kinetic data as regards the hydrolysis of four polymer models are well described provided that $k_2 = 3.7 \times 10^{-4}\,\text{min}^{-1}$ and $k_0:k_1:k_2 = 1:2.5:3.4$. Platé et al. [105] employed Klesper's data [100] on peak assignment in the α-methyl region of the NMR spectrum for the analysis of the triad composition. However, the method of determining the rate constants was essentially refined in reference [105].

To determine the fractions of triads, the poorly resolved α-methyl region of the NMR spectrum was simulated with a computer using the sum of six Lorentz components, the parameters of which were found by means of a special search program minimizing the sum of square deviations of the simulated spectral functions from experimental functions [107, 108] (cf. also Chapter 8). The search program was likewise employed to find the most probable values of rate constants that describe changes in the triad composition of polymeric models in the course of hydrolysis. This approach provides a substantially higher reliability of quantitative interpretation of poorly resolved NMR spectra as well as a better accuracy of individual rate constants determination.

Figure 4.12 illustrates that for the model of composition $P(A) = 0.89$, variation of the triad composition is well described by the constants ratio $k_0:k_1:k_2 = 1:3:5$. This ratio, which also describes the triad composition of hydrolysis products for other polymeric models, is in agreement with the kinetic method data.

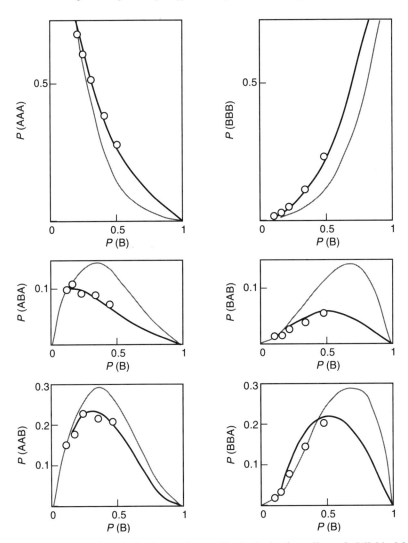

Fig. 4.12 Contents of triads in the products of hydrolysis of syndiotactic MMA–MAA copolymer containing 89% MMA (hydrolysis conditions as in Fig. 4.10). Dots, NMR data; curves, calculated for ratios $k_0:k_1:k_2$ equal to 1:1:1 (thin lines) and 1:3:5 (solid lines). (Reproduced by permission of Nauka, Moscow, Russia, from *Dokl. Akad. Nauk. SSSR*, 1975, **223**, 396)

As can be seen from the foregoing discussion, for a syndiotactic polymer the accelerating neighbor effect is substantially below that observed for an isotactic polymer. Moreover, the fact that the ratio $k_2:k_1$ is small (about 1.5) most probably points to the independent effect of each of the two reacted neighboring groups in the BAB triads on the reactivity of the central unit. Hence, the syndiotactic configuration of the MMA–MAA copolymer does hinder the formation of a cyclic intermediate compound required for the onset of anchimeric assistance, while bifunctional catalysis in BAB syndiotactic triads is totally inoperative.

In the case of DPMMA having bulky substituents in the ester groups, steric hindrances apparently become so significant that in the hydrolysis of syndiotactic DPMMA–MAA models the neighbor effect is absent altogether ($k_0 = k_1 = k_2 = 0.5 \times 10^{-4} \text{min}^{-1}$).

The influence exerted by stereochemical configurations on the kinetics of PPMA hydrolysis has been studied by Yun [31] who investigated the hydrolysis of isotactic and atactic polymeric models in dioxane–phosphate buffer (92:8 by volume) at 80 °C. The isotactic models were prepared as described earlier, while the atactic models were obtained by radical copolymerization with benzoyl peroxide at 60 °C in benzene [109]. The stereochemical microstructure of polymers subjected to quantitative hydrolysis and subsequent methylation with diazomethane has been established by NMR. The microstructure of models was found to be practically constant (*ca.* 53% syndio-, 37% hetero- and 10% iso-triads) in the composition range $0.24 \leq P(A) \leq 0.77$. The values of N_0, N_1 and N_2 for atactic models were calculated using Eqs. (4.14) and the reactivity ratios r_A and r_B determined in reference [109].

For isotactic PMA–MAA models the following data were obtained: $k_2 = 15 \times 10^{-4} \text{min}^{-1}$ and $k_0 : k_1 : k_2 = 1 : 55 : 100$. The high $k_1 : k_0$ ratio is apparently explained by strong anchimeric assistance of the reacted group. The constant k_2 is only about twice as large as the constant k_1. Under these conditions, in contrast to the hydrolysis of PMA–MAA copolymers in the pyridine–water medium at 145 °C, bifunctional catalysis appears to be inoperative. Among the causes of this dissimilarity, the interaction of undissociated MAA groups and dioxane may be mentioned: hydrogen bonding between the carboxyl hydrogen and the oxygen atom of the solvent hinders the electrophilic attack of the —COOH group on the neighboring PMA unit (an analogous interaction of the carboxyl groups with pyridine at a higher temperature, viz. 145 °C, is evidently ineffective).

If this hypothesis is sound, the $k_2 : k_1$ ratio should grow with temperature elevation which causes weakening of the bond between the carboxyl group and solvent. Yun found the ratio of constants at 100 °C to be as follows: $k_0 : k_1 : k_2 = 1 : 18 : 65$. Hence, temperature elevation resulted in increasing the $k_2 : k_1$ ratio from 2 to 4, this being consistent with the suggested hypothesis.

For atactic radical copolymers at 80 °C, $k_2 = 50 \times 10^{-4} \text{min}^{-1}$ and $k_0 : k_1 : k_2 = 1 : 2 : 10$. It will be noted that here the rate constants are averaged quantities, e.g. $k_0 = \gamma^i k_0^i + \gamma^s k_0^s + \gamma^h k_0^h$, where γ^i and k_0^i are the fraction of iso-triads and the respective individual rate constant, etc.

In so far as the microstructure of radical PMA–MAA copolymers is practically independent of their composition, in the initial models the contribution of true constants k_j^i, k_j^s and k_j^h to the averaged constants k_j ($j = 0, 1, 2$) is practically one and the same for each model. Thus the ratios of constants $k_0 : k_1 : k_2$ found from the initial rates of hydrolysis by solving Eqs. (4.29) should predominantly be representative of the influence exerted by the nature of neighboring units on the reactivity of ester groups, the stereochemical configuration being of secondary importance. The ratio $k_0:k_1:k_2 = 1:1.4:10$ found by solving Eqs. (4.29) is close to the ratio $k_0:k_1:k_2 = 1:2:10$ which describes satisfactorily the kinetics of hydrolysis of four models in the examined range of conversion degrees. However, the data obtained for atactic models are useful only for estimating qualitatively the configurational effect in the hydrolysis of PMA–MAA copolymers.

For the atactic copolymers, k_0 is by an order of magnitude higher than for the isotactic copolymers. Hence, under aforementioned conditions the syndiotactic and heterotactic configurations of AAA-type triads appear to be more favorable for the reaction than the isotactic configuration.

The low ratio $k_1/k_0 = 2$ for atactic polymer points to the fact that in syndio- and hetero-triads the mutual arrangement of ester and carboxyl groups is much less favorable for the anchimeric assistance than in iso-triads.

Thus, in studies of the hydrolysis of polymethacrylates, extensive application has been found a method of polymeric models described in the previous paragraph, according to which rate constants are determined using the results of investigating the reaction kinetics of stereoregular models of different composition and random units distribution. The values of k_0, k_1 and k_2 found by this method (cf. Table 4.5) are the true rate constants, provided reaction specificity under the experimental conditions used is indeed governed by the neighbor effect alone. In this case k_0, k_1 and k_2 should meet the following requirements.

To begin with, the sole set of k_0, k_1 and k_2 constants, as can be readily inferred from the method of their determination, should describe the reaction kinetics of all polymeric models. The violation of this requirement provides a direct indication of an essential role of factors other than the neighbor effect, this having been demonstrated in the case of isotactic DPMMA–MAA copolymers.

On the other hand, when the syndiotactic MMA–MAA copolymers are hydrolyzed with 0.05 M aqueous alkali, the found values of k_0, k_1 and k_2 comply with the requirements formulated above (the kinetic criterion), but do not fit the data on the units distribution (for this reason in Table 4.5 the values of k_0, k_1 and k_2 in question are bracketed, since they can not be regarded as true rate constants; also bracketed are the values of averaged rate constants for the hydrolysis of atactic PMA–MAA copolymers).

Only the accordance between kinetic data and those of the units distribution or composition heterogeneity appears to be a necessary and sufficient criterion of the validity, under the experimental conditions used, of the conventional model of a macromolecular reaction involving the neighbor effect. This accordance has been found for the hydrolysis of syndiotactic MMA–MAA copolymers both in 0.2 M KOH and in pyridine–water systems, as well as for the hydrolysis of isotactic and syndiotactic PMMA under the conditions intended for obtaining models with a random units distribution. As long as the correlation of kinetic

Table 4.5. Individual rate constants of hydrolysis of poly(methacrylic acid) esters

Polymer	Type	Medium	Temperature (°C)	k_0	k_1	k_2	$k_0:k_1:k_2$	Reference
MMA–MAA	Iso	0.2 M KOH	145	90	35	35	1:0.4:0.4	[25]
MMA–MAA	Syndio	0.2 M KOH	145	5.8	1.2	0.3	1:0.2:0.05	[25]
MMA–MAA	Syndio	0.05 M KOH	145	(1.9)	(1.3)	(1.3)	1:0.7:0.7	[25]
MMA–MAA	Iso	Pyridine–water	145	5.3	42	530	1:8:100	[25]
MMA–MAA	Syndio	Pyridine–water	145	1.1	2.8	3.7	1:2.5:3.4	[105]
DPMMA–MMA	Iso	Pyridine–water	145	0.3	7	33	1:20:100	[25]
DPMMA–MMA	Syndio	Pyridine–water	145	0.5	0.5	0.5	1:1:1	[25]
PMA–MAA	Iso	Pyridine–water	145	6	240	6000	1:40:1000	[25]
PMA–MAA	Iso	Dioxane–buffer	80	0.15	8.3	15	1:55:100	[31]
PMA–MAA	Iso	Dioxane–buffer	100	2	36	130	1:18:65	[31]
PMA–MAA	Atactic	Dioxane–buffer	80	(5)	(10)	(50)	1:2:10	[31]

The column header spanning k_0, k_1, k_2 reads: $k_i \times 10^4$ (min^{-1})

data with the data on the units distribution or on the composition heterogeneity of the reaction products is not available, the values of rate constants based on the kinetic criterion alone are necessarily somewhat inconclusive. It should be noted, however, that when the hydrolysis is carried out in a pyridine–water mixture it is hardly likely that supramolecular effects would be as manifest as in the case of hydrolysis in an aqueous alkali; thus constants found for the former conditions appear to be adequately reliable.

Klesper and Barth [33, 34] investigated in detail the alkaline hydrolysis of syndiotactic MMA–MAA copolymers by the NMR method and found that, in the presence of excess OH$^-$ ions, the reaction is retarded. Varying the ionic strength, polymer concentration, its molecular weight in the 1×10^5–6.5×10^5 range and the temperature in the range 115–175 °C does not affect the units distribution in the product, i.e. it fails to change the rate constants ratios. Like the authors of reference [25], Klesper and Barth ascribe reaction retardation to the repulsion of OH$^-$ ions only by neighboring COO$^-$ groups that are closest to the ester group being attacked. The fact that the rate constants ratios are temperature insensitive prompted a conclusion [33, 34] that the retardation effect of the neighbors has in this case an entropic nature.

Analysis of NMR spectra also made it possible to arrive at an interesting deduction that, where the content of MMA in copolymers corresponds to $P(A) > 0.6$, in aqueous alkali there occurs the contraction of macromolecular coils and diminution of their conformational mobility. Klesper and Barth therefore conclude that the neighbor effect *per se* manifests itself, under the conditions used, only in models of composition $P(A) < 0.6$ (it will be noted that this situation bears a resemblance to the hydrolysis of isotactic DPMMA–MAA copolymers referred to earlier [25]).

From the results of hydrolysis in excess alkali, Klesper and Barth estimated both relative [33] and absolute [34] values of rate constants and their ratio $k_0:k_1:k_2 = 1:0.5:0.17$. In both cases, they used the intersection method (see Section 4.2); however, they did not employ an optimization of these preliminary values by comparing the calculated and experimental data.

It is noteworthy that Klesper and Barth [33], contrary to the opinion of many investigators in this field, ascribe the accelerating neighbor effect in hydrolysis of polymethacrylates to the influence of undissociated carboxyl groups —COOH, and not to that of carboxylate anions —COO⁻. To corroborate this contention, the authors cite the inability of low-molecular carboxylate anions to act as nucleophilic catalysts in the hydrolysis of monocarboxylic acid esters [110]. Indeed, low-molecular carboxylate anions are much weaker nucleophilic agents than hydroxyl ions, but it would be improper to conclude, on the strength of these data, that the —COO⁻ ions in the polymer chain are incapable of anchimeric assistance, since polymeric effects, which are specific for macromolecular reactions, should not be ignored.

To begin with, the fact that both an ester and a carboxylate group are located in the neighboring units of a macromolecule signifies a tremendous increase in the 'local' concentration of carboxylate anions (as compared to the average concentration of these anions in the solution volume) and, therefore, a respectively greater frequency of collisions between the substrate and the catalytic agent. The configurational factors should also be accounted for: indeed, the anchimeric assistance of carboxylate anions is substantially facilitated due to the feasibility of forming a stable five- or six-membered intermediate cyclic compound or a transient state, as is the case when the esters of poly(methacrylic acid) or poly(acrylic acid) are subjected to hydrolysis. In contrast, for poly(vinyl pyrrolidone) hydrolysis, no such possibility exists and the accelerating neighbor effect is absent in this case [111]. No acceleration is likewise observed for the hydrolysis of poly(ethyl acrylate-co-butadiene) [112], in which the ethyl acrylate units are separated from one another by butadiene units, although the hydrolysis of homopolymer poly(ethyl acrylate) proceeds with acceleration.

The structures of effective individual rate constants may be presented in the following forms:

$$k_0 = k_0^{OH}[OH^-]$$

$$k_1 = k_1^{OH}[OH^-] + k_1^{COO^-}$$
$$k_2 = k_2^{OH}[OH^-] + {}^{COO^-}k_2{}^{COO^-}$$

or

$$k_2 = k_2^{OH}[OH^-] + {}^{COOH}k_2{}^{COO^-}$$

Here the superscripts denote the catalyzing agent nature, and for k_2 two situations are practicable: nucleophilic attack by any one of the two neighboring carboxylate anions or bifunctional catalysis.

In view of the above considerations as regards the role of concentration and configurational factors, it can be expected that, under certain conditions, the constant $k_1^{COO^-}$ reflecting the accelerating effect of the neighboring carboxylate anion may become comparable in magnitude to the product $k_1^{OH}[OH^-]$ (an analogous situation is also likely for the constants ${}^{COO^-}k_2{}^{COO^-}$ or ${}^{COOH}k_2{}^{COO^-}$). This possibility (assumed by many investigators and proved experimentally by Gaetjens and Morawetz [49]) has been substantiated in a convincing manner in an interesting paper by Polowinski and Bortnowska-Barela [113] for the alkaline hydrolysis of poly(phenyl methacrylate) in a dimethyl sulphoxide–water mixture (25:1); the reaction in this medium proceeds homogeneously in the entire range of conversions. The authors found that k_0 and k_1 depended upon the alkali concentration whereas k_2 did not. Accordingly, the $k_0:k_1:k_2$ ratios were 1:0.24:0.24 and 1:0.46:0.12 at KOH concentrations 4.8 and 9.6×10^{-3} mol/l respectively. The retardation has been explained in reference [113] by ordinary electrostatic repulsion.

At present, kinetic studies of the hydrolysis of polymethacrylates yield, at best, the effective individual rate constants k_0, k_1 and k_2. These constants constitute very valuable information on the process, since they make it possible to describe the reaction kinetics as well as the units distribution and composition heterogeneity. Moreover, the found values, as demonstrated above, provide an insight into essential features of the hydrolysis, namely the influence of such factors as the reaction medium, the nature of neighboring units, the stereochemical configuration and conformational properties of the chain on the reactivity of polymethacrylates. To interpret in greater detail the mechanism of hydrolysis, it appears pertinent to determine the partial constants of the type k_1^{OH} and $k_1^{COO^-}$, as has been done in reference [113].

4.5 CHLORINATION OF POLYETHYLENE

The kinetics of this reaction was studied in 1962–1963 by Michail *et al.* [68] and Wielopolski *et al.* [69]. These authors showed that the reaction rate of photo-initiated chlorination is almost independent of temperature, but the reaction slows down after introduction of 60 wt% Cl into polyethylene (PE) and practically ceases once 73 wt% Cl is introduced. This effect evidently reflects a deactivating action of substituents (Cl atoms) on neighboring C—H bonds [114–117]. However, in references [68] and [69] no quantitative analysis of kinetic data was carried out and individual rate constants for chlorination were not determined. Meanwhile, PE chlorination appears to be of particular value for quantitative studies of the effect of neighboring units [118].

Gusev *et al.* [119] investigated the products of chlorinated C_{20}–C_{25} n-paraffins related to PE by IR spectroscopy and found that, because of the retardation effect of chlorine atoms introduced into the molecule, the CCl_2 groups are not practically formed until a high degree of chlorination (about 73 wt% Cl) is attained and the chlorination of CH_2 groups adjacent to CHCl groups is inhibited. Oswald and Kubu [120] studied IR spectra of PE of different degrees of chlorination and established that at the Cl/C ratio of 0.7 the polymer does not contain any CCl_2 groups. It is true that later the presence of CCl_2 groups in chlorinated PE (CPE) was detected by NMR spectroscopy (cf. reference [121]). It should be noted, however, that in the solvents generally used for PE chlorination (chlorinated paraffins) the solubility of the polymer depends significantly on the degree of conversion and decreases markedly at the Cl/C ratios of 0 and 0.5 [118], so that at commencement of the process and at approximately 50% of conversion of CH_2 groups, chlorination may proceed as a quasi-heterogeneous reaction. Under these conditions, the CPE composition heterogeneity increases and components having a greater degree of conversion naturally contain appreciable amounts of CCl_2 groups.

Lukás *et al.* [122] studied poly(vinyl chloride) (PVC) chlorination in suspension (concentrated HCl, 20 °C, UV irradiation) and in tetrachloroethane solution (115 °C, thermal initiation). At the ratio Cl/C = 0.76–0.77, the ratios of converted CHCl and CH_2 groups in suspension equaled 13:1 and in solution 0.32:1. It should be borne in mind that temperature elevation (from 20 to 115 °C) is accompanied by selectivity diminution. It is, therefore, reasonable to assume that in a good solvent and at a moderate temperature and reaction rate the CPE may be regarded, up to a significantly high degree of conversion, as a copolymer composed of —CH_2— and —CHCl— units.

Insofar as the deactivating effect of chlorine rapidly falls off along the chain [115–117], it is feasible to assume in first approximation that CHCl groups retard chlorination of neighboring CH_2 groups only. Three types of triads with the unreacted CH_2 unit in the middle should be distinguished, viz. —CH_2—CH_2—CH_2, —CH_2—CH_2—CHCl— (or —CHCl—CH_2—CH_2—) and —CHCl—CH_2—CHCl—. In the latter triad Cl atoms may be either in the iso- or in the syndio-position in relation to each other, so the influence of configurational factors on the reaction kinetics could be expected. However, according to Allen and Young [123] and Kolinský *et al.* [124], the reactivity of CH_2 groups in PVC is independent of the polymer microstructure. Although Millan [125] using ^{13}C NMR spectroscopy showed the chlorination of hetero-sequences in PVC to proceed somewhat more easily than that of syndio-sequences, it was demonstrated by Quenum *et al.* [126] that PE chlorinated in solution and characterized by the ratio $Cl/C = 0.79$ contains in the chain short sequences of syndiotactic and isotactic triads. Hence, for PE, the rate of CH_2 group chlorination in the —$CHCl$—CH_2—$CHCl$— triad can be expected to be only insignificantly dependent on the stereochemical configuration of the triad. It is therefore reasonable to assume that the reactivity of CH_2 units in PE chlorination depends solely on the number of reacted neighboring groups; i.e. the central CH_2 groups in the —CH_2—CH_2—CH_2—, —CH_2—CH_2—$CHCl$— (or —$CHCl$—CH_2—CH_2—) and —$CHCl$—CH_2—$CHCl$— triads are chlorinated with the rate constants k_0, k_1 and k_2 respectively.

Krentsel *et al.* [29] employed polymeric models for separate determination of individual rate constants in PE chlorination (it should be noted that the kinetic method of rate constants estimation involving the use of polymeric models was first described in reference [29]). To estimate k_0 and k_2, the polymer models used were PE and PVC respectively, and the authors of reference [29] reasoned that inasmuch as at the initial stage of PE chlorination only centered CH_2 groups in —CH_2—CH_2—CH_2— triads are involved practically, the initial rate of PE chlorination might be employed for calculating the constant k_0. Similarly, the initial rate of PVC chlorination is used for k_2 determination (centered CH_2 groups in —$CHCl$—CH_2—$CHCl$— triads transform during the initial stage of the reaction). Then the constant k_1 can be selected using Eq. (4.31) to describe the experimental data on PE chlorination to relatively high degrees of conversions.

As emphasized in Section 4.1, in investigating the neighbor effect it is preferable to carry out a macromolecular reaction under homogeneous conditions in order to avoid the influence of intermolecular interactions on the process kinetics. In reference [29] the solvent used was chlorobenzene in which PE, PVC and their chlorination products are dissolved. However, pure PE dissolves in chlorobenzene at high temperatures only ($\sim 100\,°C$), when the reaction studies are complicated by the feasibility of reaction product dehydrochlorination. Accordingly, in kinetic experiments the starting material was the partially chlorinated PE prepared by subjecting PE (molecular weight 12 000, 1.4 CH_3 groups per 100 C atoms, according to IR spectra) to chlorination with gaseous chlorine at 80 °C in chlorobenzene (PE concentration in solution of 1%) for a period of 3.5 hours. The target PE contained 6 wt% Cl and dissolved in chlorobenzene at as low a temperature as 40 °C. IR spectroscopy demonstrated the absence of unsaturation and the presence of isolated CHCl groups solely in the polymer, the content of these groups being accounted for in the calculation of the constant k_0.

Krentsel *et al.* [29] investigated the kinetics of PE chlorination at 50 °C, illumination intensity 230 lux, polymer concentration 0.1–0.4 g/100 ml, and under chlorine partial pressures from 120 to 530 torr. Specially designed experiments demonstrated the absence of polymer degradation under the conditions used: the molecular weight of the product with almost maximal chlorine content of 62 wt% obtained in this work was found to be 34 000, this being in good agreement with the calculated value.

At the initial stage of chlorination under the conditions used, the reaction was found to be first order as regards both PE and chlorine. Hence the chlorination rate for CH_2 groups of the same type is described by the equation

$$-\frac{d[CH_2]}{dt} = \text{constant}\,[CH_2][Cl_2]$$

(This equation can be obtained by the conventional quasi-stationary method, if chlorine consumption for solvent chlorination under experimental conditions employed is accounted for and the chain termination step is assumed to occur as a result of reaction of Cl atoms with radicals formed from the solvent.)

In kinetic experiments the starting sample (6 wt% Cl) contained 97.7 mol% of CH_2 groups and 2.3 mol% of isolated CHCl groups. The sample therefore contained 4.6 mol% of type N_1 CH_2 groups disposed at the center of $-CH_2-CH_2-CHCl-$ triads and 93.1 mol% of type N_0 CH_2 groups located at the center of $-CH_2-CH_2-CH_2-$ triads. During the initial reaction stage, predominant transformation of type N_0 CH_2 groups occurs, because the content of these groups is 20 times as great as that of type N_1 CH_2 groups and also owing to the higher reactivity of the former groups.

From the initial rate of chlorination (with due regard to the considerations set forth above), the constant k_0 was found to be 6.1×10^{-4} l/mol s. The constant k_2 was estimated from the kinetic data of PVC chlorination under the same conditions as for PE. According to IR data [118], the $-CCl_2-$ groups were absent in chlorinated PVC containing 60 w% of chlorine maximum. Thus, PVC chlorination to low degrees of conversion involves H replacement with Cl in CH_2 groups solely (this finding is consistent with Kolinský *et al.* data [124]) and it is possible to find k_2 from the initial rate of PVC chlorination, the value found being $k_2 = 0.48 \times 10^{-4}$ l/mol s. Therefore k_2 is of an order of magnitude less than k_0. The found values of k_0 and k_2 were inserted into Eq. (4.31) and, varying the k_1 values, kinetic curves were calculated and compared with the experimental data of PE chlorination to high degrees of conversion (see Fig. 4.13).

It is deemed appropriate to discuss specific details of calculations. The type of equations of (4.31) are derived assuming the constancy of low-molecular reagent concentration in the course of polymer transformation. However, during PE chlorination the solvent (chlorobenzene) likewise undergoes chlorination, so that chlorine solubility in the solvent varies. Since during a kinetic experiment the partial pressure of chlorine was maintained constant in the reactor, chlorine concentration in the solvent was time dependent.

Shown in Fig. 4.14 is the dependence of chlorine concentration in reaction mixture on time, the curve being plotted using the data on chlorine solubility in pure chlorobenzene [127] and in chlorobenzene chlorinated under the conditions of kinetic experiments [29]. In order to account for this dependence, it is

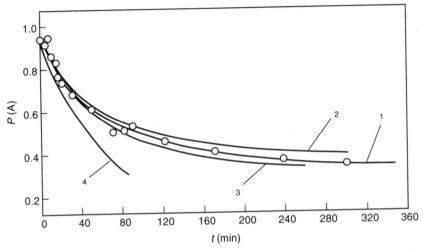

Fig. 4.13 Calculated kinetic curves for the ratios $k_0:k_1:k_2$ equal to $1:0.24:0.08$ (2), $1:0.35:0.08$ (1), $1:0.5:0.08$ (3) and $1:1:1$ (4). Dots, experimental data on polyethylene chlorination in chlorobenzene ($50\,°C$, PE concentration $0.4\,\mathrm{wt\%}$, Cl_2 partial pressure $530\,\mathrm{torr}$). (Reproduced by permission of Nauka, Moscow, Russia, from *Vysokomol. Soed. A*, 1971, **13**, 2489)

pertinent to introduce the so-called transformed time

$$t' = \int_0^t [Cl_2]\, dt$$

In practice, t' could be determined by graphic integration, i.e. calculating the area under the curve shown in Fig. 4.14. Equation (4.31) may now be used assuming $\tau = k_0 t'$.

Curves for different $k_0:k_1:k_2$ ratios are plotted in Fig. 4.13. Curve 4 corresponds to $k_0:k_1:k_2 = 1:1:1$ (the absence of a neighbor effect) and explicitly fails to fit the experimental data. The kinetic data, particularly for high degrees of conversion (in the region of pronounced divergence of the calculated curves), adequately fit curve 2, plotted for $k_0:k_1:k_2 = 1:0.35:0.08$. Hence, the most probable value of k_1 is $2.1 \times 10^{-4}/\mathrm{mol\,s}$, and the error of k_1 estimation, as can be seen from Fig. 4.13, equals $\pm 50\%$.

It would now be appropriate to correlate, in compliance with the concept stated in Section 4.1, the kinetic data with the results of investigations of units distribution and composition heterogeneity of CPE. Oswald and Kubu [120] suggested that the values of N_0, N_1 and N_2 in CPE should be estimated from the intensity of 1468, 1447 and $1428\,\mathrm{cm^{-1}}$ bands respectively. However, Krentsel [118] and later Quenum *et al.* [128] showed that this interpretation of the IR spectrum of CPE was apparently incorrect.

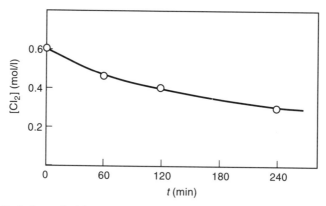

Fig. 4.14 Variation of chlorine concentration in solution during chlorination of polyethylene (50 °C, Cl_2 partial pressure 530 torr). (Reproduced by permission of Nauka, Moscow, Russia, from *Vysokomol. Soed. A*, 1971, **13**, 2489)

The distribution of $-CH_2-$ and $-CHCl-$ units in CPE has been extensively studied by NMR [70, 121, 129–132]. Saito *et al.* [121] measured spectra at a resonance frequency of 100 MHz, but Brame [70] claimed that a sufficiently accurate NMR spectrum interpretation could be obtained at 220 MHz only.

In general, Brame's approach to the problem of NMR spectrum interpretation is highly instructive. From the NMR spectrum of CPE, Brame found the relative amount of CH_2 groups that are in α-, β-, and γ- (or more remote) positions to CHCl groups. For the sample containing 42 wt% Cl, the conventional peak resolution yielded the following values: $\alpha = 0.62$, $\beta = 0.16$ and $\gamma = 0.22$ (cf. column 2 of Table 4.6). Brame also used the method of Frensdorff and Ekiner [133] (cf. Chapter 3) to calculate the fractions of α-, β- and γ-methylene groups for the ratios of constants $k_0:k_1:k_2 = 1:1:1$ and 1:0.6:0.36 (cf. columns 4 and 5 of Table 4.6). Calculations showed that variations in the constants ratios affect the value of β, but insignificantly, and this situation prompted Brame to introduce corrections into the initial resolution of the peaks of α- and β-CH_2 groups (the peak of the γ-CH_2 group is well resolved and the value of γ was assumed in

Table 4.6. Distribution of CH_2 groups in chlorinated polyethylene containing 42 wt% Cl

Group type	NMR data [70]		Calculated data for $k_0:k_1:k_2$ ratio		
	Initial	Corrected	1:1:1	1:0.6:0.36	1:0.35:0.08
α	0.62	0.53	0.48	0.53	0.56
β	0.16	0.25	0.25	0.24	0.23
γ	0.22	0.22	0.27	0.23	0.21

reference [70] to be reliable), as listed in column 3 of Table 4.6. As can be seen, Brame made practically no attempt to estimate precisely the constants ratio directly from the NMR spectrum. On the contrary, he resorted to correcting the resolution of peaks using the ratios $k_0:k_1:k_2$ adopted in reference [133]. It is therefore safe only to assert that the ratio 1:0.6:0.36 does not contradict the NMR measurements.

Such spectrum correction could have been conducted for other constants ratios as well, in particular for the ratio $k_0:k_1:k_2 = 1:0.35:0.08$ found in reference [29], inasmuch as the calculated values for α, β and γ (cf. column 6 in Table 4.6) are close to those for the ratio 1:0.6:0.36. Thus, Brame's data do not permit an unambiguous choice of the rate constants. This result points to the necessity to take the utmost caution in tackling the problem of determining the rate constants from the units distribution data, even when such an effective method as NMR is employed.

Keller [132] examined the structure of CPE by ^{13}C NMR spectroscopy. Using the $k_0:k_1:k_2 = 1:0.33:0.11$ ratio, which is very close to the ratios found in reference [29] for PE chlorination and in reference [134] for the chlorination of model compounds (linear and cyclic paraffins), Keller calculated the content of various triads and pentads in CPE, the results obtained being in good agreement with the ^{13}C NMR data. Hence, there exists satisfactory agreement between the results of investigating the units distribution in CPE and the data on the kinetics of PE chlorination.

Chlorination kinetics may also be correlated with the composition heterogeneity of the reaction product. CPE composition heterogeneity was examined [135] by the cross-fractionation technique (see Chapter 8). The dependence of precipitation threshold γ^* on the CPE composition x (x denotes the mole fraction of CHCl groups in polymer) in different solvent–precipitant systems has been studied. In compliance with requirements of the procedure, for these studies CPE samples of various composition were used, but having the same chain length. Such samples were prepared by chlorinating the starting PE to different conversion degrees; as mentioned before, polymer chain length does not change during chlorination. The γ^* versus x curves therefore represent the dependence of CPE solubility on the sample composition in the examined solvent–precipitant systems.

The chlorobenzene–cyclohexane and cyclohexanone–methanol systems were selected for fractionation of CPE of an average composition $x \approx 0.415$, since, in the range $0.35 < x < 0.45$, in the first system the γ^* value drops fairly sharply with increasing x, while in the second system γ^* grows. It should be noted that this composition range is adapted very well for investigations of composition heterogeneity since at the ratio $k_0:k_1:k_2 = 1:0.3:0.3$, which is close to the ratios found for PE chlorination, the dispersion of composition distribution almost does not change in the interval $0.3 < x < 0.7$ [136].

Fractionation was carried out at 40 °C and comprised separating the initial sample into four fractions in the cyclohexanone–methanol system, followed by

Table 4.7. Composition (mole fraction of CHCl groups) of fractions separated by cross-fractionation of chlorinated polyethylene

	n			
m	1	2	3	4
a	0.40	0.42	0.43	0.435
b	0.46	0.42	0.37	0.43
c	0.46	0.43	0.42	0.40
d	0.43	0.39	0.39	0.405

separating each intermediate fraction also into four fractions in the chloroben-zene–cyclohexane system. The results of fractionation are listed in Table 4.7 and in Fig. 4.15 (points).

The modified first-order Markovian approximation (cf. Chapter 3) has been used to calculate the theoretical composition distribution of the CPE test sample for various rate constants ratios. In Fig. 4.15 the integral distribution functions $\Phi(x)$ are plotted (with due regard to analytical errors of fractions composition

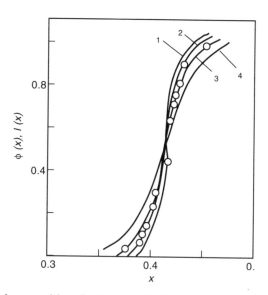

Fig. 4.15 Integral composition distribution of chlorinated polyethylene. Dots, data of cross-fractionation; curves, calculated for ratios $k_0:k_1:k_2$ equal to 1:0.1:0.08 (1), 1:0.35:0.08 (2), 1:1:1 (3) and 1:5:5 (4). (Reproduced by permission of Nauka, Moscow, Russia, from *Vysokomol. Soed. B*, 1974, **16**, 372)

determined by the Schoniger method [137, 138]). As can be seen, curve 4, which corresponds to a weak acceleration, displays significant divergence from the experimental points. The other three curves differ only slightly from one another and show fairly good agreement with the fractionation data.

Let us compare Figs. 4.13 and 4.15. The differences displayed by curves 3 and 2 in Fig. 4.15, corresponding to the absence of the neighbor effect ($k_0:k_1:k_2 = 1:1:1$) and to the retardation ($k_0:k_1:k_2 = 1:0.35:0.08$) respectively, are within the experimental error limits in an investigation of composition heterogeneity. On the other hand, the divergence of kinetic curves 4 and 2 in Fig. 4.13, which correspond to the same constants ratios, extend far beyond the limits of error of kinetic measurements. Therefore, for retardation, when composition heterogeneity is slight, the kinetic measurements may serve as a much more sensitive criterion for the quantitative evaluation of $k_0:k_1:k_2$ ratios than the data on composition heterogeneity of the reaction product. Nonetheless, the pattern presented in Fig. 4.15 provides proof that the data on CPE composition distribution are not inconsistent with the kinetic regularities of PE chlorination which characterized by ratio $k_0:k_1:k_2 = 1:0.35:0.08$ found in reference [29].

Agreement between kinetic, units distribution and composition heterogeneity data points to confidence in the ratio $k_0:k_1:k_2 = 1:0.35:0.08$, such ratios being employed to interpret the mechanism of the neighbor effect in chlorination of PE and related compounds.

The influence exerted by substituents on rate constants is interpreted in terms of correlation equations which are applied in organic chemistry both to ionic and to free-radical reactions [139] (the chlorination of PE and related paraffins is a free-radical reaction [140]). Wautier and Bruylants [141] and also Gusev [142], in interpreting the data on the chlorination of derivatives of C_3–C_5 fatty acids and monochloroparaffins, employed Taft's correlation equation [143] to account for the contribution of induction and hyperconjugation effects:

$$\log\left(\frac{k_i}{k_s}\right) = \rho^*\Delta\sigma^* + h\Delta n \tag{4.33}$$

where k_s and k_i are the rate constants for reactions of a standard and a substituted aliphatic compound respectively, ρ^* is the constant characterizing the sensitivity of a given reaction series to the induction effect of substituents, $\Delta\sigma^*$ is the algebraic sum of induction constants of substituents at the reaction site, Δn is the difference between the number of atoms participating in hyperconjugation in the case of a given and a standard substituent, and h denotes the hyperconjugation parameter of a reaction series. The values $\rho^* = -0.20$ and $h = 0.11$–0.12 were found [141, 142] for the chlorination of monochloroparaffins. Interpretation of data on the chlorination of paraffins in references [141] and [142] is of unquestionable interest, but is open to criticism.

To begin with, it is appropriate to recall that, according to Dewar [144], the replacement of the C—H bond with C—C or another bond hardly affects radical stabilization (with an unpaired electron at the C atom in the β-position relative to the bond in question). Dewar also showed that many effects formerly assumed to be caused by hyperconjugation do lend themselves to an alternative interpretation based on the role of other factors. As far as the chlorination of paraffins is concerned, the available experimental data, despite their inconsistency, are indicative of a minor role played by hyperconjugation in this reaction [142, 145, 146].

Next, using the correlation equation for evaluating the data on the chlorination of 1-chloroparaffins the authors Wautier and Bruylants [141] and Gusev [142] selected the 1-C atom as the active center. However, chlorine bonded to this carbon atom exerts on the reactivity of the α-CH bond not only an induction but also a direct mesomeric effect, the latter effect being capable of substantially compensating the induction effect [118]. Correlation of reactivities of CH_2 and $CHCl$ groups therefore appears unjustified when use is made of correlation equations that fail to account for the mesomeric interaction of Cl with the α-CH bond. In this case, it is necessary to compare the reactivities of CH_2 groups disposed at different distances from the Cl atom.

The data of the chlorination of 1-chlorohexane and 1-chlorooctane used by Gusev [142] can be recalculated (for comparison purposes, the centers at the 2-C and 4-C atoms are selected) using the Taft equation to account for the inductive effect only:

$$\log\left(\frac{k_i}{k_s}\right) = \rho^* \Delta\sigma^* \tag{4.34}$$

As these results have been obtained under the conditions when the ratios of formation rates for various dichlorides remain constant, it is feasible to substitute the ratio of the yields (R) of respective isomers for the ratio of rate constants. Then Eq. (4.34) may be rewritten as follows:

$$\log\left(\frac{R_i}{R_s}\right) = \rho^* \Delta\sigma^*$$

where the subscripts s and i denote the standard and examined CH_2 group respectively. The results of such a calculation are listed in Table 4.8, the values of σ^* being adopted from reference [147].

Figure 4.16 presents the $\log(R_i/R_s)$ versus $\rho^* \Delta\sigma^*$ plot and also the data of PE chlorination, for which the central CH_2 groups in $CH_2CH_2CH_2$ or $ClCHCH_2CH_2$ triads were selected as the standard center and use was made of the values of rate constants k_0, k_1 and k_2 found in reference [29]. As can be seen, the data on the chlorination of monochloroparaffins and PE adequately fit the common straight line or, in other words, obey Eq. (4.34), while the value of parameter $\rho^* = -0.46$ is in good agreement with the data cited in the literature. For the reaction of H atom scission from the methyl group of toluene derivatives with atomic chlorine at 80 °C in chlorobenzene $\rho^* = -0.485$ [148], while for an analogous reaction involving cumene derivatives at 70 °C in chlorobenzene

Macromolecular Reactions

Table 4.8. Recalculation of data of 1-chlorohexane and 1-chlorooctane chlorination [142] relative to active centers at 2-C and 4-C atoms

Compound	Correlated isomers	$\Delta\sigma^*$	$\log(R_i/R_s)$
1-chlorohexane	1,2/1,4	0.88	−0.439
	1,3/1,4	0.22	−0.212
	1,5/1,4	0.01	0.022
	1,3/1,2	−0.66	0.224
	1,4/1,2	−0.88	0.441
	1,5/1,2	−0.87	0.462
1-chlorooctane	1,2/1,4	0.86	−0.378
	1,3/1,4	0.20	−0.118
	1,5/1,4	−0.07	0.024
	1,6/1.4	−0.09	−0.008
	1,7/1,4	0	0.088
	1,3/1,2	−0.66	0.260
	1,4/1,2	−0.86	0.378
	1,5/1,2	−0.93	0.402
	1,6/1,2	−0.95	0.370
	1,7/1,2	−0.86	0.466

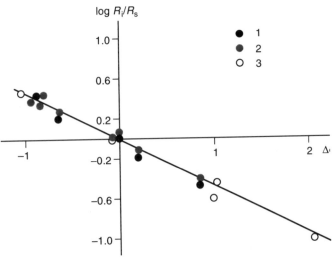

Fig. 4.16 Results of treatment of the data on chlorination of 1-chlorohexane (1), 1-chlorooctane (2) [142] and polyethylene (3) [29] in accordance with the Taft equation (4.34) (cf. Table 4.8)

$\rho^* = -0.44$ [149] (the low value found in references [141] and [142], viz. $\rho^* = -0.22$ appeared because of injudicious selection of the standard active center).

The ratio of individual rate constants for PE chlorination therefore obeys the correlation Eq. (4.34) accounting for the induction effect of substituent, the values of parameters being the same as in the case of chlorinating the paraffins. This situation stimulated investigations of the neighbor effect in the chlorination of paraffins as low-molecular analogs of PE.

Usmanov, Litmanovich, Platé and co-workers [32, 134, 150] studied the neighbor effect in the chlorination of PE and its analogs (n-paraffins and cycloparaffins) under identical conditions. The kinetics was investigated photocolorimetrically at 50 °C in CCl_4 solution. The chlorination of cyclooctane was first studied, and the rate constants were estimated using two different methods, viz. the kinetic method and ^{13}C NMR analysis of the reaction products composition.

As can be seen in Table 4.9, at early steps of chlorocyclooctane chlorination (10 and 18% conversions), when the formation of higher chlorides has as yet not commenced, the relative yields of dichlorides are practically independent of conversion. The ratios of rates of their formation at these early stages are constant and may be regarded as a measure of the relative reactivity of respective C—H bonds. With due regard to the transannular effect [151], the ratio of reactivities of $\alpha:\beta:\gamma:\delta:\varepsilon$ C—H bonds in chlorocyclooctane chlorination equals 0.22:0.43:0.96:1.0:1.0. Hence, Cl atom incorporation into the cyclooctane molecule causes the retardation of chlorination in the α- and β-positions, but γ- and more remote C—H bonds remain practically unaffected.

The ratio of β- and ε-bonds may be taken as k_1/k_0. To estimate the value of k_2/k_0, in reference [134] use was made of Eq. (4.34). Then $k_2/k_0 = (k_1/k_0)^2$, and since $k_1/k_0 = 0.43$, $k_2/k_0 \approx 0.18$. Hence, an analysis of the mixture of chlorocyclooctane chlorination products yields the ratio $k_0:k_1:k_2 = 1:0.43:0.18$.

Table 4.9 Composition of chlorooctane chlorination products (mol%) depending on the degree of conversion (50 °C, CCl_4) [150]

Conversion (%)	1,1	1,2- trans	1,2- cis	1,3- trans	1,3- cis	1,4- trans	1,4- cis	1,5- trans	1,5- cis	Higher chlorides
10	1.6	7.8	3.1	14.3	18.7	19.0	18.4	7.2	9.6	—
18	2.0	7.4	3.0	14.7	17.7	19.5	17.9	7.7	9.0	—
30	Traces	3.8	2.2	8.7	13.4	17.9	19.0	8.0	10.4	16.9
50	Traces	5.4	2.7	8.6	12.6	17.5	18.5	6.9	9.8	18.0

(Reproduced by permission of Nauka, Moscow, Russia, from *Dokl. Akad. Nauk. SSSR*, 1973, **210**, 114)

The data of Table 4.9 are likewise of interest when considering the role of stereoselective effects in chlorination. Russell and Ito [152] observed high stereoselectivity in the chlorination of monochlorosubstituted small cycles in position 2. The ratios of *trans/cis* isomers for 1,2-dichlorides were found to be 62:1 and 11:1 in chlorinating chlorocyclopentane and chlorocyclohexane respectively. For chlorocyclooctane this ratio equals 2.5:1 (see Table 4.9). Stereoselectivity is likely to be associated with the conformational characteristics of cycloparaffins [32, 152]. A pronounced diminution of stereoselectivity as the size of cycle grows makes it possible to assume that, in chlorination of linear paraffins and PE, the role of stereoselective effects is insignificant and the influence of CHCl groups on the reactivity of neighboring CH_2 units appears to be only slightly dependent on the stereochemical configuration of a chlorinated linear hydrocarbon.

Usmanov and co-workers [32, 134] investigated chlorination kinetics using the unit shown in Fig. 4.17. The photocolorimetric technique was used with continuous recording of the chlorine concentration in the solution (CCl_4 has been used as a solvent) in the course of the reaction.

In cyclooctane and cyclododecane chlorination, determining the initial reaction rate showed the reaction to be first order for the hydrocarbon and an order 1/2 for chlorine. Conversely, in the case of cyclooctacosane, n-hexadecane and PE, the reaction has an order 1/2 for hydrocarbon and first order for chlorine.

These findings necessitated a revision of the available description of the chlorination kinetics of PE and its analogs [153]. The point is that the kinetic equations discussed in Chapter 3 pertain exclusively to the reactions of the first order for the macromolecular reagent, let alone the fact that the theory of polymer-analogous transformations did not embrace chain reactions. Let us consider how Keller's approach [154] (see Chapter 3) may be modified to describe PE chlorination as a chain reaction having an order of 1 or 1/2 for the polymer [153], provided the chlorination is carried out under conditions where the formation of CCl_2 groups is practically absent.

In a chlorination, the first order for hydrocarbon may be interpreted on the basis of the following scheme:

$$Cl_2 \xrightarrow{k^{(0)}} 2\,Cl^{\cdot}$$

$$Cl^{\cdot} + RH \xrightarrow{k^{(1)}} R^{\cdot} + HCl$$

$$R^{\cdot} + Cl_2 \xrightarrow{k^{(2)}} Cl^{\cdot} + RCl$$

$$Cl^{\cdot} + Cl^{\cdot} \xrightarrow{k^{(3)}} Cl_2$$

Scheme 4.8

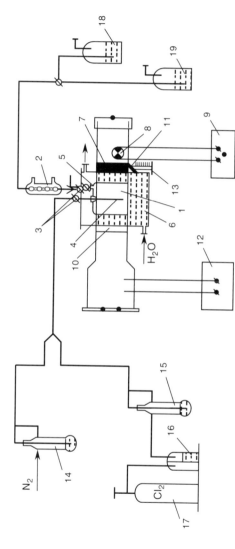

Fig. 4.17 Layout of a unit for the study of hydrocarbon chlorination kinetics: 1, reactor; 2, reflux condenser; 3, valves; 4, capillary; 5, sample inlet; 6, thermostated chamber; 7, light filter; 8, incandescent lamp; 9, rectifier; 10, 11, photocells; 12, potentiometer; 13, slit diaphragm adjuster; 14, 15, rheometers; 16, concentrated H_2SO_4; 17, chlorine cylinder; 18, KOH solution; 19, 10% solution of KI. (Reproduced by permission of Nauka, Moscow, Russia, from *Kinetika i Kataliz*, 1977, **18**, 337)

Macromolecular Reactions

where $k^{(0)}$ is the initiation constant, $k^{(1)}$ and $k^{(2)}$ are the chain propagation constants and $k^{(3)}$ is the chain termination constant.

The neighbor effect may manifest itself at both chain propagation steps. It is therefore necessary to introduce the rate constants for the reaction of Cl atoms with CH_2 groups having 0, 1 and 2 reacted neighbors: $k_0^{(1)}$, $k_1^{(1)}$ and $k_2^{(1)}$ respectively.

Denoting the concentrations of such CH_2 groups as RH_0, $2RH_1$ (factor 2 appeared because after Keller [154] RH_1 denotes CH_2 groups having a reacted neighbor on one side only) and RH_2, step 2 in Scheme 4.8 may be rewritten as follows:

$$Cl^{\cdot} + RH_0 \xrightarrow{k_0^{(1)}} R_0^{\cdot} + HCl$$

$$Cl^{\cdot} + RH_1 \xrightarrow{k_1^{(1)}} R_1^{\cdot} + HCl$$

$$Cl^{\cdot} + RH_2 \xrightarrow{k_2^{(1)}} R_2^{\cdot} + HCl$$

Scheme 4.9

Similarly, for the reaction of Cl_2 with corresponding radicals:

$$R_0^{\cdot} + Cl_2 \xrightarrow{k_0^{(2)}} Cl^{\cdot} + RCl$$

$$R_1^{\cdot} + Cl_2 \xrightarrow{k_1^{(2)}} Cl^{\cdot} + RCl$$

$$R_2^{\cdot} + Cl_2 \xrightarrow{k_2^{(2)}} Cl^{\cdot} + RCl$$

Scheme 4.10

Inserting Schemes 4.9 and 4.10 into Scheme 4.8 instead of steps 2 and 3 respectively, and resorting to the method of quasi-stationary approximation, the following reaction rate equation is obtained:

$$-\frac{dRH}{dt} = \left(\frac{2k^{(0)}}{k^{(3)}}\right)^{1/2} Cl_2^{1/2}(k_0^{(1)}RH_0 + 2k_1^{(1)}RH_1 + k_2^{(1)}RH_2) \qquad (4.35)$$

Hence, chlorination proceeding according to Scheme 4.8 yields the kinetic equation which is first order for the polymer, so that Eq. (4.35) may be solved by the Keller method [154], i.e. by solving the following equations:

$$\frac{dN_0}{dt'} = -(k_0 + 2\bar{k})N_0$$

$$\frac{dN_1}{dt'} = -(k_1 + \bar{k})N_1 + \bar{k}N_0 \tag{4.36}$$

$$\frac{dN_2}{dt'} = -k_2 N_2 + 2\bar{k}N_1$$

where N_0, N_1 and N_2 are the mole fractions of RH_0, RH_1 and RH_2 respectively, $k_i = (2k^{(0)}/k^{(3)})^{1/2} k_i^{(1)} (i = 0, 1, 2)$, $\bar{k} = (k_0 N_0 + k_1 N_1)/(N_0 + N_1)$ and $dt' = (Cl_2)^{1/2} dt$.

It should be noted that for a reaction of first order for the polymer the Keller equations (4.36) and the McQuarrie equation (4.31) are equivalent (see Chapter 3), so the above presentations of PE chlorination in chlorobenzene based on Eq. (4.31) is quite correct.

In the case of an order 1/2 for hydrocarbon, the following scheme of chlorination is usually offered:

$$Cl_2 \xrightarrow{k^{(0)}} 2Cl^{\cdot}$$

$$Cl^{\cdot} + RH \xrightarrow{k^{(1)}} R^{\cdot} + HCl$$

$$R^{\cdot} + Cl_2 \xrightarrow{k^{(2)}} Cl^{\cdot} + RCl$$

$$R^{\cdot} + Cl^{\cdot} \xrightarrow{k^{(3)}} RCl$$

Scheme 4.11

Here, in contrast to Scheme 4.8, chain termination occurs due to the recombination of Cl atoms with alkyl radicals.

Assuming the neighbor effect to manifest itself, as before, at both chain propagation steps, we believe that the $R^{\cdot} + Cl^{\cdot}$ recombination rate constant $k^{(3)}$ is the same for $R_0^{\cdot}$, $R_1^{\cdot}$ and $R_2^{\cdot}$ [155]. A further simplifying assumption is that at both chain propagation steps the neighbor effect is one and the same or, in other words, $k_0^{(1)}/k_0^{(2)} = k_1^{(1)}/k_1^{(2)} = k_2^{(1)}/k_2^{(2)}$.

Applying the quasi-stationary approximation method to Scheme 4.11 yields the following equation:

$$-\frac{dRH}{dt} = \left(\frac{k^{(0)}}{k^{(3)}}\right)^{1/2} \frac{Cl_2}{RH^{1/2}} \left[(k_0^{(1)}k_0^{(2)})^{1/2}RH_0 + 2(k_1^{(1)}k_1^{(2)})^{1/2}RH_1 + (k_2^{(1)}k_2^{(2)})^{1/2}RH_2\right]$$

(4.37)

Hence, under conditions of determination of kinetic parameters (using the initial rate when $RH = RH_0$), the reaction indeed has the order 0.5 for RH and 1 for Cl_2.

To employ Keller's approach to the case of an order 1/2 for the polymer, it is also pertinent to introduce, apart from N_0, N_1 and N_2, the following quantities: $\bar{N}_0, \bar{N}_1$ and $\bar{N}_2$ which denote the mole fractions of $R_0^{\cdot}$, $R_1^{\cdot}$ and $R_2^{\cdot}$ radicals, as well as $\tilde{N}_0$ and $\tilde{N}_1$ which stand for the mole fractions of the type $-CH_2CH_2\dot{C}H-$ and $-CHClCH_2\dot{C}H-$ radicals respectively. As a result, the following set of equations is obtained:

$$\frac{dN_0}{dt'} = -\frac{(k_0 + 2\bar{k})N_0}{(N_0 + 2N_1 + N_2)^{1/2}}$$

$$\frac{dN_1}{dt'} = -\frac{(k_1 + \bar{k})N_1 + \bar{k}N_0}{(N_0 + 2N_1 + N_2)^{1/2}}$$

(4.38)

$$\frac{dN_2}{dt'} = -\frac{k_2N_2 + 2\bar{k}N_1}{(N_0 + 2N_1 + N_2)^{1/2}}$$

where $k_i = (k^{(0)}k_i^{(1)}k_i^{(2)}/k^{(3)})^{1/2}$ $(i = 0, 1, 2)$, $\bar{k} = (k_0N_0 + k_1N_1)/(N_0 + N_1)$ and $dt' = [Cl_2]dt$.

As can be seen from Eqs. (4.35) or (4.37), it is feasible to determine the constant k_0 from the initial rate of chlorination. Next, the most probable values of k_1 and k_2 can be found for which the kinetic curve calculated from Eqs. [4.36] or (4.38) displays superior agreement with the experimental data. The values of transformed time t' are derived from the data on Cl_2 concentration and change with time in the solution: $t' = \int_0^t [Cl_2] \, dt$ and $t' = \int_0^t [Cl_2]^{1/2} \, dt$ for the reaction order of 1 and 1/2 for chlorine respectively.

It is deemed pertinent to ascertain whether the equations derived for infinite chains are valid for cycles of relatively small size in the case of the retarding neighbor effect. The situation is exemplified by cyclooctane, the smallest of the hydrocarbons studied by Usmanov *et al.*

The equations suggested by Alfrey and Lloyd [156] (see Chapter 3) are adapted for a closed loop consisting of eight units to describe the kinetics of cyclooctane chlorination:

$$\frac{dN_8}{dt} = -8k_0 N_8$$

$$\frac{dN_7}{dt} = 8k_0 N_8 - (2k_1 + 5k_0)N_7$$

$$\frac{dN_6}{dt} = 2k_1 N_7 - (2k_1 + 4k_0)N_6$$

$$\frac{dN_5}{dt} = 2k_0 N_7 + 2k_1 N_6 - (2k_1 + 3k_0)N_5$$

$$\frac{dN_4}{dt} = 2k_0(N_7 + N_6) + 2k_1 N_5 - (2k_1 + 2k_0)N_4 \qquad (4.39)$$

$$\frac{dN_3}{dt} = 2k_0(N_7 + N_6 + N_5) + 2k_1 N_4 - (2k_1 + k_0)N_3$$

$$\frac{dN_2}{dt} = 2k_0(N_7 + N_6 + N_5 + N_4) + 2k_1 N_3 - 2k_1 N_2$$

$$\frac{dN_1}{dt} = 2k_0(N_7 + N_6 + N_5 + N_4 + N_3) + 2k_1 N_2 - k_2 N_1$$

where N_i is the number of sequences comprising i CH_2 groups flanked on both sides by $CHCl$ groups calculated per one unit.

In Fig. 4.18, the solutions of Eqs. (4.39) at initial conditions $N_8(0) = 1/8$, $N_i(0) = 0$ $(i = 1, \ldots, 7)$ for various ratios of constants $k_0 : k_1 : k_2$ are correlated with results of calculations using Eq. (4.31) (the Cl_2 concentration was assumed to be constant, this limitation producing no effect on the results of comparing various methods of kinetic curve calculation, since in either case a variation of Cl_2 concentration would be accounted for in a similar manner). Good agreement between the calculation according to Eqs. (4.39) and (4.31) provides grounds to employ Eqs. (4.36) or (4.38) (depending on the reaction order for hydrocarbon) for interpreting kinetic data on the chlorination of cycloparaffins containing a minimum of eight CH_2 groups in the cycle, and also of unbranched paraffins with a sufficiently long chain (in order to disregard the influence of end groups).

A comparison of curves calculated by solving Eq. (4.31) and experimental data on the kinetics of chlorocyclooctane chlorination [32, 134] (see Fig. 4.19) shows that the best agreement is observed at the ratio $k_0 : k_1 : k_2 = 1 : 0.43 : 0.18$, i.e. the kinetic method yielded the same ratios as that obtained from NMR analysis of the reaction products. This coincidence demonstrates the efficiency of the kinetic method which found an application to estimate the rate constants of chlorination of cyclododecane, cyclooctacosane and n-hexadecane (in the latter case, in calculations of k_0 the presence of CH_3 groups was accounted for, and since the

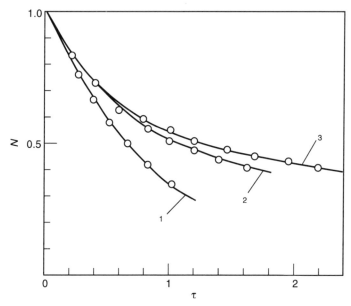

Fig. 4.18   Kinetics of 'mathematical chlorination' of cyclooctane (N-molar fraction of CH_2 groups). Curves, calculations for infinite chain according to Eq. (4.12); dots, calculations for an eight-membered cycle according to Eq. (4.39); for ratios $k_0{:}k_1{:}k_2$ equal to 1:1:1 (1), 1:0.43:0.18 (2) and 1:0.35:0.08 (3). (Reproduced by permission of Nauka, Moscow, Russia, from *Vysokomol. Soed. A*, 1977, **19**, 1211)

ratio of reactivities of primary and secondary C—H bonds in the chlorination of paraffins equals 1:3 [146], two CH_3 groups were assumed to be approximately equivalent to one CH_2 group as regards the rate of chlorine consumption).

Usmanov *et al.* [134] evaluated the rate constants of PE chlorination in CCl_4 by the method of polymeric models using chlorinated PE samples containing 6, 17.4 and 57 wt% Cl as models. The samples were prepared by Krentsel *et al.* [29] as a result of PE chlorination in chlorobenzene as described above. It should be noted that the solubility of CPE in CCl_4 depends largely on the chlorine content in the sample, so that reaction system homogeneity is not preserved at high degrees of polymeric model conversion. In this case, studies were necessarily restricted to the determination of approximate values of k_0, k_1 and k_2 from the initial rates of chlorination of three polymeric models by solving simultaneously the corresponding Eq. (4.31). The requisite values of N_0, N_1 and N_2 for each model were calculated by methods described in Chapter 3 using the ratio $k_0{:}k_1{:}k_2 = 1{:}0.35{:}0.08$.

The results concerning the chlorination kinetics of various compounds are presented in Table 4.10. The found reaction orders correspond to a dissimilar

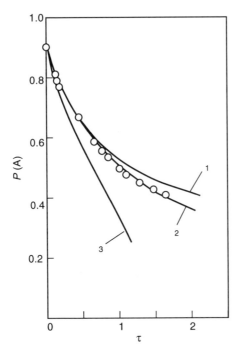

Fig. 4.19 Kinetics of chlorocyclooctane chlorination ($50\,^\circ$C, $[Cl_2]_0 = 19 \times 10^{-2}$ mol/l, $[CH_2]_0 = 0.252$ mol/l). Dots, experimental data; curves, calculated for ratios $k_0:k_1:k_2$ equal to 1:0.33:0.10 (1), 1:0.43:0.18 (2) and 1:1:1 (3) [134]. (Reproduced by permission of Nauka, Moscow, Russia, from *Kinetika i Kataliz*, 1977, **18**, 337–41)

Table 4.10. Reaction orders and rate constants for the chlorination of polyethylene and its analogs ($50\,^\circ$C, solvent CCl_4) [134]

| Compound | Reaction order | | $k_0 \times 10^3$ $(l^{1/2}/mol^{1/2}\,s)$ | $k_0:k_1:k_2$ |
	For RH	For Cl_2		
Cyclooctane	1	0.5	0.7	1:0.43:0.18
Cyclododecane	1	0.5	0.8	1:0.43:0.18
Cyclooctacosane	0.5	1	1.0	1:0.43:0.18
n-Hexadecane	0.5	1	0.7	1:0.35:0.08
Polyethylene	0.5	1	1.8	1:0.38:0.11

(Reproduced by permission of Nauka, Moscow, Russia, from *Kinetika i Kataliz*, 1977, **18**, 337)

mechanism of the chain termination step, viz. $Cl^. + Cl^. \rightarrow Cl_2$ for cyclooctane and cyclododecane, and $R^. + Cl^. \rightarrow RCl$ for the remaining compounds. Insofar as in all cases (except PE) in kinetic studies practically the same concentration ranges of hydrocarbon and chlorine were used, the dissimilarity of the chain termination mechanism should presumably be ascribed to conformational particularities of the compounds studied.

In the alkyl or cycloalkyl radical, the C atom with an unpaired electron is characterized by the coordination number of 3 and by the planar position of C—H bonds. The termination step $R^. + Cl^. \rightarrow RCl$ causes the coordination number to change from 3 to 4 and active center transition from the planar to the tetrahedral configuration.

In medium-sized C_8–C_{12} cycles, this transition is accompanied by an increase in the number of shielded C—H bonds and, therefore, brings about torsional strain enhancement. Owing to this situation, for cyclooctane and cyclododecane the reaction $R^. + Cl^.$ is somewhat hindered, and chain termination appears to proceed mainly according to the $Cl^. + Cl^.$ mechanism.

No such hindrances to the reaction $R^. + Cl^.$ exist in linear hydrocarbons. With regard to macrocycles, it should be noted that their conformations are close to that of linear hydrocarbons [157]. Therefore, for n-hexadecane, PE and cyclooctacosane, chain termination may proceed according to the $R^. + Cl^.$ reaction.

The $k_0:k_1:k_2$ ratios for all compounds are very close to one another, so the retarding action of chlorine substituent on the neighboring CH_2 groups is practically the same in both PE and its linear and cyclic analogs. This means that the conformational peculiarities of cycloparaffins having eight or more atoms in the cycle do not influence the magnitude of the neighbor effect in chlorination. Moreover, the found rate constants obey the ratio $k_2/k_0 \approx (k_1/k_0)^2$ derived from the correlation Eq. (4.34), as demonstrated previously for cyclooctane. On these grounds, it was assumed [134] that the contribution of steric factors to the retardation mechanism in chlorination of PE and its analogs could be disregarded and that retardation is primarily caused by the induction effect of substituents (Cl atoms). It is noteworthy that in the case when precisely this mechanism of the neighbor effect is operative, it would be natural to employ low-molecular analogs of the polymer of interest as adequate models for studying the reaction kinetics.

As can be concluded from the material presented in this section, the chemical behavior of polyethylene under homogeneous reaction conditions is described with good approximation within the framework of the neighbor effect, is not complicated in any appreciable manner by the conformational or other effects associated with macromolecular reagent and lends itself to modeling by means of linear (having a long chain of about sixteen carbon atoms) and cyclic paraffins, provided the conformational characteristics of the latter are close to those of linear hydrocarbons (for an estimation of the neighbor effect proper, i.e. rate constants ratios, medium-size cycles are useful as well).

4.6 QUATERNIZATION OF POLYVINYLPYRIDINES

This reaction was one of the first objects of research into the effect of neighboring units in polymer-analogous transformations. Fuoss and co-workers [88, 158] studied in detail the kinetics of quaternization of poly-4-vinylpyridine (P4VP) and a series of low-molecular models using n-butyl bromide and found that in the polymer the alkylated pyridine group retarded the reaction of neighboring units. Later Morcellet-Sauvage and Loucheux [159] carried out a further detailed investigation of the kinetic behavior of P4VP, poly-2-vinylpyridine (P2VP), poly-2-methyl-5-vinylpyridine (P2M5VP) and model compounds in quaternization with alkyl bromides in tetramethylene sulfone. The quaternization of low-molecular models obeys the second-order kinetics. The second-order kinetics is also obeyed for statistical copolymers of both 4-vinylpyridine and 2-methyl-5-vinylpyridine with styrene containing less than 20 mol% of pyridine units. Moreover, even the rate constants for copolymers are the same as for related model compounds. However, the quaternization of homopolymers proceeds with retardation. The authors interpreted their results in terms of the neighbor effect and used some semiempirical equations to estimate individual rate constants. They found $k_0 : k_1 : k_2$ ratios of 1:0.67:0.10 and 1:0.63:0.09 at 70 °C for P4VP and P2M5VP respectively. (For P2VP the steric hindrances caused by the very close N atom position to the polymer backbone are so great that it is practically impossible to reach high degrees of conversion.)

As pointed out in Section 4.3, a precise kinetic equation for the reaction involving the neighbor effect was initially employed by Arends [87] to evaluate the three rate constants just in the case of P4VP quaternization. Arends, however, had at his disposal only the kinetic data of Fuoss and co-workers [88, 158] while no experimental data on the units distribution or composition heterogeneity in the quaternized P4VP were available at the time (let alone the fact that the theoretical methods of describing such functions have as yet not been elaborated). Noah *et al.* [26] correlated the results of studies on the kinetics of P4VP quaternization with benzyl chloride and the data on the composition heterogeneity of the reaction product.

Torchilin [160] fractionated P4VP (prepared by anionic polymerization) by a successive precipitation technique from methanol with ethyl acetate followed by repeated fractionation of the middle fraction on a Sephadex G-100 column. The quaternization of the thus-isolated narrow fraction ($M_w = 48 \times 10^3$) with a tenfold molar excess of benzyl chloride was carried out at 60 °C in a nitromethane–methanol mixture (6:1 by volume). The kinetic data are presented in Fig. 4.20(a).

The constant $k_0 = 5.8 \times 10^{-4}$ l/mol s was found from the initial slope of the kinetic curve. The data reported in references [160] to [162] provide grounds for believing that at a conversion of about 85% the polymer undergoes conformational transition and the quaternization rate decreases sharply (as can also be

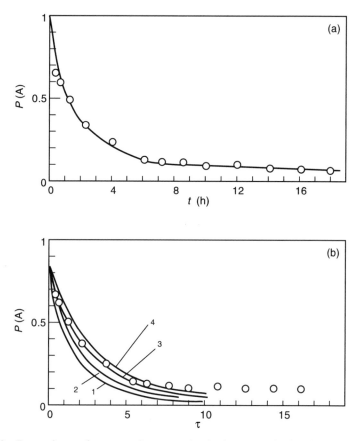

Fig. 4.20 Dependence of content of unreacted units in quaternized poly-4-vinylpyridine (with benzyl chloride, at 60 °C, in nitromethane–methanol mixture 6:1 by volume) on (a) time and (b) dimensionless time ($\tau = k_0 t$) for ratios $k_0:k_1:k_2$ equal to 1:1:0.3 (1), 1:0.6:0.3 (2), 1:0.3:0.3 (3) and 1:0.2:03 (4). Dots, experimental data. (Reproduced by permission of Nauka, Moscow, Russia, from *Vysokomol. Soed. A*, 1974, **16**, 668)

seen in Fig. 4.20a). Assuming the neighbor effect up to conversion of 85% to be the sole factor governing the reaction kinetics, Noah *et al.* [26] estimated k_2 according to Arends [87] and found that $k' = k_2/k_0 = 0.3$. It follows from Fig. 4.20(b) that the best agreement with the experimental data (at conversions of up to about 85%) is displayed by the kinetic curve calculated for $k = k_1/k_0 = 0.3$. In this figure, the fraction of unreacted units is represented as a function of dimensionless time $\tau = k_0 t'$, where $t' = \int_0^t c(t) \, dt$, $c(t)$ being the time-dependent concentration of benzyl chloride.

The composition heterogeneity of polymer samples having 48, 85.5 and 92.5% conversion has been studied by the GPC technique on a Sephadex G-100 column.

This technique provides fractionation of prepared samples on composition, since the macromolecules of equal chain length but of dissimilar benzylation extent display different conformations and sizes [163].

To calculate the composition distribution functions at the found ratio $k_0:k_1:k_2 = 1:0.3:0.3$ for the samples having the mean degrees of conversion mentioned use was made of Monte Carlo simulation [136] (cf. Chapter 3). It follows from Fig. 4.21 that the calculated curves fit the experimental data for 48 and 85.5% conversion. In the case of 92.5% conversion, experimental data point to a more narrow composition distribution than the calculated one. Discrepancy is consistent with sharp retardation of the reaction in this conversion range (cf. Fig. 4.20). This finding provides further corroboration of the assumption that the course of the reaction is affected, apart from the neighbor effect, by some new factor. A sharp drop in the reduced viscosity and also in electrophoretic mobility of the polymer observed by Torchilin [160] made it possible to assume the existence of conformational transition which results in essentially hindering further P4VP benzylation.

On the other hand, the feasibility of describing by a single set of rate constants both the kinetics of quaternization and the composition heterogeneity of benzylated P4VP (at conversions of up to 85%) is indicative of the fact that the

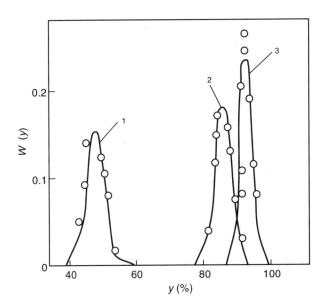

Fig. 4.21 Composition distribution functions $w(y)$ (y is degree of benzylation) of P4VP quaternization products at y equal to 48 (1), 85.5 (2) and 92.5% (3). Dots, experimental data; curves, calculated by Monte Carlo simulation. (Reproduced by permission of Nauka, Moscow, Russia, from *Vysokomol. Soed. A*, 1974, **16**, 668)

reaction regularities are governed, in the range of conversion mentioned, just by the effect of neighboring units.

It should be noted, however, that P4VP quaternization has been investigated using atactic samples. In particular, according to reference [26] the examined P4VP fraction contained 59% syndio-, 5% iso- and 36% hetero-triads. Hence, in studies concerned with the quantitative treatment of P4VP quaternization, the polymer stereochemical configuration is tacitly assumed to exert no effect on the reaction kinetics. True, Morcellet-Sauvage and Loucheux [159] found no difference in the reaction kinetics for atactic and isotactic P2VP. Whether this is correct or not for P4VP should be verified experimentally once the stereoregular samples of the polymer become available.

Nevertheless, the agreement between kinetic and composition distribution data mentioned above provides a basis for employing the obtained values of rate constants for predicting the units distribution in the chain of benzylated P4VP, i.e. for gaining the information which is at present unobtainable experimentally for the polymer.

In general, the fact that in PVP quaternization just the neighbor effect plays a determining role has been substantiated by cited Morcellet-Sauvage and Loucheux studies [159]: for copolymer of 4VP and styrene (about 20 mol% 4VP), not only the standard second-order kinetics was observed but the rate constants for copolymer and low-molecular analogs of P4VP were close to one another. It means that a bonding of the pyridine ring to the macromolecular chain *per se* failed to affect pyridine ring reactivity. Conversely, Fuoss *et al.* [88] showed the quaternization of one pyridine group in 1,3-di(4- pyridyl)propane to diminish by a factor of 3 the reactivity of the second pyridine group. These results seem to prove that the retardation in P4VP quaternization is primerly caused by the neighbor effect.

As to the mechanism proper of the neighbor effect, Fuoss *et al.* [88] believed that reaction retardation was caused by the inductive effect or by electrostatic interaction of quaternized groups with either the intermediate complex formed from a neighboring vinylpyridine unit and an alkylating agent (heterolytic decomposition of the complex is hindered) or polar solvent (decreased accessibility of the neighboring groups to the attack of an alkylating agent as a result of increased solvent density in the strong field of an alkylpyridinium ion).

Arends [87] and Noah *et al.* [26] suggested that retardation is associated with steric hindrances, which at the time was a speculative hypothesis. Conclusive proofs of the determining role of steric factors in the retardation of PVP quaternization have been done by Boucher and co-workers [164–165].

Boucher and Mollett [165] discussed the results of P4VP quaternization with alkyl and arylalkyl bromides, and also allyl bromide in sulpholane at 305–349 K. Figure 4.22 shows $K = k_1/k_0$ and $L = k_2/k_0$ plotted against the extended lengths of the molecules relative to ethyl bromide as obtained from space-filling scale models. The results for all bromides tested fall well on respective straight

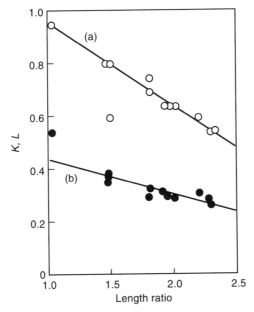

Fig. 4.22 Dependence of (a) K and (b) L on the relative extended length of the organic groups of the bromide reagents used in P4VP quaternization in sulpholane. (Reproduced by permission of The Royal Society of Chemistry from *J. Chem. Soc. Faraday Trans. 1*, 1982, **78**, 77)

lines (for K and L), the retardation being more pronounced with an increase in the bromide extended length.

Boucher and Mollett [165] pointed out, rightly, that local electrostatic repulsion would not depend on alkyl size. Besides, for P4VP quaternization in propylene carbonate [164] also proceeded with retardation no primary salt effect on the kinetics has been found. The inductive effect was rejected because the same values of $K = 0.75$ have been found for both n-propyl and allyl bromides. Meanwhile, retardation would be expected to be less for reaction with allyl bromide compared with n-propyl bromide. Therefore Boucher and Mollett concluded that steric hindrance in the transition state could be the sole factor in the neighboring group effect. The meaning of the ratio $K/L \approx 2$ found for the twelve reagents is in accordance with such a mechanism of retardation.

Quite an original interpretation has recently been given by Frere and Gramain [166]. They suggested an empirical equation to describe the P4VP quaternization in sulpholane:

$$F(t) = (\beta b - a)^{-1} \ln \left[\frac{1 - (a/b)x/a}{1 - \beta x/a} \right] = k_0 t \qquad (4.40)$$

where x is the concentration of halogen ion at time t, a and b are initial concentrations of pyridine groups and alkyl halide respectively, k_0 is the sole true rate constant and β is a macromolecular steric coefficient that describes the effect of the size and the nature of the substituents on the accessibility of the pyridine groups. Frere and Gramain [166] underlined the fact that β reflects not a local but a global steric effect related to the macromolecular coil as a whole, namely β related not only to the geometric size of the reactant but also to the induced change of the coil conformation due to the interactions of pendant substituents with the backbone and solvent caused the coil contraction and, in the case of voluminous substituent, e.g. ω-[(4′-methoxybiphenyl-4-yl)dodecyl] bromide, even internal phase separation. Using Eq. (4.40) Frere and Gramain succeeded in describing both their own and published data related to P4VP quaternization in sulpholane.

Thus, both groups of authors suppose steric hindrance to be the main factor affecting retardation. The principal difference is that Boucher and co-workers [164, 165] interpreted retardation in terms of the local neighbor effect while Frere and Gramain [166] interpreted it in terms of the global coil state.

Remember now that the kinetics of P4VP quaternization depends strongly upon experimental conditions. In particular, as mentioned above, the reaction proceeds with retardation in sulpholane and without one in dimethylformamide [67]. (The solvent nature also affects the reaction kinetics of model compounds: quaternization of 1,3-di(4-pyridyl)propane proceeds with retardation in propylene carbonate [88] but without one in sulpholane [167].) The reacting polymer–solvent interaction may cause a sharp coil contraction, so that the reaction almost ceases; this was observed in P4VP quaternization with benzyl chloride in nitromethane–methanol [160]. In such a case the polymer–solvent interaction might contribute in a local neighbor effect before the coil collapse and produces a global retarding effect after that. Therefore, any interpretation of the retarding effect in P4VP quaternization should be connected closely with experimental conditions.

In this connection, let us note that the local retardation would lead to a units distribution in quaternized P4PV characterized by a tendency to alternation, whereas the global effect would lead to a Bernoullian units distribution. Therefore the question as to whether the local or global factor contributes mainly to retardation (under concrete experimental conditions) could be solved by means of a crucial experiment: an investigation of a units distribution in the chain of the reaction product.

4.7 MISCELLANEOUS REACTIONS

4.7.1 EPOXYDATION OF POLYISOPRENE

The reaction of polydienes with peracids results in the formation of epoxy groups in the polymer chain:

$$\diagdown CH{=}CH\diagup \;+\; R{-}\underset{O}{\overset{\text{||}}{C}}{-}O{-}O{-}H \;\longrightarrow\; \overset{\diagup}{\underset{O}{CH{-}CH}}\diagup \;+\; R{-}\underset{O}{\overset{\text{||}}{C}}{-}O{-}H$$

Scheme 4.12

Tutorsky, Khodzhaeva and co-workers [30, 168] studied the kinetics of the polyisoprene (PI) reaction with perbenzoic acid (PBA) and established that the reaction proceeded with retardation. The constant k_0 was determined from the initial rate of epoxydation followed by calculating the kinetic curves for different values of k_1 and k_2 using the equation analogous to Eq. (4.12). The reaction is first order with respect to PI and PBA, the variation of PBA concentration during epoxydation being accounted for by the introduction of the transformed time $t' = \int_0^t [\text{PBA}]\, dt$. As can be seen in Fig. 4.23, the experimental data on PI (natural rubber) epoxydation at 25 °C (solvent CCl_4) fit the ratio $k_0{:}k_1{:}k_2 = 1{:}0.60{:}0.30$. For PI epoxydation in benzene at 25 °C, the found ratios were close to these ones, viz. $k_0{:}k_1{:}k_2 = 1{:}0.67{:}0.42$. Next, using the turbidimetric technique, it was shown

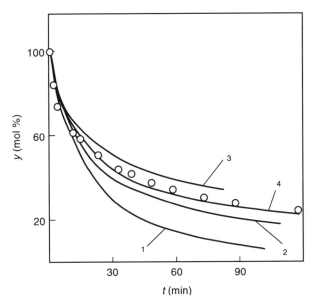

Fig. 4.23 Kinetics of natural rubber reaction with perbenzoic acid at 25 °C (solvent CCl_4): y, content of double bonds; dots, experimental data; curves, calculated for ratios $k_0{:}k_1{:}k_2$ equal to 1:1:1 (1), 1:0.67:0.42 (2), 1:0.4:0.2 (3) and 1:0.6:0.3 (4). (Reproduced by permission of Nauka, Moscow, Russia, from *Vysokomol. Soed. A*, 1974, **16**, 157)

that at conversions of 10, 20, 30 and 50% the samples of partially epoxydized PI displayed a relatively narrow composition distribution. Moreover, NMR spectroscopy showed PI epoxydized to the 38% conversion to contain epoxy groups disposed predominantly between unreacted double bonds. These results agree (if qualitatively) with the kinetic data reflecting the retarding neighbor effect.

Of major interest are the data obtained by Tutorsky and co-workers [30, 168] on the PBA reaction with the low-molecular analog of PI, viz. squalene (isoprene hexamer). For epoxydation of internal double bonds (the terminal double bonds display enhanced reactivity) in CCl_4 at 25 °C, the ratios $k_0:k_1:k_2 = 1:0.60:0.37$ were found from the kinetic data. These ratios are very close to that found for PI epoxydation under the same conditions. The authors of references [30] and [168] showed that the rate constants ratios for epoxydation of both PI and squalene obey the Taft correlation Eq. (4.34). (Apparently, this was the first example of using the correlation equations for interpreting the mechanism of the neighbor effect in macromolecular reactions.) Therefore they regard the inductive electronegative influence of epoxy groups on neighboring double bonds as the principal cause of reaction retardation.

It is appropriate to recall that in the case of polyethylene chlorination the ratios of constants $k_0:k_1:k_2$ were found to be very close to those for low-molecular analogs of the polymer. As can be seen from the foregoing, in the case of PI epoxydation the constants ratios are likewise close to one another for the polymer and squalene. (This similarity of constants ratios was established by Khodzhaeva [168] before an analogous pattern was shown to exist in the case of polyethylene chlorination [118].) It can therefore be expected that, in case the inductive mechanism causes mainly the neighbor effect and steric factors exert a relatively small influence on the magnitude of this effect, low-molecular models for determining the individual rate constants for macromolecular reactions can be employed.

4.7.2   REACTION OF POLY(METHACRYLOYL CHLORIDE) WITH AMINES

Turaev et al. [169] studied the reaction kinetics of atactic poly(methacryloyl chloride) (PMAC) with secondary amines, such as piperidine and 6,7-dimethoxy-1,2,3,4-tetrahydroisoquinoline (THQ). Interaction between PMAC and THQ proceeds with pronounced retardation and is characterized by limiting conversion to about 50%. The reaction kinetics is described by the constants ratio $k_0:k_1 = 1:0.1$ and $k_2 = 0$.

The formation of a six-membered cyclic complex between the neighboring substituted and unsubstituted units is regarded by Turaev et al. [169] as a cause of reaction retardation:

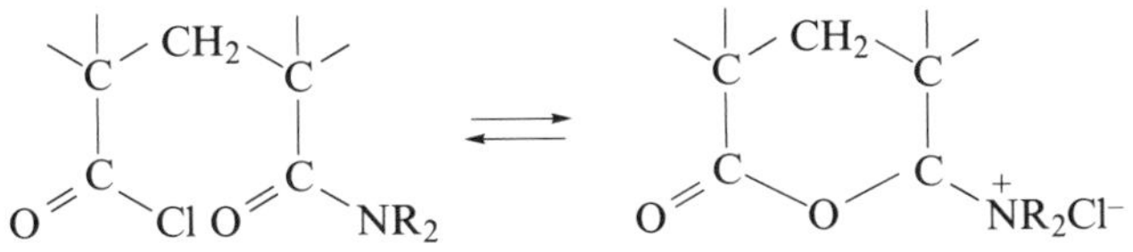

Scheme 4.13

The validity of the suggested mechanism is also favored by the effect exerted by the solvent nature on the reaction kinetics: in alkyl halides (a methylene chloride–chloroform mixture, 2:1 by volume) the initial rate of interaction between PMAC and THQ is 5 times as great as that in dimethyl formamide, which forms complexes with carboxylic acid chloranhydrides [170] analogous to the complexes shown in Scheme 4.13, thereby preventing the transformation of polymer functional groups bound in the complex.

However, complexing according to Scheme 4.13 is reversible and as such should fail to prevent the attainment of high conversions. Indeed, in piperidine the limiting conversion equals about 93%. Moreover, it was demonstrated [169] that, upon attaining the limiting conversion of the reaction with THQ, the unsubstituted groups in PMAC retain the ability to react with piperidine. In view of this observation, Turaev *et al.* regard the restriction of the reaction in question as being associated with the fact that THQ is a bulky reactant, i.e. with the steric hindrances involved. The existence of such steric hindrances does not mean, however, that it would be impossible to obtain a completely substituted polymer. Turaev *et al.* [169] provided a proof of this contention by resorting to the radical polymerization of an appropriate monomer, viz. methacryloylsalsolidine (the product of the THQ reaction with methacryloyl chloride).

Polymerization of this compounds yielded polymethacryloylsalsolidine having a molecular weight of up to 25×10^3, and this result enabled Turaev *et al.* [169] to interpret the retarding neighbor effect in the PMAC and THQ reaction in terms of the joint action of two factors, viz. complexing and steric hindrances.

4.7.3 NUCLEOPHILIC SUBSTITUTION ON POLY(METHYL METHACRYLATE)

Bourguigon *et al.* [171] studied the reaction of PMMA with RCH_2Li, where R is preferably—$SO_2N(CH_3)_2$. The reaction proceeds as follows:

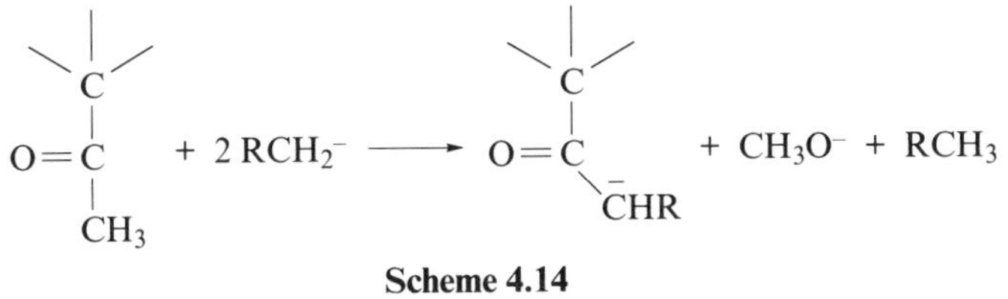

Scheme 4.14

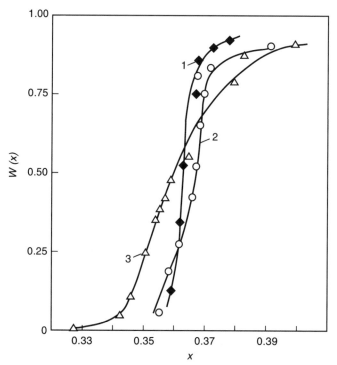

Fig. 4.24 Composition heterogeneity of the product of PMMA reaction with $(CH_3)_2NSO_2CH_2Li$. Mean substitution degree $x = 0.366$. Results of successive precipitation in solvent–nonsolvent systems: chloroform–diethyl ether (1), dimethylformamide–(water + 0.5% NH_4Cl) (2) and the data on cross-fractionation in the same systems (3). (Reproduced from *Polymer*, 1977, **18**, 937, by permission of the publishers, Butterworth Heinemann Ltd ©)

The authors investigated the composition heterogeneity of the polymer having a substitution degree of 0.366 by the cross-fractionation, the selection of solvent–nonsolvent systems for fractionation being performed in accordance with the procedure described in Chapter 8. As can be seen in Fig. 4.24, cross-fractionation was found to be substantially more effective than successive precipitation. It follows from the fractionation results that the composition distribution dispersion for the examined sample $\sigma^2 = 22 \times 10^{-5}$. In reference [171] this value was compared with the calculations of the σ^2 value [136] for various rate constants ratios. For $k_0:k_1:k_2 = 1:1:1$, $\sigma^2 = 31 \times 10^{-5}$, while for $k_0:k_1 = 1:0.2$ and $k_2 = 0$, $\sigma^2 = 10 \times 10^{-5}$. This comparison enabled the authors of reference [171] to conclude that there exists a qualitative agreement between the results of studying the composition heterogeneity and the kinetic data, pointing to the retarding neighbor effect.

4.7.4 ESTERIFICATION OF SYNDIOTACTIC POLY(METHACRYLIC ACID)

As noted in Section 4.3, Klesper *et al.* [66, 172] employed the NMR method to show that the esterification of syndiotactic poly(methacrylic acid) (PMA) with methanol, ethanol or 2,2,2-trifluoroethanol in concentrated sulfuric acid proceeds to a high degree of conversion ($\geqslant 0.85$) and yields copolymers characterized by Bernoullian units distribution.

However, when PMA esterification with methanol, 2,2,2-trifluoroethanol or benzyl alcohol is carried out by action of dicyclohexylcarbodiimide (DCC), the limiting conversions do not exceed 0.61 and a pronounced tendency to units alternation in the chain of the thus-prepared copolymers manifests itself. Formally, it is feasible to obtain such a distribution on the basis of a conventional reaction model involving a retarding neighbor effect, the constants ratio being $k_0 : k_1 : k_2 = 1 : 0.1 : 0$. In references [66] and [172] the authors believe another interpretation to be more plausible, viz. that the molecule of DCC reacts consecutively with two neighboring carboxyl groups in PMA, so that a cyclic anhydride is formed in the chain, followed by the reaction of this anhydride with alcohol and anhydride transformation into an ester and a carboxyl vicinal group:

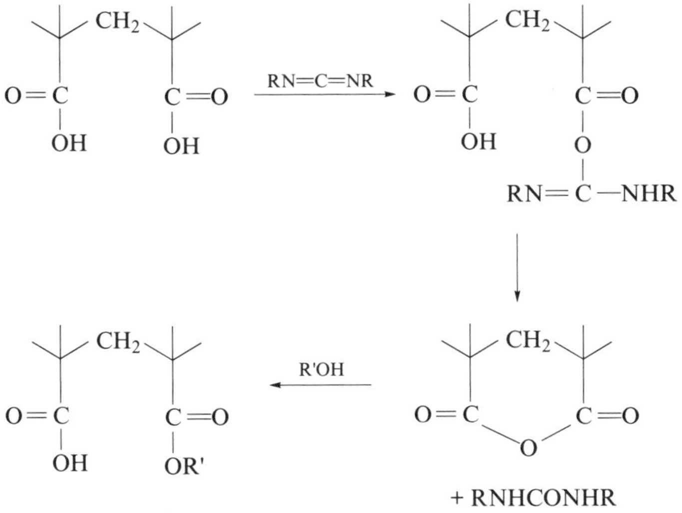

Scheme 4.15

The validity of Scheme 4.15 is substantiated by IR spectra which point to the presence of cyclic anhydride groups in the polymer.

These results are of utmost interest and demonstrate the fact that in interpreting experimental data relevant theoretical models should be employed in the correct manner. Unfortunately, no kinetic measurements are cited in references

[66] and [172], although it would be of interest to check whether the data in question fit Eq. (4.12) for the $k_0:k_1:k_2$ ratios of 1:1:1 and 1:0.1:0 in the case of PMA esterification by the action of H_2SO_4 and DCC respectively. It is, however, reasonable to expect such an equation to be inadequate, inasmuch as the simplest model of a reaction involving the nearest neighbor effect fails to embrace the mechanism of esterification in the presence of DCC suggested by Klesper. To account for this mechanism, relevant equations describing the reaction kinetics and the units distribution should be modified accordingly.

4.7.5   HYDROLYSIS OF POLYACRYLAMIDE

The basic hydrolysis of polyacrylamide (PAAm) proceeds in homogeneous conditions up to a high degree of conversion. Sawant and Morawetz [173, 174] tried to elucidate whether the reaction kinetics obeys the model of a neighbor effect. They rejected a computer search for the k_0, k_1 and k_2 values which best fit the kinetic data concerning only homopolymer PAAm transformation [175] and used polymeric models prepared by copolymerization of acrylamide and acrylic acid. The units distribution for the models of various compositions was calculated using comonomer reactivity ratios. Then k_0, k_1 and k_2 were estimated from the initial rates of the hydrolysis of three models in 0.2N NaOH at 53 °C, in

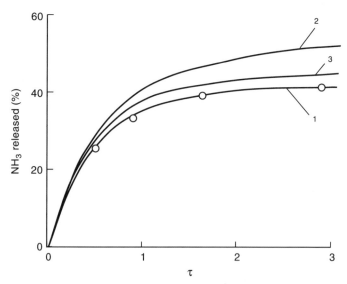

Fig. 4.25   Hydrolysis of polyacrylamide in 0.2 N NaOH at 53 °C. Experimental data (1) and calculations for $k_1:k_0 = 0.11$, $k_2:k_0 = 0.013$ (2) and for $k_1 = k_2 = 0$(3). (Reprinted with permission from *Chemical Reactions on Polymers*, Eds. J. L. Benham and J. F. Kinstle, ACS Symposium Series 364, ACS, Washington, D.C., 1988, p. 315. Copyright American Chemical Society)

a similar way to the method described in Section 4.3: Eqs. (4.13) and (4.14). The constants ratios found were $k_0:k_1:k_2 = 1:0.11:0.013$. However, the kinetic curve calculated using these ratios does not fit experimental data on PAAm hydrolysis (cf. Fig. 4.25). Moreover, experimental conversions lay below the curve calculated for the case where a single reacted nearest neighbor completely inhibits hydrolysis: $k_1 = k_2 = 0$. Therefore Sawant and Morawetz [173, 174] assumed that not only the nearest but also the non-neighboring carboxylate groups take part in the repulsion of the catalyzing hydroxyl ions from amide residues, thus retarding the hydrolysis.

The data presented above show the modern methods of quantitative investigation of the neighbor effect to be applicable to a broad range of macromolecular reactions proceeding without polymer chain alteration. (It will be noted in this connection that the epoxydation of polydienes cannot be regarded as a polymer-analogous transformation proper [176] because in this case the nature of bonds in the backbone undergoes alterations during the reaction.)

4.8 ON THE EMPIRICAL KINETIC EQUATIONS

The kinetics of macromolecular reactions was the object of numerous investigations, but very often in evaluating the effect of neighboring units they do not use precise kinetic equations of the type of Eq. (4.12) and the methods for determining the constants k_0, k_1 and k_2 based on such equations. Polyvinylpyridine quaternization with alkyl halides provides an example of such an approach. For alkylpyridines as model compounds the kinetics of the well-known Menshutkin reaction [177] is described by the equation

$$(a-b)^{-1}\ln\left(\frac{a-x}{b-x}\right) = kt \tag{4.41}$$

where a and b are the initial concentrations of alkyl halide and alkylpyridine, x is the concentration of halide ions at time t and k denotes the rate constant.

Having detected the retarding effect in the quaternization of P4VP Fuoss *et al.* [88] suggested the following semiempirical equation:

$$\frac{f(x)}{t} = k_2 + (k_0 - k_2)\phi(\alpha t) \tag{4.42}$$

where $f(x)$ stands for the left-hand side of Eq. (4.41), k_0 and k_2 are the values of the effective rate constant at the reaction initial and final stages respectively, α is the parameter depending on the magnitude of rate constant, temperature and the initial concentrations of the reactants, and $\phi(\alpha t) = (1 - e^{-\alpha t})/\alpha t$. To find the rate constants, the values of $f(x)/t$ are calculated using Eq. (4.41), resorting to the trial-and-error method to select the magnitude of α so as to obtain the linear relationship of $f(x)/t$ versus $\phi(\alpha t)$. Then, in compliance with Eq. (4.42), the constants k_0 and k_2 can readily be found from the parameters of the thus-obtained straight line. In the specific case of P4VP quaternization with butyl bromide in sulpholane at 75 °C, the application of this method yielded the constants ratio $k_0:k_2 = 1:0.14$.

Morcellet-Sauvage and Loucheux [159] employed the Fuoss method for determining all of the three rate constants for quaternization of polyvinylpyridines having different structures. These authors presented Eq. (4.42) in the following form:

$$\frac{f(x)}{t} = k_n + (k_0 - k_n)\phi(\alpha t) \tag{4.43}$$

where k_n has the meaning of k_1 or k_2 depending upon the examined range of conversion degree y. The selected parameter $\alpha \approx 1/t(y_{dev})$, where y_{dev} denotes the degree of conversion at which there is observed the deviation of experimental data from the kinetic curve corresponding to Eq. (4.41). In the case of poly(2-methyl-5-vinyl-pyridine) quaternization at 70 °C in the $0 < y < 0.5$ range at $\alpha \approx 1/t_{0.33}(y_{dev} = 0.33)$ the values of k_0 and k_1 were determined, while in the $0 < y < 1$ range at $\alpha \approx 1/t_{0.80}(y_{dev} = 0.80)$ the constants found were k_0 and k_2, the ratio being $k_0:k_1:k_2 = 1:0.6:0.1$. For P4VP at 70 °C, the ratio was $k_0:k_1:k_2 = 1:1:0.1$. The authors note that their method displays low precision and the estimation of errors presents major difficulties.

The Fuoss method was also used by Rempp [178] in a study concerned with the kinetics of methylsulphonylmethyllithium reaction with PMMA at 25 °C in the dimethylsulphoxide–benzene (5:1 by volume) mixture:

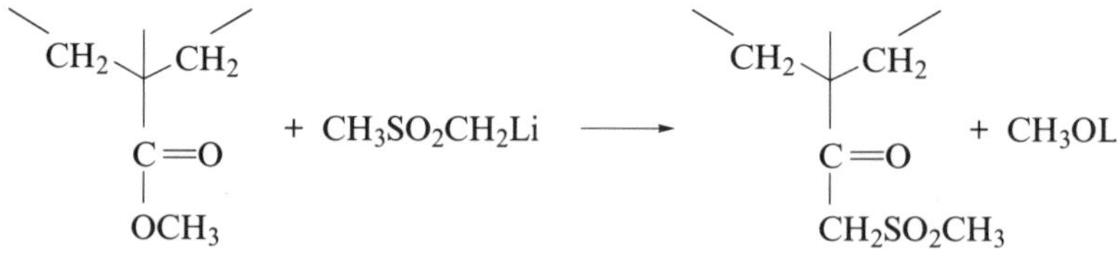

Scheme 4.16

For syndiotactic PMMA, $k_0 = 9.9 \times 10^{-2}$ l/mol min and $k_0:k_1:k_2 = 1:0.055:0$. For isotactic PMMA, $k_0 = 4.7 \times 10^{-1}$ l/mol min and $k_0:k_1:k_2 = 1:1:0.003$. Rempp, however, failed to interpret the observed marked effect of polymer microtacticity on kinetic behavior.

Kawabe and Yanagita [179] studied the kinetics of chloromethylated polystyrene amination with 2-aminobutanol and found the neighboring units to exert an accelerating effect on this reaction. To interpret the kinetic data, these authors [179] discussed the transformations of a trifunctional cyclic model and assumed the constants k_1 and k_2 to be equal. Then the constants k_0 and k_1 ($= k_2$) can be found from the equations:

$$k_0 = \frac{1}{[(3 - 2k_1/k_0)a - b]t} \ln \frac{1 - x/a}{1 - (3 - 2k_1/k_0)x/b}$$

$$k_1 = \frac{1}{(a - b)(t - \tau)} \left[\ln \frac{1 - x/a}{1 - x/b} - \ln \frac{1 - x_\tau/a}{1 - x_\tau/b} \right]$$

(4.44)

where a and b stand for the initial concentrations of 2-aminobutanol and chloromethylated styrene units respectively, x is the concentration of chloride ions at time t and τ denotes the time t at which $x = x_\tau = b/3$ (once this degree of amination has been attained, the deviation of experimental data from the kinetic curve for the standard second-order reaction commences).

In the case of a first-order macromolecular reaction $A \to B$, Stoy [180] also assumes the equality of constants $k_1 = k_2$ and suggests that k_0 and k_1 ($= k_2$) should be found from the dependence of the conversion $\beta = 1 - A/A_0$ on time and of the apparent rate constant $k_{app} = (1/t)\ln(A_0/A)$ on conversion:

$$k_0 = \left(\frac{d\beta}{dt} \right)_{t=0}$$

$$k_1 - k_0 = 0.5 \left(\frac{dk_{app}}{d\beta} \right)_{\beta=0}$$

(4.45)

Having processed experimental data cited in reference [179], Stoy inferred that the Kawabe and Yanagita equations (4.44) are applicable for the reaction involving the retarding neighbor effect, whereas in the case of acceleration, use should be made of Eqs. (4.45). For amination under the conditions specified in reference [179], the $k_1:k_0$ ratio found using Eqs. (4.44) equaled 2.6:1, whereas according to Eqs. (4.45) the ratio was 1.6:1.

The method suggested by Stoy [180] was employed by Hudecek *et al.* [181] for interpreting the data on the kinetics of acid hydrolysis of polymeric ketal poly(2,2-dimethyl-1,3-dioxolan-4-yl)methyl methacrylate]. In this reaction the ratio $k_1:k_0$ equals 3:1 at 25 °C and 1.5:1 at 40 °C.

The empirical approach is likewise employed in the studies of poly(vinyl acetate) (PVAc) hydrolysis (solvolysis). In the early 1940s both the alkaline and the acidic hydrolysis of this polymer were found to proceed with autoacceleration [182, 183]. Sakurada [184] suggested that the kinetics of PVAc alkaline hydrolysis could be described by the equation

$$\frac{dx}{dt} = k_0 \left(1 - \frac{mx}{a} \right) \frac{a - x}{b - x}$$

(4.46)

where a and b are the initial concentrations of the acetate groups and alkali respectively, x is the concentration of hydrolyzed units at time t, k_0 is the rate constant at process commencement and m is the acceleration coefficient.

Fujii [185] used Eq. (4.46) to determine the values of k_0 and m for alkaline hydrolysis of highly isotactic and atactic PVAc samples, whereas Arranz and Tahir [186] made use of Eq. (4.46) in their study of the kinetics of poly(vinyl isopropionate) alkaline hydrolysis.

Pichot *et al.* [187] studied methanolysis of PVAc in a methanol–tetrahydrofuran solution using HCl as the catalyst. The reaction proceeded with an autoacceleration similar to that of alkaline hydrolysis. Using Eq. (4.46) the authors found m to depend linearly upon the solvent composition, decreasing from 36 (in pure methanol) to 0 (in pure THF). To elucidate the mechanism of acceleration, Pichot *et al.* investigated methanolysis kinetics for model compounds, and the results were compared with the data on the alkaline hydrolysis of PVAc. The rate constant k_0 values were found to be lower in the case of acidic solvolysis, presumably because methanol is a poor solvent for PVAc, so that macromolecular coils are more contracted and acetate groups are less accessible. Nevertheless, the accelerating effect is more pronounced in methanol. To explain the fact Pichot *et al.* [187] assumed the acceleration to be caused by an attack of the carbonyl group by $CH_3 \overset{+}{-} \overset{+}{C}(OH)(OCH_3)$ ions formed on neighboring units. The contracted conformation of the coil prevents these ions from diffusing into solution and retains them near the chain.

Pichot *et al.* [188] also studied acidic methanolysis of poly(vinyl acetate-co-vinyl chloride) and established the dependence of the initial reaction rate and the m value upon the copolymer composition and the units distribution.

Prior to the estimation of the aforementioned approximate methods of describing the kinetics of macromolecular reactions, it was deemed appropriate to elucidate how the methods in question originated. Of particular importance in this context is the paper by Fuoss *et al.* [88] who disclosed the procedure of deriving an accurate kinetic equation that accounts for the neighbor effect. These authors cite no final expression for the dependence of unreacted units concentration on time, but nonetheless note that this relationship may be presented as incomplete Γ-functions of the constants k_0, k_1 and k_2, this being the solution of Eq. (4.12) according to reference [189]. The authors further contend that the analysis of P4VP quaternization kinetics by means of an accurate equation is excessively sophisticated and therefore suggest an empirical relationship (4.42) for carrying out such an analysis.

Hence, Fuoss *et al.* [88] were actually the first who derived an equation of the type of (4.12), but at the time of their studies (1960) failed to visualize how to employ the equation in question for determining the constants k_0, k_1 and k_2 or else to arrive at a conclusion that this procedure for determining the constants would involve prohibitively great experimental difficulties. Therefore they decided to resort to an approximate approach.

The correct methods of determining the individual rate constants do involve the necessity of overcoming major experimental difficulties associated with the preparation of polymeric models, as well as with examination of the units distribution or of reaction product composition heterogeneity. Because of this situation many investigators, even at present, give preference to approximate approaches in kinetic studies and, in doing so, generally restrict their tasks to the kinetics of initial homopolymer transformations and describe the results obtained in terms of semiempirical equations based on not quite correct assumptions (it is, for example, evident that an assumption regarding the equality of constants $k_1 = k_2$ would be valid in specific cases only). The validity of thus-obtained information is, therefore, questioned.

Insofar as approximate equations account for macromolecule reactivity variation with the degree of conversion, it is practicable to employ these equations for a preliminary estimate of neighboring unit effects, as well as of the relative effect in the series of polymers of different structure. These equations are, however, unsuited for quantitative investigations.

For this purpose, it is primarily necessary to employ exact equations, in which the parameters have a lucid physical meaning, viz. that of individual rate constants k_0, k_1 and k_2, and next to determine the constants by an adequately reliable method. In kinetic studies *per se* it is therefore desirable to employ as a starting sample not only a homopolymer but also polymeric models of various composition (for a single kinetic curve representing homopolymer transformation it is easy to select a suitable mathematical function, so that the temptation to resort to a relatively unsophisticated approximation arises). Moreover, it is essential to make sure that the kinetic data are in agreement with the results of estimating the parameters of the units distribution or the composition heterogeneity of reaction products. Only the constants k_0, k_1 and k_2 found by this method can be employed with certainty for calculating the distribution of units and the composition heterogeneity of reaction products and, hence, for interpreting the physical and chemical properties of modified polymers. Such constants solely are of true value as far as the elucidation of the detailed mechanism of a macromolecular reaction is concerned.

REFERENCES

1. Gaylord, N. G., *J. Polymer Sci. Part C*, 1968, **24**, 1–5.
2. Harwood, H. J., *ACS Polymer Prepr.*, 1971, **2**, 46–51.
3. Platé, N. A. and Litmanovich, A. D., 23rd International Congress on *Pure Applied Chemistry*, Boston, 1971, Vol. 18, pp. 123–50.
4. Platé, N. A., Litmanovich, A. D. and Noah, O.V., *Makromoleculiarnye Reaktsii* (*Macromolecular Reactions*), Khimiya, Moscow, 1977.
5. Ravens, D. A., *Polymer*, 1960, **1**, 375–83.
6. Dzhagatspanyan, R. V., Kolbasov, V. I., Bardenshtein, S. B., *et al.*, *Vysokomol. Soed*, 1965, **7**, 1959–63.

7. Winslow, F. H. and Matreyek, W., *ACS Div. Polymer Chem. Prepr.*, 1962, **3**(1), 229–33.
8. Fawcett, E. W., Pat. USA 2183356 19/XI—1939; *C.A.*, 1940, **34**, 2500³.
9. Mnatsakanov, S. S., Thesis, Leningrad, 1974.
10. Harwood, H. J. and Robertson, A. B., International Symposium on *Macromolecular Chemistry*, Budapest, 1969, Preprint 10/9.
11. Hvidt, A. and Corett, R., *J. Am. Chem. Soc.*, 1970, **92**, 5546–50.
12. Goodman, N. and Morawetz, H., *J. Polymer Sci. Part C*, 1970, **31**, 177–92.
13. Morawetz, H. and Goodman, N., *Macromolecules*, 1970, **3**, 699–700.
14. Goodman, N. and Morawetz, H., *J. Polymer Sci. A-2*, 1971, **9**, 1657–69.
15. Marcus, R. A. *J. Phys. Chem.*, 1954, **58**, 621–3.
16. Katchalsky, A. and Feitelson, J., *J. Polymer Sci.*, 1954, **13**, 385–92.
17. Morawetz, H. and Gaetjens, E., *J. Polymer Sci.*, 1958, **32**, 526–8.
18. Smets, G. and De Loecker, W., *J. Polymer Sci.*, 1960, **45**, 461–7.
19. Smets, G. and Van Hambeeck, W., *J. Polymer Sci. A*, 1963, **1**, 1227–38.
20. Gaetjens, E. and Morawetz, H., *J. Am. Chem. Soc.*, 1961, **83**, 1738–42.
21. Bender, M. L., *J. Am. Chem. Soc.*, 1957, **79**, 1258–9.
22. Smets, G. and Hesbain, A. M., *J. Polymer Sci.*, 1959, **40**, 217–26.
23. Arcus, C. L., *J. Chem. Soc. (Lond.)*, 1949, 2732–6.
24. Pinner, S. N., *J. Polymer Sci.*, 1953, **10**, 379–84.
25. Litmanovich, A. D., Platé, N. A., Agasandyan, V. A., *et al.*, *Vysokomol. Soed A*, 1975, **17**, 1112–22.
26. Noah, O. V., Torchilin, V. P., Litmanovich, A. D. and Platé, N. A., *Vysokomol. Soed A*, 1974, **16**, 668–71.
27. Morawetz, H. and Zimmering, P. E., *J. Phys. Chem.*, 1954, **58**, 753–6.
28. Zimmering, P. E., Westhead, E. W. and Morawetz, H., *Biochim. Biophys. Acta*, 1957, **25**, 376–81.
29. Krentsel, L. B., Litmanovich, A. D., Pastukhova, I. V. and Agasandyan, V. A., *Vysokomol. Soed A*, 1971, **13**, 2489–95.
30. Tutorsky, I. A., Khodzhayeva, I. D. and Dogadkin, B. A., *Vysokomol. Soed A*, 1974, **16**, 157–68.
31. Yun, E., Thesis, Moscow, 1973.
32. Usmanov, T. I., Thesis, Moscow, 1974.
33. Barth, V. and Klesper, E., *Polymer*, 1976, **17**, 777–86.
34. Klesper, E. and Barth, V., *Polymer*, 1976, **17**, 787–94.
35. Harwood, H. J., Kemp, K. G. and Landoll, L. M., *J. Polymer Sci. Polymer Lett. Ed.*, 1978, **16**, 109–114.
36. Harwood, H. J., Landoll, L. M. and Kemp, K. G., *J. Polymer Sci. Polymer Lett. Ed.*, 1978, **16**, 91–94.
37. Bauer, B. J., *Macromolecules*, 1979, **12**, 704–8.
38. Merle, Y., *J. Polymer Sci. Polymer Phys. Ed.*, 1984, **22**, 525–7.
39. Stroganov, L. B., Thesis, Moscow, 1972.
40. Olonovskii, A. N., Stroganov, L. B., Noah, O. V. and Platé, N. A., *Vysokomol. Soed A*, 1983, **25**, 882–8.
41. Bender, M. L., Chow, I. L. and Chloupek, F., *J. Am. Chem. Soc.*, 1958, **80**, 5380–4.
42. Bender, M. L., Chloupek, F. and Neveu, M. C., *J. Am. Chem. Soc.*, 1958, **80**, 5384–7.
43. Bender, M. L. and Neveu, M. C., *J. Am. Chem. Soc.*, 1958, **80**, 5388–91.
44. Ruzicka, L., Brugger, W., Pfeiffer, M., Schinz, H. and Stoll, M., *Helv. Chim. Acta*, 1926, **9**, 499–520.
45. Sisido, M., *Macromolecules*, 1971, **4**, 737–42.
46. Stoll, M. and Rouvé, A., *Helv. Chim. Acta*, 1935, **18**, 1087–125.

47. Morawetz, H., *Macromolecules in Solution*, Interscience, New York, London, Sydney, 1965, Chap. 7.
48. Ryabov, A. V., Semchikov, Yu. D. and Slavnitskaya, N. I., *Vysokomol. Soed A*, 1970, **12**, 553–60.
49. Gaetjens, E. and Morawetz, H., *J. Am. Chem. Soc.*, 1960, **82**, 5328–35.
50. Bruice, T. C. and Pandit, U. K., *J. Am. Chem. Soc.* 1960, **82**, 5858–65.
51. Sakurada, I., Sakaguchi, Y. and Kagai, M., *Kobunshi Kagaku*, 1960, **17**, 87–94.
52. Westhead, E. W. and Morawetz, H., *J. Am. Chem. Soc.*, 1958, **80**, 237–42.
53. Shimanouchi, T., Tasumi, M. and Abe, Y., *Makromol. Chem.*, 1965, **86**, 43–63.
54. Doskocilová, D., Stokr, J., Schneider, B., Pivková, H., Kolinský, M. and Petránek, J., *J. Polymer Sci. Part C*, 1967, **16**, 215–28.
55. Doskocilová, D., Stokr, J., Votavová E., Schneider, B. and Lim, D., *J. Polymer Sci. Part C*, 1967, **16**, 2225–37.
56. Stokr, J., Schneider, B. and Vodnasky, J., *J. Polymer Sci.*, 1964, **2**, 783–8.
57. Matsuzaki, K., Uryu, T., Ishida, A. and Takeuchi, M., *J. Polymer Sci. Part C*, 1967, **16**, 2099–110.
58. Shibatani, K. and Fujii, K., *J. Polymer Sci. A-1*, 1970, **8**, 1647–56.
59. Skorokhodov, S. S., Thesis, Leningrad, 1972.
60. Goldstein, B. N., Goryunov, A. N., Gotlib, Yu. Ya., *et al.*, *J. Polymer Sci. A-2*, 1971, **9**, 769–77.
61. Platé, N. A., Noah, O. V. and Stroganov, L. B., *Vysokomol. Soed A*, 1983, **25**, 2243–66.
62. Platé, N. A., Stroganov, L. B. and Noah, O. V., *Polymer J.*, 1987, **19**, 613–22.
63. Ito, K. and Yamashita, Y., *J. Polymer Sci. A*, 1965, **3**, 2165–87.
64. Smets, G. and De Loecker, W., *J. Polymer Sci.*, 1959, **41**, 375–80.
65. Klesper, E., Strasilla, D. and Regel, W., *Makromol. Chem.*, 1974, **175**, 523–34.
66. Klesper, E., Strasilla, D. and Berg, M. C., *Europ. Polymer J.*, 1979, **15**, 593–601.
67. Arcus, C. L. and Hall, W. A., *J. Chem. Soc. (Lond.)*, 1964, 5995–9.
68. Michail, R., Gherghel, F., Stanescu, M. and Kornbaum, S., *Plaste und Kautschuk*, 1962, **9**, 397–400.
69. Wielopolski, A., Krajewski, J. and Swierkot, J., *Plaste und Kautschuk*, 1963, **10**, 467–9.
70. Brame, E. G., *J. Polymer Sci. A-1*, 1971, **9**, 2051–61.
71. Klesper, E., Gronski, W. and Barth, V., *Makromol. Chem.*, 1970, **139**, 1–16.
72. Klesper, E., *J. Polymer Sci. B*, 1968, **6**, 663–72.
73. Litmanovich, A. D. and Agasandyan, V. A., *Kinetika i Kataliz*, 1966, **7**, 309–18.
74. Agasandyan, V. A., Kudryavtseva, L. G., Litmanovich, A. D. and Shtern, V. Ya., *Vysokomol. Soed A*, 1967, **9**, 2634–6.
75. Harwood, H. J., *Angew. Chem.*, 1965, **77**, 405–13, 1124–34.
76. Wall, F. T., *J. Am. Chem. Soc.*, 1941, **63**, 1862–6.
77. Skeist, I., *J. Am. Chem. Soc.*, 1946, **68**, 1781–4.
78. Spinner, I. H., Lu, B. C.-Y. and Graydon, W. F., *J. Am. Chem. Soc.*, 1955, **77**, 2198–200.
79. Meyer, V. E. and Lowry, G. G., *J. Polymer Sci. A*, 1965, **3**, 2843–51.
80. Markert, G., *Makromol. Chem.*, 1967, **109**, 112–19.
81. Iziumnikov, A. L. and Vyrsky, Yu. P., *Vysokomol. Soed A*, 1967, **9**, 1996–2000.
82. Myagchenkov, V. A. and Frenkel, S. Ya., *Vysokomol. Soed A*, 1969, **11**, 2348–50.
83. Myagchenkov, V. A., Frenkel, S. Ya., Tsentovsky, V. M. and Tsentovskaya, V. S., *Vysokomol. Soed B*, 1972, **14**, 693–6.
84. Harwood, H. J., *J. Polymer Sci. C*, 1968, **25**, 37–46.
85. Johnston, N. W. and Harwood, H. J., *J. Polymer Sci. C*, 1969, **22**, 591–610.

86. Johnston, N. W., *ACS Polymer Preprints*, 1969, **10**, 608–13.
87. Arends, C. B., *J. Chem. Phys.*, 1963, **39**, 1903–4.
88. Fuoss, R., Watanabe, M. and Coleman, B. D., *J. Polymer Sci.*, 1960, **48**, 5–15.
89. Klesper, E., Barth, V. and Johnsen, A. 23rd IUPAC International Congress on *Macromolecular Preparations*, Boston, 1971, Vol. 8, pp. 151–65.
90. Robertson, A. B. and Harwood, H. J., *ACS Polymer Preprints*, 1971, **12**(1), 620–7.
91. Semen, J. and Lando, J. B., *Macromolecules*, 1969, **2**, 570–5.
92. Selegny, E. and Segain, P., *J. Macromol. Sci. Chem. A*, 1971, **5**, 603–9,
93. Matsuzaki, K., Uryu, T. and Okada, M., *Makromol. Chem.*, 1970, **140**, 295–7.
94. Platé, N. A. and Litmanovich, A. D., *Vysokomol. Soed A*, 1972, **14**, 2508–17.
95. Kotake, Y. and Ide, F., *Chem. High Polymers Japan*, 1969, **26**, 126–33.
96. Kryshtob, V. I., Litmanovich, A. D. and Platé, N. A., *Vysokomol. Soed B*, 1972, **14**, 326.
97. Yun, E., Stroganov, L. B., Agasandyan, V. A., Litmanovich, A. D. and Platé, N. A., *Vysokomol. Soed B*, 1972, **14**, 292–5.
98. Yuki, H., Hatada, K., Ninomi, T. and Kikuchi, Y., *Polymer J.*, 1970, **1**, 36–45.
99. Klesper, E., Gronski, W. and Barth, V., *Makromol. Chem.*, 1971, **150**, 223–49.
100. Klesper, E., *J. Polymer Sci. B*, 1968, **6**, 313–21.
101. Klesper, E., IUPAC International Symposium on *Macromolecular Chemistry*, Budapest, 1969, Vol. 5, p. 91.
102. Klesper, E. and Gronski, W., *J. Polymer Sci. B*, 1969, **7**, 727–38.
103. Benson, S. W., *The Foundation of Chemical Kinetics*, McGraw-Hill, New York, Toronto, London, 1960, Chap. 4.
104. Morawetz, H. and Oreskes, I., *J. Am. Chem. Soc.*, 1958, **80**, 2591–2.
105. Platé, N. A., Stroganov, L. B., Seifert, T. and Noah, O. V., *Dokl. Akad. Nauk. SSSR*, 1975, **223**, 396–9.
106. Seifert, T., Thesis, Moscow, 1974.
107. Stroganov, L. B., Taran, Yu. A., Platé, N. A. and Seifert, T., *Vysokomol. Soed A*, 1974, **16**, 2147–53.
108. Stroganov, L. B., Taran, Yu. A. and Platé, N. A., *Zh. Fiz. Khim.*, 1975, **49**, 2696–701.
109. Yun, E., Lobanova, L. B., Litmanovich, A. D., Platé, N. A., Shishkina, M. V. and Polikarpova, T. A., *Vysokomol. Soed A*, 1970, **12**, 2488–93.
110. Schwetlick, K., *Kinetische Methoden zur Untersuchung von Reaktionmechanismen*, VEB Deutsch Verlag Wissensch, Berlin, 1971, Chap. 6.
111. Konix, A. and Smets, G., *J. Polymer Sci.*, 1955, **15**, 221–9.
112. Cooper, W., *Chem. and Ind.*, 1958, 263–4.
113. Polowinski, S. and Bortnowska-Barela, B., *J. Polymer Sci. Polymer Chem. Ed.*, 1981, **19**, 51–5.
114. Krentsel, B. A., *Khlorirovanie Parafinovykh Uglevodorodov (Chlorination of Paraffins)*, Nauka, Moscow, 1964, Chap. 2.
115. Bratolyubov, A. S., *Uspekhi Khim;* 1961, **30**, 1391–409.
116. Huyser, E. S., in *Advances in Free-Radical Chemistry*, Ed. G. H. Williams, Logos Press, London 1965, Vol. 1, pp. 77–135.
117. Soumillion, J. Ph., *Ind. Chim. Belge*, 1970, **35**, 851–77.
118. Krentsel K. B., Thesis, Moscow, 1972.
119. Gusev, M. N. Kissin Yu. V., Voronovitsky, M. M. and Berlin, A. A., *Neftekhimiya*, 1968, **8**, 435–41.
120. Oswald, H. J. and Kubu, E. T., *SPE Trans.*, 1963, **3**, 168–75.
121. Saito, T., Matsumura, Y. and Hayashi, S., *Polymer J.*, 1970, **1**, 639–55.
122. Lukás, R., Paleckova, V., Svétly, J., Kolinský, M. and Doskocilová, D., *J. Polymer Sci. Polymer. Chem. Ed.*, 1979, **17**, 2263–7.

123. Allen, V. R. and Young, R. D., *J. Polymer Sci. A-1*, 1970, **8**, 3123–33.
124. Kolinský, M., Doskocilová, D., Schneider, B., Stokr, J., Drahorádová, E. and Kuska, V., *J. Polymer Sci. A-1*, 1971, **9**, 791–800.
125. Millan, J., *J. Macromol. Sci. Chem. A*, 1978, **12**, 315–21.
126. Quenum, B. M., Berticat, Ph. and Vallet, G., *Polymer J.*, 1975, **7**, 287–99.
127. Krentsel, L. B., Litmanovich, A. D. and Shtern, V. Ya., *Zh. Prikl. Khim.*, 1972, **45**, 1875–6.
128. Quenum, B. M., Berticat, Ph. and Vallet, G., *Polymer J.*, 1975, **7**, 277–86.
129. Abu-Isa, I. A. and Myers Jr, M. E., *J. Polymer Sci. Polymer Chem. Ed.*, 1973, **11**, 225–31.
130. Quenum, B. M., Berticat, Ph. and Pham, Q. T., *Europ. Polymer J.*, 1971, **7**, 1527–36; 1973, **9**, 777–87.
131. Humbert, G., Quenum, B. M., Pham, Q. T., Berticat, Ph. and Vallet, G., *Makromol. Chem.*, 1974, **175**, 1597–609.
132. Keller, F., *Plaste und Kautschuk*, 1979, **26**, 80–2, 136–42.
133. Frensdorff, N. K. and Ekiner, O., *J. Polymer Sci. A*, 1967, **2**, 1157–75.
134. Usmanov, T. I., Krentsel, L. B., Litmanovich, A. D. and Platé, N. A., *Kinetika i Kataliz*, 1977, **18**, 337–41.
135. Krentsel, L. B. and Litmanovich, A. D., *Vysokomol. Soed B*, 1974, **16**, 372–4.
136. Noah, O. V., Litmanovich, A. D. and Platé, N. A., *J. Polymer Sci. Polymer Phys. Ed.*, 1974, **12**, 1711–25.
137. Schoniger, W., *Microchim. Acta*, 1955, 123–9.
138. Sokolova, N. V., Orestova, V. A. and Nikolayeva, N. A., *Zh. Analit. Khim.*, 1959, **14**, 472–7.
139. Afanasyev, I. B., *Usp. Khim.*, 1971, **40**, 385–416.
140. Semyonov, N. N., *Tsepnye Reaktsii* (*Chain Reactions*), Goskhimizdat, Leningrad, 1934, Part 2, Chap. 1.
141. Wautier, J. and Bruylants, A., *Bull. Soc. Chim. Belge*, 1963, **72**, 222–38.
142. Gusev, M. N., Thesis, Moscow, 1970.
143. Taft Jr, R. V., in *Steric Effects in Organic Chemistry*, Ed. M. S. Newman, Wiley, New York, and Chapman & Hall, London, 1956, Chap. 13.
144. Dewar, M. J. S., *Hyperconjugation*, Ronald Press, New York, 1962, Chap. 1.
145. Colebourn, N. and Stern, E. S., *J. Chem. Soc. (Lond.)*, 1965, 3599–605.
146. Fredricls, P. S. and Tedder, J. M., *Chem. and Ind.*, 1959, 490–1.
147. Zhdanov, Yu. A. and Minkin, V. I., *Korrelyatsionny Analiz v Organicheskoy Khimii* (*Correlation Analysis in Organic Chemistry*), Rostov University Press, Rostov on Don, 1966, Sec. Y, Chap. 2.
148. Hradil, J. and Chvalovsky, V., *Collect Czechoslov Chem. Commun.*, 1968, **33**, 2029–40.
149. Harvey, L., Gleicher, G. J. and Tetherow, W. D., *Tetrahedron*, 1969, **25**, 5019–26.
150. Litmanovich, A. D., Platé, N. A., Sergeyev, N. M., Subbotin, O. A. and Usmanov, T. I., *Dokl. Akad. Nauk. SSSR*, 1973, **210**(1), 114–17.
151. Usmanov, T. I., Subbotin, O. A., Litmanovich, A. D. and Sergeyev, N. M., *Zh. Org. Khim.*, 1973, **9**, 428–9.
152. Russell, G. A. and Ito, A., *J. Am. Chem. Soc.*, 1963, **85**, 2983–8, 2988–91.
153. Noah, O. V. and Litmanovich, A. D., *Vysokomol. Soed A*, 1977, **19**, 1211–17.
154. Keller, J. B., *J. Chem. Phys.*, 1962, **57**, 2584–6.
155. Chilts, G., Goldfinger, P., Huybrechts, G., Martens, G. and Verbeke, G., *Chem. Rev.*, 1963, **63**, 355–72.
156. Alfrey Jr, T. and Lloyd, W. G., *J. Chem. Phys.*, 1963, **38**, 318–21.
157. Eliel, E. L., Allinger, N. L., Angyal, S. J. and Morrison, G. A., *Conformational Analysis*, Interscience–Wiley, New York, London, Sydney, 1965, Chap. 4.

158. Coleman, B. D. and Fuoss, R. M., *J. Am. Chem. Soc.*, 1955, **77**, 5472–6.
159. Morcellet-Sauvage, J. and Loucheux, C., *Makromol. Chem.*, 1975, **176**, 315–31.
160. Torchilin, V. P., Thesis, Moscow, 1972.
161. Torchilin, V. P., Papisov, I. M. and Kirsh, Yu. E., 3rd Symposium on *Physiologically Active Polymers*, Riga, 1971, Abstracts 10.
162. Torchilin, V. P., Maklakova, T. A., Papisov, I. M., Kirsh, Yu. E. and Kabanov, V. A., *Vysokomol. Soed B*, 1972, **14**, 5.
163. Flodin, P., *J. Chromatography*, 1961, **5**, 103–15.
164. Boucher, E. A., Khosravi-Babadi, E. and Mollet, C. C., *J. Chem. Soc. Faraday Trans. 1*, 1979, **75**, 1728–35.
165. Boucher, E. A. and Mollet, C. C., *J. Chem. Soc. Faraday Trans. 1*, 1982, **78**, 75–88.
166. Frere, Y. and Gramain, Ph., *Macromolecules*, 1992, **25**, 3184–9.
167. Boucher, E. A. and Khosravi-Babadi, E., *J. Chem. Soc. Faraday Trans. 1*, 1981, **77**, 2259–63.
168. Khodzhaeva, I. D., Thesis, Moscow, 1970.
169. Turaev, A. S., Nadzhimutdinov, Sh. and Usmanov, Kh. U., *Vysokomol. Soed A*, 1977, **19**, 1347–56.
170. Nadzhimutdinov, Sh., Khalikov, T. and Usmanov, Kh. U., *Dokl. Akad. Nauk. SSSR*, 1973, **211**, 642–5.
171. Bourguignon, J. J., Bellissent, H. and Galin, J. C., *Polymer*, 1977, **18**, 937–44.
172. Klesper, E., Strasilla, D. and Berg, M. C., *Europ. Polymer J.*, 1979, **15**, 587–91.
173. Sawant, S. and Morawetz, H., *Macromolecules*, 1984, **17**, 2427–30.
174. Morawetz, H., in *Chemical Reactions on Polymers*, Eds J. L. Benham and J. F. Kinstle, ACS Symposium Series 364, ACS, Washington D.C., 1988, Chap. 23.
175. Boucher, E. A., *Progr. Polymer Sci.*, 1978, **6**, 63–120.
176. *Entsiklopediya Polimerov* (*Encyclopedia of Polymers*), Sovet. Entsiklopediya P. H., Moscow, 1974, Vol. 2, pp. 874–5.
177. Menshutkin, N., *Z. Phys. Chem.*, 1890, **6**, 41–57.
178. Rempp, P., *Pure Appl. Chem.*, 1976, **46**(1), 9–17.
179. Kawabe, H. and Yanagita, M., *Bull. Chem. Soc. Japan*, 1971, **44**, 896–901.
180. Stoy, V., *J. Polymer Sci. Polymer Chem. Ed.*, 1975, **13**, 1175–82.
181. Hudecek, S., Otoupalova, J., Ryska, M. and Svetlik, J., 4th International Conference *on Modified Polymers*, Bratislava 1975, Vol. 1, p. 30.
182. Minsk, L. M., Priest, J. W. and Kenyon, W. O., *J. Am. Chem. Soc.*, 1941, **63**, 2715–21.
183. Sakurada, I., *Kobunshi Tembo*, 1951, **5**, 64–7.
184. Sakurada, I. *Pure Appl. Chem.*, 1968, **16**, 263–83.
185. Fujii, K., *J. Polymer Sci. D*, 1971, **5**, 431–40.
186. Arranz, F. and Tahir, M. A., *Rev. Plast. Mod.*, 1970, **21**, 643–9.
187. Pichot, C., Guillot, J. and Guyot, A., *J. Macromol. Sci. Chem. A*, 1974, **8**, 1073–86.
188. Pichot, C., Guillot, J. and Guyot, A., *J. Macromol. Sci. Chem. A*, 1974, **8**, 1087–98.
189. McQuarrie, D. A., *J. Appl. Probability*, 1967, **4**, 413–78.

CHAPTER 5

Configurational and Conformational Effects: Quantitative Approaches

5.1 DESCRIPTION OF AN INFLUENCE OF MICROTACTICITY

The reactivity of functional groups in macromolecular reactions can be affected by the stereochemistry of the polymer chain. There are three possible types of triads for vinyl polymers:

Isotactic

Heterotactic

Syndiotactic

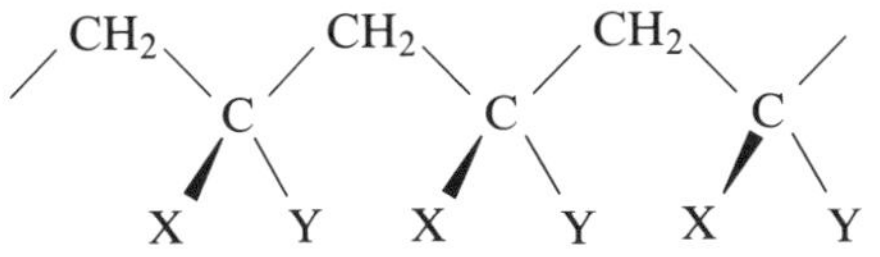

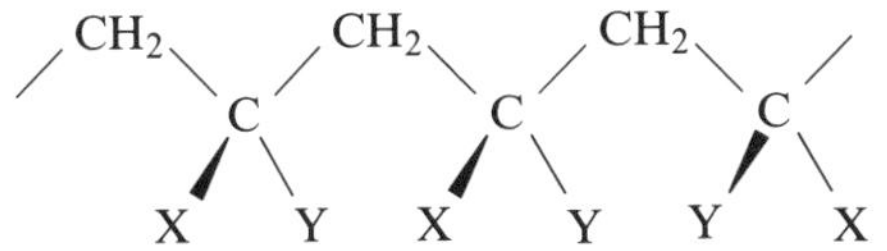

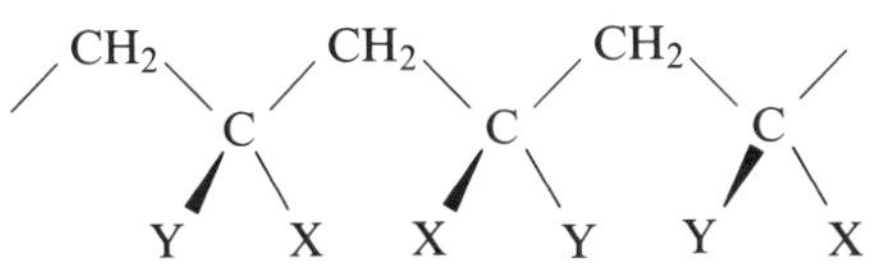

When the configurational effect is not accompanied by the neighboring groups effect, the process can be considered to include three independent reactions with rate constants k_i, k_h and k_s. The calculation of the kinetics of such a reaction is a rather simple task. However, when the reactivity depends on both of these effects (which is more probable), instead of the problem with three parameters (k_0, k_1, k_2; see Chapter 3) we have the problem with ten parameters, as an unreacted A unit can be in the center of any one of the ten triads, and therefore can be converted into B with one of the ten rate constants [1]:

$k_0^i N_0^i$ $k_0^h N_0^h$ $k_0^s N_0^s$

$k_1^i N_1^i$ $k_1^h N_1^h$ $k_1^{h'} N_1^{h'}$

$k_1^s N_1^s$

$k_2^i N_2^i$ $k_2^h N_2^h$ $k_2^s N_2^s$

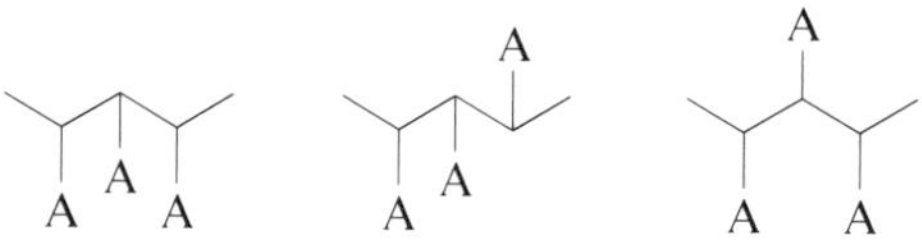

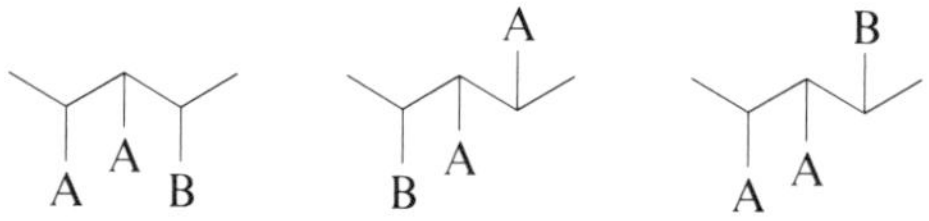

where N^i, N^h, N^s are the mole fractions of iso-, hetero- and syndio-triads and k^i, k^h, k^s are the rate constants of A units being in the center of the corresponding triads (the subscripts denote the number of reacted units).

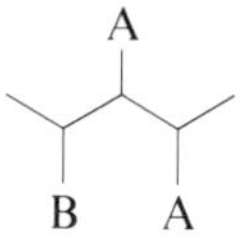

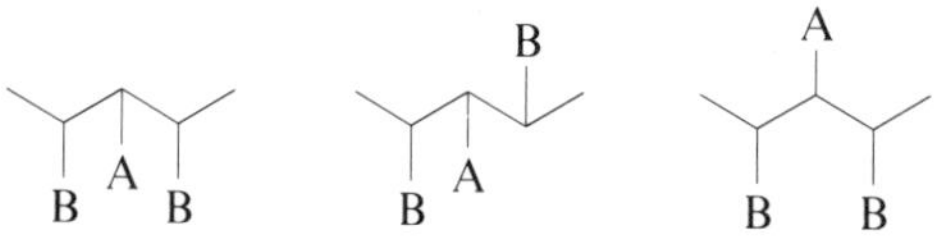

In principle this problem is very similar to the one described in Chapter 3. Therefore, for example, the kinetics of such a reaction can be described by the system of ten equations analogous to the Keller system (3.22):

$$-\frac{dN_0^i}{dt} = k_0^i N_0^i + 2N_0^i \frac{k_0^i N_0^i + k_0^h N_0^h + k_1^i N_1^i + k_1^h N_1^h}{N_0^i + N_0^h + N_1^i + N_1^h}$$

$$-\frac{dN_0^h}{dt} = k_0^h N_0^h + N_0^h \left(\frac{k_0^i N_0^i + k_0^h N_0^h + k_1^i N_1^i + k_1^h N_1^h}{N_0^i + N_0^h + N_1^i + N_1^h} \right.$$

$$+ \frac{k_0^s N_0^s + k_0^h N_0^h + k_1^s N_1^s + k_1^{h'} N_1^{h'}}{N_0^s + N_0^h + N_1^s + N_1^{h'}} \Bigg)$$

$$-\frac{dN_0^s}{dt} = k_0^s N_0^s + 2N_0^s \frac{k_0^s N_0^s + k_0^h N_0^h + k_1^s N_1^s + k_1^{h'} N_1^{h'}}{N_0^s + N_0^h + N_1^s + N_1^{h'}} \tag{5.1}$$

and so on.

In the same way one can transform equations describing the sequence length distribution and the compositional heterogeneity.

The only difficulty in the description of macromolecular reactions proceeding with the configurational effect is related to the experimental difficulties in evaluating the rate constants. This problem is rather complicated even for the stereoregular samples (see Chapter 4). The application of the same approaches to atactic polymers includes the synthesis of isotactic and syndiotactic models of high regularity, which is rather difficult itself.

It would appear that the calculation of kinetics and statistics of reactions in atactic chains could be simplified if some of the ten constants were equal or if there were some other restrictions. Pis'men [2] calculated the kinetics of a polymer-analogous reaction in atactic chains for the particular case: $k_0^i = k_0^h = k_0^s = k_0, k_1^h = k_1^s = k_1, k_1^i = k_1^{h'} = k_2, k_2^i = 2k_2 - k_0, k_2^h = k_1 + k_2 + k_0, k_2^s = 2k_1 - k_0$ and $k_2 > k_1 > k_0$ (autoaccelerating reaction). In this case one can distinguish three types of 'n-clusters' (i.e. sequences of n unreacted units bordered by reacted ones): open, semiopen and closed, depending on the number of ultimate unreacted units in the iso-position to the reacted neighbor (open (2), semiopen (1) and closed (0) respectively).

Denoting the open, semiopen and closed n-clusters as A_n^o, A_n^s and A_n^c one can represent the transformations of ultimate unreacted units as follows:

$$A_n^0 \xrightarrow{\;2\chi k_2\;} A_{n-1}^0 \qquad A_n^0 \xrightarrow{\;2(1-\chi)k_2\;} A_{n-1}^p$$

$$A_n^s \xrightarrow{\;2\chi k_1\;} A_{n-1}^p \qquad A_n^s \xrightarrow{\;2(1-\chi)k_1\;} A_{n-1}^s$$

$$A_n^p \xrightarrow{\;\chi k_1\;} A_{n-1}^0 \qquad A_n^p \xrightarrow{\;(1-\chi)k_1 + \chi k_2\;} A_{n-1}^p$$

$$A_n^p \xrightarrow{\;(1-\chi)k_2\;} A_{n-1}^s$$

where χ is the probability of finding an iso-diad in the chain ($1 - \chi$ is the probability to find a syndio-diad).

The rate of the decay of a cluster by the reaction of internal units

$$A_n \xrightarrow{\;k_2\;} A_m + A_{n-m-1}$$

is the same for all clusters. In this case the open cluster is transformed into two open ones with the probability χ^2 into open and semiopen with the probability $2\chi(1-\chi)$ and into two semiopen with the probability $(1-\chi)^2$. The probabilities of the transformation of a semiopen cluster into open and semiopen (1), two semiopen (2), open and closed (3) and semiopen and closed (4) clusters are equal to χ^2 (1), $\chi(1-\chi)$ (2 and 3) and $(1-\chi)^2$ (4) respectively. For the closed cluster the probability of the transformation into two semiopen ones is equal to χ^2, into semiopen and closed clusters to $2\chi(1-\chi)$ and into two closed ones to $(1-\chi)^2$.

The solution of the kinetic equations for the concentrations of n-clusters of three types C_n^o, C_n^s, C_n^c with the condition $k_2 \gg k_1, k_0$ gives the fraction of unreacted units $P(A)$ as

$$P(A) = \left(\frac{1-\chi e^{-k_0 t}}{1-\chi}\right)^{(2/\chi)[(k_1/k_0)(1-\chi)-1]} e_0^{-(2k_1-k_0)t} \tag{5.2}$$

For $\chi \to 0$ (stereoregular polymer) this expression coincides with the expression (3.14) for $j=1$ (it is true for $k_2 = 2k_1 - k_0$).

In general the kinetics of the polymer-analogous reaction in atactic chains can be calculated.

5.2 INTRACHAIN CATALYSIS AND CYCLIZATION

The reactions of intramolecular catalysis depend essentially on the conformation of a macromolecule because the reacting groups can be isolated from each other along the chain, and the probability of their interaction is related with the flexibility of the polymer chain. Another peculiarity of these reactions is the relative stability of the macromolecule conformation in the course of the reaction.

Theoretical aspects of the intrachain catalysis were considered for the first time by Morawetz and co-workers [3–8]. The simplest case of reactions between functional groups which are situated apart one from another are the reactions of cyclization, i.e. the reactions of end groups of the same macromolecule. Another case of the intramolecular reaction not accompanied by the change of the conformation is the interaction between reactive and catalytic groups. The interaction between such groups is possible only if they are close to one another. The probability of such proximity for the end groups as well as for the groups situated at random in the chain can be calculated through the function of end-to-end distance distribution [5–6]. For the Gaussian model of the polymer chain of Z segments of length b the probability to have a distance between the ends being equal to h is determined as

$$W(h)dh = \left(\frac{2\pi Z b^2}{3}\right)^{-3/2} \exp\left(\frac{-3h^2}{2\pi Z b^2}\right) 4\pi h^2 \, dh \tag{5.3}$$

where $W(h)/4\pi h^2$ denotes the concentration and trends to some limit for

$h^2 \ll Zb^2$. This limit value $\lim [W(h)/4\pi h^2] \equiv c_{ef}^0$ can be regarded as the average concentration of one end of the polymer chain in the vicinity of the other end. In the mol/1 unities c_{ef}^0 is expressed as

$$c_{ef}^0 = \left(\frac{1000}{N_A}\right)\left(\frac{3}{2\pi\overline{h^2}}\right)^{3/2} \qquad (5.4)$$

where $\overline{h^2}$ is a mean-square distance between the ends and N_A is the Avogadro number. If k_2 is the rate constant of the second-order reaction of functional groups not fixed on the chain, the rate constant of the first-order reaction between the end groups of the chain is determined as

$$k_1 = k_2 c_{ef}^0 = k_2 \left(\frac{1000}{N_A}\right)\left(\frac{3}{2\pi\overline{h^2}}\right)^{3/2} \qquad (5.5)$$

When the effect of the excluded volume can be neglected, the probability of the reaction between the group situated on the nth site of the polymer chain and catalytic group situated on the jth site can be estimated in the same way as for the reaction of end groups of the chain of $(j - n)$ segments. Then the mean-square distance between the nth and jth units can be expressed through the mean-square end-to-end distance as

$$\overline{h_{nj}^2} = \overline{h^2}\,|j - n|/Z$$

for $|j - n| \geq x'$ (x' is the minimum number of monomer units allowing the interaction).

The effective local concentration of the catalyst near the nth reactive segment of the chain of Z segments with random distribution of catalytic groups can be written as

$$c_{ef} = \left(\frac{1000}{N_A}\right)\left(\frac{3Z}{2\pi\overline{h^2}}\right)^{3/2} \left(\sum_{x=1-n}^{-x'} |P_j x|^{-3/2} + \sum_{x=x'}^{Z-n} |P_j x|^{-3/2}\right) + c_{ef}' \qquad (5.6)$$

where $x = j - n$, $P_j = 1$ or 0 depending on the presence of catalytic groups on the jth site and c_{ef}' is a correction parameter which accounts for the probability of the interaction of groups being on the nearer distance than j monomer units.

The rate constant is given by $k_1 = k_2 c_{ef}$. If there are little reactive groups in the chain and all other segments carry the catalytic substituents, $P_j = 1$ for all j in Eq. (5.6) and $k_1 = k_2 c_{ef}^{max}$. If the fraction of catalytic groups is equal to ω, then $k_1 = \omega k_2 c_{ef}^{max}$. After integration of Eq. (5.6) and substitution of $\overline{h^2} = Z(K_\theta/\Phi)^{2/3} M_0/2$ (where $K_\theta = [\eta]_\theta M^{-1/2}$, $\Phi = 2.6 \times 10^{21}$ is the Flory constant and M_0 is the molecular mass of a monomer unit), one obtains [6]

$$k_1 = \omega k_2 \left(\frac{4000\Phi}{N_A K_\theta}\right)\left(\frac{3}{2\pi M_0}\right)^{3/2} (x')^{-1/2} \qquad (5.7)$$

All these calculations correspond to θ-conditions, while taking account of the excluded volume results in a decrease in the probability of cyclic conformations. In that case $k_1 \sim Z^{-a}$ where $a > 3/2$.

The experimental study of hydrolysis of ternary copolymer of acrylamide with small amounts of reactive monomers I or II and catalitically active III [6] shows that there is an intramolecular catalysis in such a reaction, but that the rate constant decreases with a decrease in the thermodynamic quality of a solvent, although its increase seems to be more reasonable:

$$
\begin{array}{ccc}
\text{CH}_3 & \text{CH}=\text{CH}_2 & \text{CH}=\text{CH}_2 \\
| & | & | \\
\text{C}=\text{CH}_2 & \text{CO} & \text{CO} \\
| & | & | \\
\text{CO} & \text{NH} & \text{NH} \\
| & | & | \\
\text{O} & (\text{CH}_2)\text{n} & \text{CH}_2 \\
\end{array}
$$

I   II   III

Goodman and Morawetz [5] also used a computer to simulate the kinetics of intramolecular reactions for the chains with some fraction ω of catalytic substituents and only one reactive groups in the middle of the chain. The interaction between groups separated by more than ten units was allowed. The rate constant for the case of 100 chains of 1000 units was calculated as follows:

$$
k_n = c \left[\sum_{j=1}^{490} P_{nj}(500 - j)^{-a} + \sum_{j=510}^{1000} P_{nj}(j - 500)^{-a} \right] \tag{5.8}
$$

where $P_{nj} = 1$ if there is a reactive group in the jth site and $P_{nj} = 0$ if there is no reactive groups in the jth site. The fraction of reacted groups was calculated as

$$
y(t) = \frac{1}{1000} \sum_n - k_n t \tag{5.9}
$$

The experimental dependences of $\log y$ on t were compared with curves calculated for various a. The kinetics of the reaction in a good solvent was shown to be described by the expression with $a = 2$, but close to the θ-conditions a is equal to 1.6 (this value is more than for the model with zero excluded volume).

The set of papers of Sisido and co-workers [9–21] is also devoted to the theoretical and experimental study of intramolecular catalysis reactions. These authors used the thermodynamic approach to calculate the probability of the proximity of reactive and catalytic groups [9], assuming that

$$W(r < r_0) = \frac{Z(r < r_0)}{Z_t}$$

$$Z(r < r_0) = \sum_{r < r_0} \exp\left(\frac{-E_i}{RT}\right)$$

$$Z_t = \sum_{\text{over all states}} \exp\left(\frac{-E_i}{RT}\right)$$

where $W(r < r_0)$ is the probability of groups coming to one another down to the distance less than some limiting value r_0, $Z(r < r_0)$ is the partition function for this probability and Z_t is the partition function over all states.

The thermodynamic functions Z are calculated by the Monte Carlo method for chains disposed on a lattice with some energetic restrictions. Using this procedure the absolute values of constants, of the conformational energy and of the activation entropy are obtained.

Sisido [10] also studied the influence of catalytic groups distribution in the chain on the rate of the reaction. Three cases were considered:

(a) The catalytic groups distribution is averaged for all molecules and can be neglected in the kinetic calculation.
(b) The catalytic groups are fixed in the polymer chain, and the probability of finding the catalytic group on a given site independent on the state of neighboring groups is given.
(c) The probability of finding the catalytic group on a given site depends on the state of neighboring groups.

The equations derived permit the kinetics of the reaction to be calculated, taking into account the type of the units distribution in a chain and the distances between reactive groups. The results obtained were compared with the experimental data of Goodman and Morawetz [6], the good coincidence of calculated and experimental values being shown.

Later Sisido and co-workers published some papers about the experimental studies of intramolecular reactions. They studied the hydrolysis of the terminal *p*-nitrophenyl ester group catalyzed by the terminal 4-pyridyl groups attached to a polysarcosine [12] and to a polyoxyethylene [17] chain, the formation of charge transfer complexes between terminal groups [13, 19, 20], the formation of disulfide linkages in the air oxidation of sulfhydryl groups attached to the ends of a polysarcosine chain [16], photoselective cyclization of azobenzene-containing oligosarcosines [21] and other systems. In these papers the dependences of the rate constant on the chain length and flexibility are given, some thermodynamic

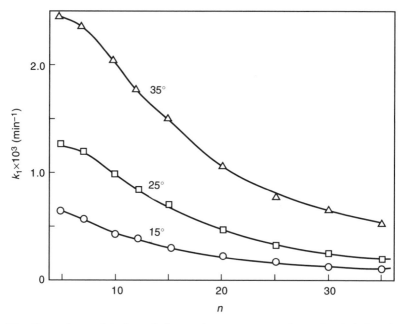

Fig. 5.1   Dependence of the intrachain reaction rate constant k_1 on the chain length n at three different temperatures. (Reprinted with permission from *Macromolecules*, 1976, **9**, 316. Copyright 1976 American Chemical Society)

parameters are calculated and the experimental data are compared with results of the theoretical calculation and of the Monte Carlo simulation.

In Fig. 5.1 the rather typical dependence of the intrachain reaction rate constant k on the chain length n at different temperatures is presented [12]. In Fig. 5.2 an example of the comparison of the theoretical calculation and the Monte Carlo simulation is given for the rate constant of the hydrolysis of the terminal p-nitrophenyl ester group catalyzed by the terminal 4-pyridyl group in the polysarcosine chain [12].

It should be mentioned that in many experimental works on intramolecular interactions the photochemically interacting systems are described [21–23]. The elementary act in such reactions is very fast; therefore these processes are diffusion controlled. In this chapter we consider only the chemically controlled reactions and are not concerned with the very large domain of diffusion-controlled processes which is successfully developed by Wilemski and Fixman [24, 25], Doi [26, 27] and other authors [28–34]. Kozlov [28] described a model of the chemically controlled reaction of terminal groups of the same chain taking into consideration the excluded volume effect and the selective sorption.

In this paper the ratio of the effective rate constant k_{ef} to the real one k_r is given

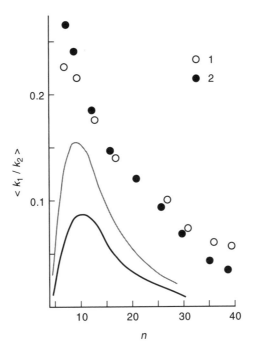

Fig. 5.2 $\langle k_1/k_2 \rangle$ ratio versus the chain length n. Curves, the calculation for the poly-sarcosine chain with Poisson molecular mass distribution (solid curve, $r_0 = 4\,\text{Å}$; broken curve, $r_0 = 6\,\text{Å}$); points, experimental data obtained at 15 (1) and 35 °C (2). (Reprinted by permission from *Macromolecules*, 1976, **9**, 316. Copyright 1976 American Chemical Society)

by the expression

$$\frac{k_{\text{ef}}}{k_{\text{r}}} = \frac{\left(\int_V \rho_{12} I_{\text{r}}\, \mathrm{d}V\right)}{\overline{\rho^2}} \tag{5.10}$$

where ρ_{12} is the concentration of coils being at the r distance from molecule 1 under consideration, $\overline{\rho^2}$ is the average concentration of molecules 2, I_{r} is the overlap integral for the functions of the reactive groups distribution relatively to the mass center of reacting molecules:

$$I_{\text{r}} = \left(\frac{\beta_1 \beta_2}{\pi^{1/2}(\beta_1^2 + \beta_2^2)^{1/2}}\right)^3 \exp\left(\frac{\beta_1^2 \beta_2^2}{\beta_1^2 + \beta_2^2} r^2\right) \tag{5.11}$$

where β_1 and β_2 are the parameters of the Gaussian functions of the reactive groups distribution relatively to the mass center and r is the distance between the mass centers.

The distribution function for the distances of the kth segment from the mass center (s_k) is defined as

$$P(s_k) = \left\{ \frac{9}{2\pi R^2 [u^3 + (1-u)^3]} \right\}^{3/2} \exp \left\{ \frac{-9s_k^2}{2R^2 [u^3 + (1-u)^3]} \right\} \tag{5.12}$$

where $u = k/N$ is the number of segments in a chain. It follows from Eqs. (5.11) and (5.12) that

$$\beta^2 = \frac{\beta_1^2 \beta_2^2}{\beta_1^2 + \beta_2^2} = \frac{9}{2\{R_1^3 [u^3 + (1-u)^3] + R_2^3 [v^3 + (1-v)^3]\}} \tag{5.13}$$

The condition of the equilibrium for the spatial distribution of reactive groups can be written as

$$\rho_{12} = \rho_2(\infty) \exp \left(\frac{-U_r}{kT} \right) \tag{5.14}$$

where U_r is the free energy of the interaction of two molecules with the distance between mass centers being equal to r:

$$U_r = kTXe^{-\beta^2 \rho r^2} \tag{5.15}$$

where $X = 2(\alpha^2 - 1)$ is the Flory–Fox parameter, taking into account the dimensions rise in a good solvent. Then

$$\beta_\rho = \frac{\beta_{\rho_1} \beta_{\rho_2}}{(\beta_{\rho_1}^2 + \beta_{\rho_2}^2)^{1/2}} = \frac{3}{(R_1^2 + R_2^2)^{1/2}} \tag{5.16}$$

Introducing

$$t = \left(\frac{\beta_\rho}{\beta} \right)^2 = \frac{2\{R_1^2 [u^3 + (1-u)^3] + R_2^2 [v^3 + (1-v)^3]\}}{R_1^2 + R_2^2} \tag{5.17}$$

one obtains, after the substitution of Eqs. (5.11), (5.14), (5.15) and (5.17) into (5.10),

$$\frac{k_{ef}}{k_r} = \frac{4}{\pi^{1/2}} \int_0^\infty \exp(-Xe^{-tx^2} - x^2)x^2 \, dx \qquad (x = \beta r) \tag{5.18}$$

For the terminal groups $u = v = 0$, $t = 2$; for the groups situated in the middle of a chain $u = v = 0$, $t = 0.5$; for other groups $0.5 < t < 2$.

If $X \leq 1$,

$$\frac{k_{ef}}{k_r} = \sum_{n=0}^\infty (-1)^n \frac{1}{(tn+1)^{3/2}} \frac{X^n}{n!} \tag{5.19}$$

The rate constant of the intramolecular reaction is decreased with an increase in the solubility and approaching to the middle of a chain.

The calculation of k_{ef}/k_r with parameters $\alpha = 1.27$ and $X = 2.12$ corresponding to the experimental data obtained by Goodman and Morawetz [6] gives the 0.74

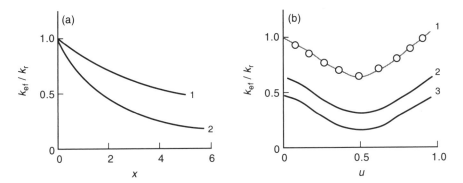

Fig. 5.3 k_{ef}/k_r dependence on the solvent quality (a) for $u = v = 0$ (1) and $u = v = 0.5$ (2) and on the groups position (b) for $X = 1$ (1), 3 (2) and 6 (3) (points, the calculation following Eq. 5.19. (Reproduced by permission of Nauka, Moscow, Russia, from *Dokl. Akad. Nauk. SSSR*, 1978, **243**, 410)

value, being much closer to the experimental result (0.79) than the results of the theoretical calculation of the authors [6] (Fig. 5.3).

Friedman and O'Shaughnessy [35–40] used the scaling approach [41] for the solution of the intramolecular cyclization problem assuming that the reaction rate is determined by the exponent θ:

$$\theta \equiv \frac{d + g}{z} \tag{5.20}$$

Here $d = 3$, z is the dynamical exponent ($\tau \sim R$, where τ is the longest relaxation time, R is the root-mean-square end-to-end distance of a single polymer chain) and g is the 'correlation hole exponent' which quantifies the diminished probability that the chain ends meet one another. When excluded-volume interactions are not screened, the relevant end-to-end equilibrium probability distribution $p(r)$ is that describing self-avoiding walks:

$$p(r) = R^{-d} f\left(\frac{r}{R}\right)$$

where $R = aN^v$ and v is the Flory exponent relating chain size to the number of segments N each of the size a. For long chains and small r/R, $f(x) \sim xg$ for $x \ll 1$. In θ-solvents Gaussian statistics are recovered, $p(0)$ equals a nonzero constant, g vanishes and the reaction exponent (5.20) reduces to $q = d/z$. Using the renormalization group approach Friedman and O'Shaughnessy [35–37] obtained $g = 0.27$ and derived for the rate constant the expression:

$$k(t) \sim \left(\frac{1}{s}\right)^{3/2} [\log s \text{ of } s \text{ and } t] \tag{5.21}$$

where s is the number of units in the chain segment connecting the interacting groups.

In terms of $\tau_s = \tau_a s^{yz}$ (the relaxation time of the segment of polymer connecting the reactive groups), where t_a is the relaxation time of the single chain, the 'short time' k_0 and 'long time' k_∞ were derived, being equal to

$$k_0(t) = \frac{A_1}{\tau_s(t/\tau_s)^{-1/4}}$$

$$k_\infty(t) = \frac{A_1}{\tau_s} \tag{5.22}$$

where $A_1 = 16/\pi^3$ and $\tau_s = $ constant s^2. Besides the cases of θ-solvents Friedman and O'Shaughnessy [35–40] also described the cyclization reactions in melts and good solvents.

5.3 INTRAMOLECULAR CROSSLINKING

The reactions of intramolecular crosslinking also depend essentially on the chain conformation, but in addition the formation of each crosslink results in the change of the conformation of a macromolecular coil. As a result of this complication such processes are an area badly suited to theoretical analysis. That is why despite interest both from the viewpoint of the theory of macromolecular reactions and the theory of the network formation there are very few papers written on a theoretical study of this problem.

5.3.1 DIMENSIONS OF CROSSLINKED MACROMOLECULES

The first paper dealing with the calculation of dimensions of macromolecular coils containing crosslinks is probably that of Zimm and Stockmayer [42]. In 1949 these authors calculated the mean-square radius of gyration for the model of a free-rotational cyclic chain [42]. It was shown that

$$\overline{R_{cycl}^2} = \frac{Nb^2}{12} \tag{5.23}$$

while for the linear chain

$$\overline{R_{lin}^2} = \frac{Nb^2}{6} \tag{5.24}$$

is valid (N is the number of segments in the chain and b is the length of the segment).

Then the authors of reference [42] assumed that the contribution of the linear and cyclic parts into the value of the overall radius of gyration is proportional to the respective number of units in them, i.e.

$$\overline{R^2} = \frac{Z}{N}\frac{Zb^2}{12} + \frac{N-Z}{N}\frac{(N-Z)b^2}{6} \qquad (5.25)$$

where Z is the number of units in the cycle.

The approach of Zimm and Stockmayer allows dimensions of macromolecules having one cycle, i.e. with one crosslinkage, to be calculated. It is impossible to apply this approach to chains with large numbers of crosslinks because of the rapid increase in the number of possible topological structures accompanied by the complication of the algorithm of the calculation of $\overline{R^2}$ for each of them. There are therefore three possible structures for the chain with two crosslinkages, for three crosslinkages there are eight structures, and so on [43]:

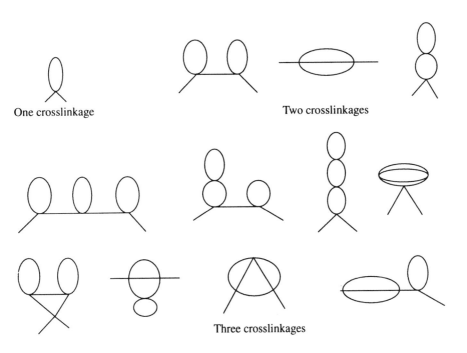

One crosslinkage   Two crosslinkages

Three crosslinkages

The problem of classifying complicated cycles being formed in crosslinked chains was considered by Yerukhimovich [44], who described them as graphs and calculated their topological size distribution dependent on the only parameter—the complicity index of a cycle.

Edwards and co-workers [45, 46] used a thermodynamic approach to the problem of the calculation of dimensions of crosslinked macromolecules. In reference [45] the general thermodynamic theory of polymer chains with cross-linkages is described. The equation of state of this crosslinked polymer is derived using the method of secondary quantization which allowed the average number of crosslinks to operate in the chain neglecting the details of the chain structure. In reference [46] this theory is applied to the calculation of dimensions of chains with intramolecular crosslinkages. The model considers the isolated chains with sites of crosslinking distributed at random. It is further assumed that the chain is a free-rotational one, and the volume effects and van der Waals interactions between units are taken into account. The mean-square end-to-end distance is calculated as $\overline{h^2} = C_N N l^2$, where N is the chain length, l is the length of the unit and C_N is the so-called characteristic ratio. The volume interactions are taken into account by introducing the 'thickness' of the chain a (the square root of the area of a cross-section of the chain). The crosslinkage is considered to be an equilibrium form with limited dimensions (two crosslinked units are at a distance b_2 apart).

The calculation of the thermodynamic potential of a crosslinked chain assuming the energetic advantage of crosslinking leads to the following relationship between chain dimensions and the number of cross-linkages m:

$$\frac{1 + (m\xi^3/Nla^2)}{1 - (1 + (m\xi^3/Nla^2))(1/\tau^3)} = \frac{V}{kT}\frac{1}{2la^2}\left[1 - \frac{10\pi b_1^2}{3(Nla^2)^{2/3}\tau^2}\right]$$

$$- \frac{\pi C_N l^2 \tau}{3(Nla^2)^{2/3}} + \frac{m+1}{N}\tau^3\left[1 - \frac{4\pi b_2^2}{3(Nla^2)^{2/3}\tau^2}\right] \quad (5.26)$$

where τ is an intrinsic volume: $\tau^3 = (2R/3)^3(1/Nla^2)$, ξ^3 is the volume of one crosslink, b_1 is the distance between units interacting with the attracting force determined according to the Flory–Huggins theory and V is the change in interaction energy due to the formation of the additional polymer–polymer contact $V = V_{pp} + V_{ss} - 2V_{ps} = 2kT\chi la$ (χ is the Huggins interaction parameter).

Due to assumptions made in the course of deriving Eq. (5.26) the latter is correct only if the number of crosslinkages is not very small. Equation (5.26) can be solved graphically, and the dependence of the radius of gyration of the crosslinked chain on the average number of crosslinkages can be found for every given value of the structural and thermodynamic parameters. The authors of reference [46] solved this equation for polystyrene corsslinked with diiso-cyanates (all parameters for this system are known and can be taken from the literature).

When deriving Eq. (5.26) the considerable gain in energy and the considerable decrease of dimensions in the course of crosslinking were admitted. However, the calculation performed for polystyrene shows that the decrease of dimensions is not very great. This fact requires the introduction of the correction parameter into Eq. (5.26), although it is difficult to evaluate the degree of this correction. Simplifying Eq. (5.26) for a small number of crosslinkages one obtains

$$\overline{R_m^2} = \frac{\overline{R_0^2}}{m+1} \tag{5.27}$$

However, this simplification is not completely correct because of the assumption about a rather large number of crosslinkages taken at the derivation of Eq. (5.26).

Gordon *et al.* [47] proposed another empirical equation, also based on thermodynamic considerations:

$$\overline{R_m^2} = \frac{\overline{R_0^2}}{(m+1)^q} \tag{5.28}$$

where $q \approx 0.2$. Evidently this relation is more probable for chains with small amount of crosslinkages. Later Ross-Murphy confirmed the validity of this relation using the Monte Carlo calculation [48].

Yerukhimovich [49] derived the expressions for dimensions of moderately crosslinked chains with the number of crosslinks n satisfying the condition $1 \ll (n^2/N) \ll (a^3/v_0)^8$, where N is the number of units in the chain and a is the bond length:

$$\overline{R^2} = \frac{Na^2}{6} \frac{A}{2} \left(\frac{n^2}{\pi N} \right)^{-1/2} \tag{5.29}$$

$$\overline{h^2} = Na^2 \times 8A^2 \left(\frac{n^2}{N} \right)^{-1} = 8A^2 \left(\frac{N}{n} \right)^2 a^2 \tag{5.30}$$

For the Gaussian chain $A \approx 0.7$.

Allen *et al.* [46] compared the results of their calculations with experimental data measuring the intrinsic viscosity of intramolecularly crosslinked polystyrene and found a rather good coincidence.

The approach considered above is the only attempt of an analytical calculation of dimensions of chains with intramolecular crosslinkages. Another approach is a simulation of this process using the Monte Carlo method [50–54].

Bonetzkaya *et al.* [50] simulated so-called instantaneous crosslinking when the conformation of the chain is not changed until the end of the reaction. Macromolecules were simulated by a random walk procedure on the tetrahedral lattice, assuming that the chain can be crosslinked only in sites of self-intersection. The authors of reference [50] assumed that the probability of the reaction between two definite units is proportional to the number of conformations in which these units are close to one another. Then the ensemble of the chains with a given number of crosslinkages can be obtained from the ensemble of noncrosslinked conformations. The mean-square radius of gyration for the chain with

m crosslinkages can be calculated as follows:

$$\overline{R_m^2} = \frac{\sum_{n_i \geq m} \binom{n_i}{m} q^{n_i - m} R_i^2}{\sum_{n_i \geq m} \binom{n_i}{m} q^{n_i - m}} \tag{5.31}$$

where n_i is the number of units drawn together in a given conformation (with summation over all conformations in which the number of pairs drawn together is not less than the number of crosslinkages); $\binom{n_i}{m}$ is the number of ways of crosslinking of m pairs out of n_i possible pairs and q is the statistic weight of pairs that have been brought close to each other but are not crosslinked. During the computation the number of noncrosslinked conformations is chosen at random, and the dimensions are evaluated approximately using Eq. (5.31).

The results of calculations show the monotonic decrease of chain dimensions during crosslinking. Instantaneous crosslinking apparently corresponds to 'fast' reactions with a reaction time of the same order as the time of the change of the conformation. One can assume that usual 'slow' chemical reactions for which the reaction time is much longer than the time of the conformation change are accompanied by the more significant decrease of dimensions during crosslinking. Such a case is considered by Platé and co-workers [51, 52] who used Monte Carlo simulation.

In these papers the chains are simulated on various lattices allowing self-intersections. Each particular chain is subjected to consequent crosslinking, and then the parameters obtained are averaged over the ensemble of chains with a given number of crosslinkages. For each random conformation the number of reactive contacts is calculated (all noncrosslinked self-intersections are considered as reactive contacts), and then one of the contacts is crosslinked with a probability

$$W_j = \beta Z_{j-1} \tag{5.32}$$

where W_j is the probability of the jth crosslinkage formation, Z_{j-1} is the number of reactive contacts in the conformation with $j-1$ crosslinkages and β is a normalization coefficient.

If the chain is not crosslinked the new conformation with the same number of crosslinkages $(j-1)$ is generated and so on up to the crosslinkage formation. The procedure is repeated until a given number of crosslinkages is formed. In Fig. 5.4 the dependences of the relative average dimensions of polymer coils on the degree of crosslinking are presented. It can be seen that chain dimensions decrease essentially in the course of crosslinking and that this effect is greater with an increase in the chain length.

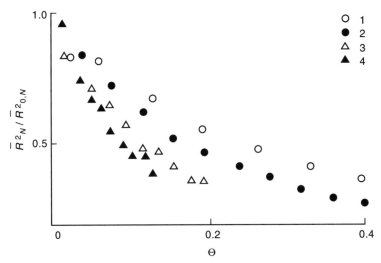

Fig. 5.4 Mean-square dimensions of crosslinked macromolecules versus the degree of crosslinking θ for model chains of the length $N = 30$ (1), 50 (2), 100 (3) and 150 units (4). (Reproduced by permission of Nauka, Moscow, Russia, from *Vysokomol. Soed. A*, 1977, **19**, 2800)

In Fig. 5.5 the results on the change of relative dimensions of the polymer coil in the course of intramolecular crosslinking are compared with experimental data on the change of intrinsic viscosity accompanying the formation of intramolecular complex between poly-N-methacryloyl-L-lysine and Cu^{II} ions in aqueous solution [53]. The rather satisfactory coincidence of calculations with experiment is observed.

5.3.2 KINETICS OF INTRAMOLECULAR CROSSLINKING

The rate of the reaction corresponding to the model described above (a relatively slow chemical reaction between any two units of the macromolecular that come close to each other because of the flexibility of the chain) should be proportional to the number of reactive contacts in a partially crosslinked coil Z_j (j is the number of crosslinkages). Assuming the independence of an average number of contacts on the crosslinkages configuration [54] one can write the following system of kinetic equations [51, 52]:

$$\frac{dC_0}{dt} = k_0 \overline{Z_0} C_0$$

$$\dots$$

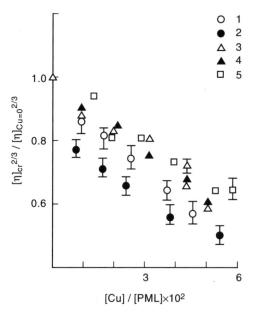

Fig. 5.5 Relative intrinsic viscosity of the complex of polymethacyloyl-L-lisine with Cu^{2+} ions as a function of the relative content of the crosslinking agent for $M_{PML} = 10^6$ (1), 83×10^4 (2), 55×10^4 (3), 52×10^4 (4) and calculated following the Monte Carlo technique (5). (Reproduced by permission of Nauka, Moscow, Russia, from *Vysokomol. Soed. A*, 1985, **27**, 1243)

$$\frac{dC_j}{dt} = k_0(\overline{Z_{j-1}C_{j-1}} - \overline{Z_jC_j}) \tag{5.33}$$

$$\cdots$$

$$\frac{dC_M}{dt} = k_0\overline{Z_{M-1}C_{M-1}}$$

where C_j is the number of chains with j crosslinkages, M is the maximum number of crosslinkages and k_0 is the rate constant of elementary crosslinking.

Then the average number of crosslinkages in the chain $\bar{n}(t) = 1/C\sum_{j=1}^{M} jC_j(t)$ (where $C = \sum_{j=0}^{M} C_j$ is the total number of chains) is determined as follows:

$$\frac{d\bar{n}}{dt} = \left(\frac{k_0}{C}\right) \sum_{j=0}^{M} \overline{Z_jC_j} \tag{5.34}$$

The problem of calculating the kinetics of intramolecular crosslinking is equivalent to finding the equilibrium average values of Z_j. The analytical approach to the estimation of Z_j is rather complicated for reasons mentioned above in connection with the calculation of chain dimensions.

Recently Friedman and O'Shaughnessy [38–40] have developed a general approach to the kinetic description of intramolecular reactions, including both cases of diffusion-controlled and 'law of mass action' regimes using scaling arguments and detailed renormalization group treatment. These authors describe three 'extreme' cases: cyclization (two end groups), reaction between one end and one deeply internal group and reaction between two deeply internal groups. The results were obtained for melts, in good solvents and in θ-solvent. The cyclization case is described above (Eqs. 5.20 to 5.22). For the general case corresponding to the model considered here—a chemically controlled reaction between two deeply internal groups in θ-conditions—the same approach results in the following expressions for 'short time' k_0 and 'long time' k_∞:

$$k_0(t) = \frac{A_3}{\tau_s(t/\tau_s)^{-1/4}}$$

$$k_\infty(t) = \frac{A_4}{\tau_s \ln(a/s)} \qquad (5.22')$$

where $A_3 = 4/\pi^3$, $A_4 = 1/\pi$, $\tau_s = $ constant s^2 and $v = 3/5$ and $g \approx 0.71$ (the variables are determined above when deriving Eqs. (5.20) to (5.22).

In references [52] and [53] the kinetics of intramolecular crosslinking has been studied by the Monte Carlo method. The procedure of the simulation is analogous to that described above with time between two consequent cross-linking acts determined as

$$t = \frac{\beta(m + \xi)}{k_0} \qquad (5.35)$$

where m is the number of conformations derived before the formation of a cross-linkage [55] and ξ is the random number evenly distributed between 0 and 1. The procedure is repeated up to a given time T_{max}.

In Fig. 5.6 the change in the number of contacts with an increase in the number of crosslinkages is shown. The increase in the number of contacts is a result of a decrease in the effective volume of a polymer coil in the course of crosslinking. The decrease in the number of contacts for large degrees of crosslinking can be explained by exhaustion of free reactive groups. The numerical solution of Eqs. (5.33) with Z_j obtained from the computer experiment gives the kinetic curve of the reaction.

In Fig. 5.7 the results of such calculation are compared with data obtained directly by simulation. The good agreement of the two approaches confirms the validity of the assumption made when deriving Eqs. (5.33) that the average number of contacts is independent on the configuration of crosslinkages. It can be seen from Fig. 5.7 that the initial part of the curve can be represented in the

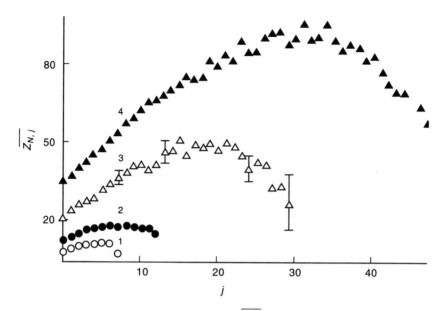

Fig. 5.6 Average number of reactive contacts $\overline{Z_{N,j}}$ versus the number of crosslinkages j for $N = 30$ (1), 50 (2), 100 (3) and 150 units (4). (Reproduced by permission of Nauka, Moscow, Russia, from *Vysokomol. Soed. A*, 1977, **19**, 2800)

linear form:

$$\overline{Z}_j = Aj + B \tag{5.36}$$

Substitution of Eq. (5.36) into Eq. (5.34) results in

$$\frac{d\bar{n}}{dt} = k_0(A\bar{n} + B) \tag{5.37}$$

The solution of Eq. (5.37) is

$$\bar{n}(t) = \frac{B}{A(e^{k_0 At} - 1)} \tag{5.38}$$

This expression with A and B values found from Fig. 5.6 also coincides well with results of the computer experiment. That is why the kinetics of intramolecular crosslinking can be initially described by the linear approximation.

Thus intramolecular crosslinking is an autoaccelerated reaction, and the initial rate and the degree of autoacceleration increase with an increase in the chain length. Another important conclusion [51, 52] is the existence of a uniform relationship between the kinetics of the reaction and the equilibrium properties of partially crosslinked chains.

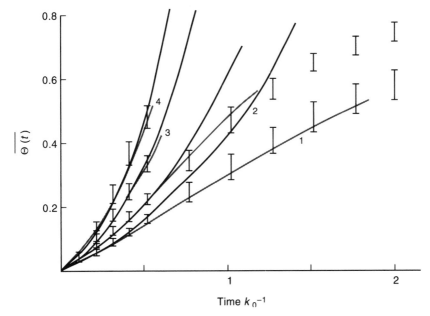

Fig. 5.7 Average degree of crosslinking θ versus time. Thin curves, the solution of Eqs. (5.33); solid curves, the solution of Eq. (5.38); points, Monte Carlo calculation for $N = 30$ (1), 50 (2), 100 (3) and 150 units (4). (Reproduced by permission of Nauka, Moscow, Russia, from *Vysokomol. Soed. A*, 1977, **19**, 2800)

5.3.3 COMPOSITION HETEROGENEITY OF CROSSLINKED MACROMOLECULES

The composition distribution of the products for various degrees of crosslinking is also calculated in reference [52] using Monte Carlo simulation. The exact distribution of the number of crosslinkages at any time is determined by the solution of Eqs. (5.33). The dispersion of this distribution is

$$D = \overline{n^2(t)} - [\overline{n(t)}]^2 \tag{5.39}$$

where $\overline{n^2(t)} = 1/C\sum_{j=1}^{M} j^2 C_j(t)$ is the mean-square number of crosslinks in the chain.

If the average number of contacts in the chain remains constant during the reaction, the process would be a random one. The number of crosslinkages would have a Poisson distribution with a dispersion

$$D_P = \overline{Z_0} k_0 t \tag{5.40}$$

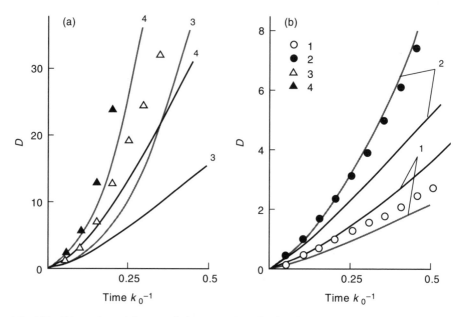

Fig. 5.8 Dispersion of the crosslinkage number distribution as a function of time. Solid curves, Poisson distribution; thin curves, linear approximation; points, Monte Carlo calculation for $N = 30$ (1), 50 (2), 100 (3) and 150 units (4). (Reproduced by permission of Nauka, Moscow, Russia, from *Vysokomol. Soed.* A, 1977, **19**, 2800)

The calculation of the dispersion in the linear approximation gives

$$D_L = e^{k_0 At}\left(\frac{B}{A}\right)(e^{k_0 At} - 1) = e^{k_0 At}\overline{n(t)} \tag{5.41}$$

In Fig. 5.8 the values of the dispersion obtained from computer simulation are compared with results of calculations using Eqs. (5.40) and (5.41). It can be seen from the figure that the true distribution is much wider than the Poisson one, and that the width increases with an increase in the chain length. In the initial stage the dispersion is described well by the linear approximation. For short chains subjected to a reaction lasting for some time the distribution becomes narrower due to the accumulation of chains with many crosslinkages (close to the maximum value) and the dispersion tends to the Poisson one.

5.3.4   EFFECT OF REACTIVE GROUPS DISTRIBUTION AND OF MMD

Platé and co-workers [55] described intramolecular crosslinking for the case when not all units of a chain have reactive groups. The reaction was simulated in

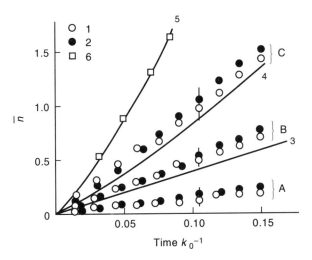

Fig. 5.9 Kinetics of crosslinking for regular (1) and Bernoullian (2) distributions of reactive groups for content of reactive groups $w = 0.3$ (A), 0.5 (B), 0.7 (C); solid curves correspond to $w = 1$ and $N = 30$ (3), 50 (4) and 100 (5) units; results for the polydisperse sample with Flory's MMD and $\overline{N} = 100$ (6). (Reproduced by permission of Nauka, Moscow, Russia, from *Vysokomol. Soed. A*, 1979, **21**, 1176)

model chains with the content of reactive groups (w) being equal to 0.3–0.7 and with even and random distributions of these groups.

In Fig. 5.9 the results of such a procedure are compared with results obtained for molecules having the same number of reactive groups, but situated in every unit ($w = 1$). It is obvious that the rate of the reaction decreases with a decrease in the number of reactive groups at a constant chain length. If the number of reactive groups is constant, the rate is higher for the case $w = 1$ ($N = 50$) than for $w < 1$ ($N = 100$, $w = 0.5$). However, the type of reactive group distribution affects very slightly the kinetics of crosslinking. In Fig. 5.9 the results obtained for a polymolecular sample with Flory's molecular mass distribution (MMD) ($N = 100$) are also shown. One can see that these results are rather close to those obtained in the simulation procedure when the chain contains 100 units.

The independence of the kinetics of intramolecular crosslinking on the type of reactive groups distribution and MMD [55] is of interest. It should be mentioned, however, that these results are obtained only for an initial part of a kinetic curve.

5.4 EFFECT OF COIL EXPANSION

In Chapter 2 we described some examples of the manifestation of conformational effect in polymer-analogous reactions related to the change in the degree of coiling of a macromolecule in the course of the reaction because of the change of the solvent quality (see references [117] to [122] in Chapter 2). It is rather difficult to describe quantitatively such an effect. The only attempt at such a quantitative description was made in references [56] to [58] on the basis of the mathematics developed for reactions of intramolecular crosslinking.

It was assumed that the reversible process of folding and unfolding macro-molecular coils related to a change in solvent quality can be simulated by another reversible process—reversible crosslinking—'uncrosslinking' accompanied by analogous conformational changes. Accepting the 'pseudocrosslinking' hypoth-esis the authors of references [56] to [58] do not assume the formation of any real physical bonds, but only the possibility of the application of mathematical apparatus developed for the description of conformational changes in the course of intramolecular crosslinking to analogous changes in polymer-analogous reactions.

As in any reversible process, the change in the chain conformation in the course of the polymer-analogous reaction can be described by the equilibrium constant $K = K_1/K_2$, where K_1 is some parameter characterizing the interactions resulting in the coil folding and K_2 characterizes the interactions resulting in the coil unfolding. Then the K value may be called the equilibrium constant of intra-molecular pseudocrosslinking.

The process including parallel concurrent reactions of functional groups with a low-molecular reagent (polymer-analogous reaction) and each with other (intramolecular crosslinking) is characterized by the value of the equilibrium constant K. The Monte Carlo simulation of such a process allows the equilibrium number of crosslinks (or 'pseudocrosslinks') to be determined, as well as the equilibrium dimensions of the coil containing this number of crosslinks and the kinetics of the polymer-analogous reactions as functions of K.

The validity of such an approach was checked in references [56 and 57] for hydrolysis of polyvinyl acetate in mixed solvent first studied by Turska and Jantas (reference [118] in Chapter 2). To simulate this process using the Monte Carlo procedure the initial K value should be defined. As any equilibrium constant K is related to the change of free energy of the system,

$$kT \ln K = \frac{-\mathrm{d}\Delta F_{cr}}{\mathrm{d}n_2 X}$$ (5.42)

where

$$\Delta F_{cr} = \Delta F^f_{mix} - \Delta F^0_{mix}$$
$$\Delta F_{max} = \Delta H_{mix} - T\Delta S_{max}$$

and X is the degree of polymerization.

For the lattice model,

$$\Delta F_{mix} = \Delta H_{mix} + kT(n_1 \ln v_1 + n_2 \ln v_2)$$ (5.43)
$$\Delta H_{mix} = \Delta H^f_{mix} - \Delta H^0_{mix}$$

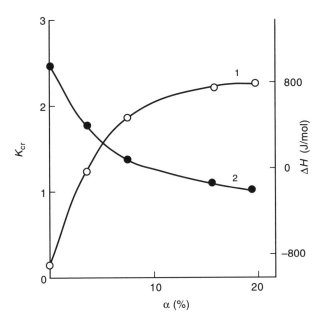

Fig. 5.10   Change of the equilibrium constant K as a function of the degree of hydrolysis of polyvinyl acetate α. (Reproduced by permission of Nauka, Moscow, Russia, from *Vysokomol. Soed. A*, 1986, **28**, 2251)

and n_1 and n_2 are amounts of the solvent and polymer molecules, v_1 and v_2 are their volume fractions, ΔH^0_{mix} is the heat of mixing initial polymer and ΔH^f_{mix} is the heat of mixing partially hydrolyzed polymer. Thus the change in the heat effect of dissolution of initial and intermediate products of the reaction permits the change in the K value to be determined.

The dependence of K on the degree of hydrolysis of partially hydrolyzed polyvinyl acetate α in mixed acetone–water solvent obtained using thermodynamic measurements [56, 57] is presented in Fig. 5.10. It is seen from Fig. 5.10 that the K value is decreased by 2.5 times in the 0–20% range of degrees of hydrolysis, pointing out the unfolding of the chain and enhancing the accessibility of functional groups for the low-molecular reagent. The results of the Monte Carlo simulation of the kinetics of hydrolysis with $K(\alpha)$ obtained as a parameter are presented in Fig. 5.11.

Figure 5.12 shows the change of relative dimensions of the macromolecular coil in the course of hydrolysis calculated in the same mathematical experiment and measured by the viscometry method.

The coincidence of calculated and experimental data both on kinetics and dimensions confirms the possibility of applying the pseudocrosslinking

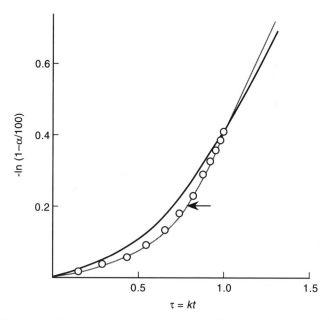

Fig. 5.11 Kinetics of hydrolysis of polyvinyl acetate. Points, experimental data; curve, calculation (an arrow shows the point $\alpha = 20\%$). (Reproduced by permission of Nauka, Moscow, Russia, from *Vysokomol. Soed. A*, 1986, **28**, 2251)

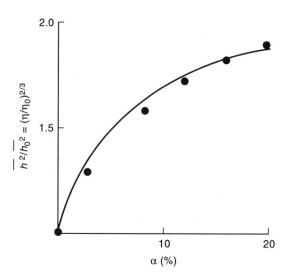

Fig. 5.12 Change of relative dimensions of polymer coils in the course of hydrolysis. Points, viscometry data; curve, calculation. (Reproduced by permission of Nauka, Moscow, Russia, from *Vysokomol. Soed. A*, 1986, **28**, 2251)

hypothesis to an evaluation of the conformational effect in polymer-analogous reactions.

5.5 INFLUENCE OF REMOTE UNITS

Let us describe one more possibility of the manifestation of the conformational effect. The flexibility of a macromolecule can result in the rapproachment of functional groups being chemically distant along the chain. If in some chemical reaction of functional groups the neighboring group effect is manifested, such approached reacted groups can also affect the reactivity of unreacted groups and as a result the structure of a product.

Such conformational effects can be significant if the reactivity is essentially affected by intermolecular interactions. In this case the effects of internal (nearest along the chain) and external (including the chemically distant units of the same chain) are distinguished [59]. For reactions proceeding in bulk, for example in polymer melt, the effect of distant groups can be neglected in the first approximation. This problem is discussed in detail in Chapter 6. However, for the case of quasi-isolated macromolecules evaluation of the possible contribution of such a conformational effect should be specially studied.

In reference [60] the simultaneous effect both of nearest neighbors and distant reacted groups on the structure of the chain of the macromolecular reaction product has been studied. The problem is solved using two approaches: the analytical solution with the aid of the body of mathematics analogous to that developed in references [61] to [63] and the Monte Carlo simulation.

The following model of the reaction was studied. An isolated macromolecule without volume interactions is subjected to the polymer-analogous reaction with the neighboring units effect: the A units transform irreversibly into B units with probabilities of the reaction per the time unity for central units of AAA, AAB (BAA), BAB triads being equal to α, β, γ respectively (the number of units in the chain $N \gg 1$; the end effects are neglected). The presence of the non-neighboring reacted B unit near the A unit changes the corresponding probabilities by the ε_α, ε_β, ε_γ values. If the rate of conformational rearrangements of the chain is much higher than the rate of the chemical reaction (as assumed further), the probability of the reaction of the ith unit being, for example, in the centre of the AAA triad in any time moment can be written as

$$\alpha_i = \alpha + \varepsilon_\alpha \sum_{j \neq i} g(|i - j|)(1 - n_j) \tag{5.44}$$

Here $g(|i - j|)$ is the equilibrium probability of the contact of the ith and jth units dependent on the physical model of a chain (freely jointed, persistent and so on), whereas n_j values denote the unit type: $n_j = 1$ for the A type and $n_j = 0$ for the B type. Expressions for β_j and γ_j probabilities are written in the same way as Eq. (5.44).

The complete probabilistic description of the structure of the reacting chain includes the set of probabilities of various sequences of A and B units, $P(n_1, n_2, \ldots, n_N; t) \equiv P(\{n_i\}; t)$ dependent on time with the obvious normalization condition

$$\sum_{\{n_i\}} P(\{n_i\}; t) = 1 \qquad (5.45)$$

where the summing is made for of all 2^N sets of $\{n_j\}$. Such standard characteristics of the chain structure as the probability of the sequence of m unreacted units (m-tuplet) $p(A_m)$ and the probability of finding unreacted units separated by m units (paired correlation function) $p(AX_mA)$ are expressed in terms of $P(\{n_i\}; t)$ as

$$p(A_m) = \sum_{\{n_k\}} n_i n_{i+1} \cdots n_{i+m-1} P(\{n_k\}; t) \qquad (5.46)$$

$$p(AX_mA) = \sum_{\{n_k\}} n_i n_{i+m+1} P(\{n_k\}; t) \qquad (5.47)$$

In the limit for the infinite long chain the right-hand parts of these expressions do not depend on i.

Evolution of the system is described by the master kinetic equation

$$\frac{\partial}{\partial t} = P(\{n_k\}; t) = \hat{K} P(\{n_k\}; t) \qquad (5.48)$$

Consider a simplified model for which

$$\alpha - \beta = \beta - \gamma \equiv u \qquad \varepsilon_\alpha = \varepsilon_\beta = \varepsilon_\gamma \equiv \varepsilon \qquad (5.49)$$

In this case the kinetic operator $\hat{K}$ can be written as

$$\hat{K} = -\gamma \sum_i (n_i - \hat{f}_i) - u \sum_i (n_i - \hat{f}_i)(n_{i-1} + n_{i+1}) - \varepsilon \sum_{i,j}{}' g(|i-j|)(n_i - \hat{f}_i)(1 - n_j)$$

$$(5.50)$$

where the operator $\hat{f}_i$ is defined as

$$\hat{f}_i P(\ldots n_i \ldots; t) = \delta_{n_i, 0} P(\ldots n_i \ldots; t)$$

The prime on the summation sign indicates the summing of the nonequal values of i and j indexes. The first two summands in the kinetic operator (5.50) describe the proper neighboring group effect, whereas the third one is related to the effect of contacts with remote reacted units.

Account of the effect of intrachain contacts essentially complicates the problem of the calculation of the characteristics of the chain structure. The accurate solution of this problem (unlike the neighboring group effect) is not available; therefore approximations are necessary. To derive various approximations the general method of references [61] to [63] can be used to apply the body of mathematics of the quantum theory, in particular the diagram technique of the parturbation theory, to study the classical systems.

The process of the irreversible polymer-analogous reaction $A \rightarrow B$ can be described as an irreversible turn over of Ising spins in the unidimensional chain. According to references [61] to [63] the set of probabilities $P(\{n_i\}; t)$ corresponds to the vector of states $|\Psi(t)\rangle$, being an analog of the wave function. The master kinetic equation (5.48) is written as the Schrödinger equation with the imaginary time

$$\frac{\partial}{\partial t} |\Psi(t)\rangle = \hat{H} |\Psi(t)\rangle$$

The $\hat{H}$ operator, being an analog of the hamiltonian, depends on the kinetic operator in the unique way of Eq. (5.50) and consists of the spin operators known from magnetism theory [60]. The averages, for example probabilities (5.46) to (5.47), are calculated following the rules analogous to quantum-mechanical ones, but not identical to them because of the nonquantum character of the problem.

The formal quantum entry develops the diagram technique of the perturbation theory in the standard way to take into account the effect of intrachain contacts on various characteristics of the structure of products of the polymer-analogous reaction. Application of the diagram technique can be used to develop various approximations and evaluate the limits of its validity.

In reference [60] with the aid of the diagram technique, the expressions for probabilities of m-tuplets (5.46) and the paired correlation function (5.47) are derived for the simplest approximation: intrachain contacts are taken into account in the first order of the perturbation theory towards the ε parameter. This limitation is acceptable only for small times and low flexibility of the chain, i.e. for

$$\varepsilon t \sum_{j=1}^{\infty} g(j) \ll 1 \qquad (5.51)$$

Besides the analytical solution in reference [60] the problem of the effect of remote units was solved using the Monte Carlo simulation of the reaction on the volume-centered lattice, with the lattice step being equal to $3^{1/2}$. Self-intersections were allowed; the reverse step was prohibited. The reactivity of the given unit was assumed to be affected by the distant unit, if these two units formed a self-intersection, i.e. were in the same lattice point.

The calculations were made for a wide range of constants ratios including the cases of retardation and slight and strong acceleration of the reaction by reacted neighbors. In particular, the case of $\alpha - \beta = \beta - \gamma$ and $\varepsilon_{\alpha} = \varepsilon_{\beta} = \varepsilon_{\gamma} = \varepsilon$ was studied to compare Monte Carlo results with those of analytical approximation.

Results of calculations of kinetic curves and dependences of some parameters of units sequence distribution on the degree of conversion $P(B)$ show that the strongest effect of remote units is observed for probabilities of alternating

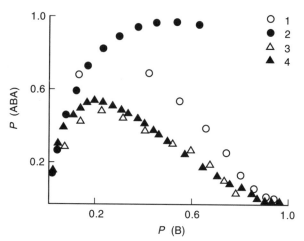

Fig. 5.13 Dependence of the triad probability $P(\text{ABA})$ on conversion $P(\text{B})$ with $(1, 2)$ and without $(3, 4)$ the conformational effect for $k_0/k_1/k_2/k_0^*/k_1^*/k_2^* = 1:5:9:5:9:13$ calculated following the Monte Carlo procedure $(1, 3)$ and analytical approach $(2, 4)$. (Reproduced by permission of Nauka, Moscow, Russia, from *Vysokomol. Soed. A*, 1993, **35**, 559)

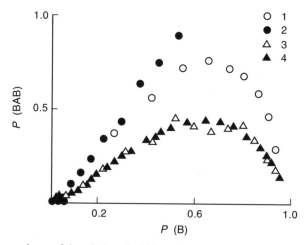

Fig. 5.14 Dependence of the triad probability $P(\text{BAB})$ on conversion $P(\text{B})$ with $(1, 2)$ and without $(3, 4)$ the conformational effect for $k_0/k_1/k_0^*/k_1^*/k_2^* = 1:5:9:5:9:13$ calculated following the Monte Carlo procedure $(1, 3)$ and analytical approach $(2, 4)$. (Reproduced by permission of Nauka, Moscow, Russia, from *Vysokomol. Soed. A*, 1993, **35**, 559)

sequences of A and B units in the case of the strong accelerating effect of distance reacted units, being of the same order as the effect of the nearest B neighbor.

In Figs. 5.13 and 5.14 the dependences of $P(\text{ABA})$ and $P(\text{BAB})$ on the degree of conversion $P(\text{B}) = 1 - P(\text{A})$ for the constants ratio $k_0/k_1/k_2/k_0^*/k_1^*/k_2^* = 1:5:9:$

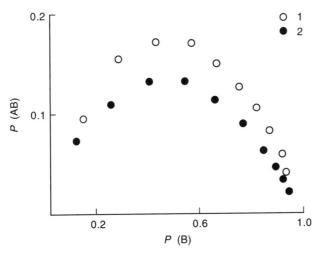

Fig. 5.15 Dependence of the run number $P(AB)$ on conversion $P(B)$ with (1) and without (2) the conformational effect for $k_0/k_1/k_2/k_0^*/k_1^*/k_2^* = 1{:}5{:}9{:}5{:}9{:}13$ calculated following the Monte Carlo procedure (Reproduced by permission of Nauka, Moscow, Russia, from *Vysokomol. Soed. A*, 1993, **35**, 559)

5:9:13 are presented. Here k_i are the rate constants of the reaction of a functional group having i nearest reacted neighbors, while k_i^* correspond to the case when there are i nearest reacted neighbors and one remote reacted unit in the same lattice point. It can be seen that under the effect of remote units the maximal values of these probabilities increase by 1.7–1.8-fold compared with the 'pure' nearest neighbor effect. This result is explained as follows. The accelerating nearest neighbor effect results in the block units distribution. Interaction with remote B units increases the total rate of transformation, and this effect is more pronounced in the long sequences of A units. As a result the average block length in the chain decreases, the content of alternating sequences increases and the probability of boundaries between blocks $P(AB)$ rises (Fig. 5.15).

In Figs. 5.16 and 5.17 the same dependences of $P(ABA)$ and $P(BAB)$ probabilities on the degree of conversion for the case of strong acceleration by two nearest reacted neighbors (1:5:100:5:100:150) are presented. In this case the conformational effect also enhances the probabilities of finding the alternating triads, but for the ABA triad the probability increases by 1.4–1.5-fold, whereas for the BAB triad a sixfold increase is observed.

This difference can be explained by comparing the rates of formation and consumption of triads. For the ABA triad assuming $k_0 = 1$ one can derive the

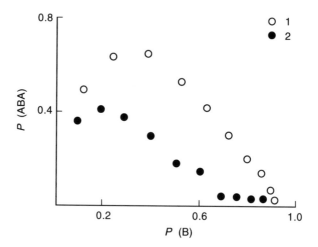

Fig. 5.16 Dependence of the triad probability $P(ABA)$ on conversion $P(B)$ with (1) and without (2) the conformational effect for $k_0/k_1/k_2/k_0^*/k_1^*/k_2^* = 1:5:100:5:100:150$ calculated following the Monte Carlo procedure. (Reproduced by permission of Nauka, Moscow, Russia, from *Vysokomol. Soed. A*, 1993, **35**, 559)

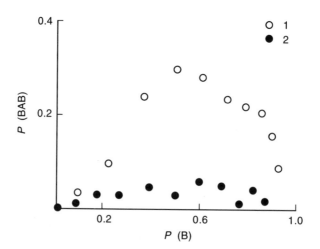

Fig. 5.17 Dependence of the triad probability $P(BAB)$ on conversion $P(B)$ with (1) and without (2) the conformational effect for $k_0/k_1/k_2/k_0^*/k_1^*/k_2^* = 1:5:100:5:100:150$ calculated following the Monte Carlo procedure. (Reproduced by permission of Nauka, Moscow, Russia, from *Vysokomol. Soed. A*, 1993, **35**, 559)

following equations:

$$\frac{dP(ABA)}{dt} = P(AAA) - 2\Phi[5P(BAA) + 100P(BAB)] \qquad (5.52)$$

$$\frac{dP(ABA)}{dt} = 5P(AAA) - 2\Phi[100P(BAA) + 150P(BAB)] \qquad (5.53)$$

for the absence and presence of the conformational effect respectively, where $\Phi = P(ABA)/[P(BAA) + P(BAB)]$.

For the BAB triad one can write, in the same manner,

$$\frac{dP(BAB)}{dt} = 2\Theta[P(AAA) + 5P(BAA)] - 100P(BAB)] \qquad (5.54)$$

$$\frac{dP(BAB)}{dt} = 2\Theta[5P(AAA) + 100P(BAA)] - 150P(BAB)] \qquad (5.55)$$

where $\Theta = P(BAA)/[P(AAA) + P(BAA)]$.

It can be seen from Eqs. (5.52) and (5.53) that the effect of distant B units results in a fivefold increase in the rate of formation of ABA triads with a simultaneous twentyfold increase in the rate of their consumption because of the fast transformation of BAA triads. As a result the maximal content of ABA triads only slightly exceeds the corresponding value obtained for the 'pure' neighboring groups effect. On the contrary, it follows from Eq. (5.54) and (5.55) that the effect of distant units results in a twentyfold increase in the rate of BAB triads formation, whereas the rate of their consumption increases only by 1.5-fold. Therefore the content of such triads in the chain can essentially increase as compared with the case of the 'pure' neighboring groups effect.

As can be seen from Figs. 5.13 and 5.14, the analytical solution (in the first-order approximation toward ε) gives results coinciding with those obtained by computer simulation up to $\sim 30\%$ conversion. The analytical description of the distant units effect in a wide range of conversions apparently requires a transition to an approximation of a higher order.

5.6 CONCLUSIONS

In concluding this chapter one can say that the quantitative approaches to configurational and conformational effects in macromolecular reactions have essentially been developed over the last 15 years. The known calculational methods permit to compare the calculated and experimental results and give grounds for further development of the theory of macromolecular reactions.

REFERENCES

1. Platé, N. A., Litmanovich, A. D. and Noah, O. V., *Makromolekulyarnye Reaktsii* (*Macromolecular Reactions*), Khimiya, Moscow, 1977.
2. Pis'men, L. M., *Vysokomol. Soed A*, 1972, **14**, 1861–8.
3. Morawetz, H. and Song, W. R., *J. Am. Chem. Soc.*, 1966, **88**, 5714–18.
4. Morawetz, H. and Goodman, N., *Macromolecules*, 1970, **3**, 699–700.
5. Goodman, N. and Morawetz, H., *J. Polymer Sci. Part C*, 1970, **31**, 177–92.
6. Goodman, N. and Morawetz, H., *J. Polymer Sci. PA-2*, 1971, **9**, 1657–68.
7. Morawetz, H., Cho, J.-R. and Gans, P. J., *Macromolecules*, 1973, **6**, 624–7.
8. Morawetz, H., *Pure Appl. Chem.*, 1974, **38**, 267–77.
9. Sisido, M., *Macromolecules*, 1971, **4**, 737–42.
10. Sisido, M., *Polymer J.*, 1972, **3**, 84–91; 1973, **4**, 534.
11. Sisido, M., *Seibutsu Butsuri*, 1974, **14**, 135–47.
12. Sisido, M., Mitamura, T., Imanishi, Y. and Higashimura, T., *Macromolecules*, 1976, **9**, 316–19, 320–4.
13. Sisido, M., Imanishi, Y. and Higashimura, T., *Macromolecules*, 1977, **10**, 125–30.
14. Tagaki, H., Sisido, M., Imanishi, Y. and Higashimura, T., *Bull. Chem. Soc. Japan*, 1977, **50**, 1807.
15. Sisido, M., *Kobunshi*, 1977, **26**, 260–3.
16. Sisido, M., Tamura, F., Imanishi, Y. and Higashimura, T., *Biopolymers*, 1977, **16**, 2723–38.
17. Sisido, M., Yoshikawa, E., Imanishi, Y. and Higashimura, T., *Bull. Chem. Soc. Japan*, 1978, **51**, 1464–8.
18. Sisido, M., Imanishi, Y. and Higashimura, T., *Bull. Chem. Soc. Japan*, 1978, **51**, 1469–72.
19. Sisido, M., Shimizu, T., Imanishi, Y. and Higashimura, T., *Biopolymers*, 1980, **19**, 701–11.
20. Sisido, M. and Imanishi, Y., *Biopolymers*, 1981, **20**, 653–64, 665–77.
21. Sisido, M. and Imanishi, Y., *Biopolymers*, 1982, **21**, 1613–21.
22. Horie, K., Schnabel, W., Mita, I. and Ushiki, H., *Macromolecules*, 1981, **14**, 1422–8.
23. Ushiki, H., Horie, K., Okamoto, A. and Mita, I., *Polymer Photochem.*, 1981, **1**, 303–14.
24. Wilemski, G. and Fixman, M., *J. Chem. Phys.*, 1973, **58**, 4009–19.
25. Wilemski, G. and Fixman, M., *J. Chem. Phys.*, 1974, **60**, 866–77, 878–90.
26. Doi, M., *Chem. Phys.*, 1975, **9**, 455–66.
27. Sunagawa, S. and Doi, M., *Polymer J.*, 1975, **7**, 604–12.
28. Kozlov, C. V., *Vysokomol. Soed B*, 1976, **18**, 642–6.
29. Perico, A. and Cuniberti, C., *J. Polymer Sci. Polymer Phys.*, 1977, **15**, 1435–50.
30. Cuniberti, C. and Perico, A., *Prog. Polymer Sci.*, 1984, **10**, 271–316.
31. Perico, A. and Beggiato, M., *Macromolecules*, 1990, **23**, 797–803.
32. Bernard, D. A. and Noolandi, J., *Phys. Rev. Lett.*, 1983, **50**, 253–4.
33. Noolandi, J., Hong, K. M. and Bernard, D. A., *Macromolecules*, 1984, **17**, 2895–901.
34. Kozlov, S. V., *Dokl. Akad. Nauk. SSSR*, 1978, **243**, 410–13.
35. Friedman, B. and O'Shaughnessy, B., *Phys. Rev. Lett.*, 1988, **60**, 64.
36. Friedman, B. and O'Shaughnessy, B., *Phys. Rev. A*, 1989, **40**, 5950.
37. Friedman, B. and O'Shaughnessy, B., *J. Phys. II (Paris)*, 1991, **1**, 471.
38. O'Shaughnessy, B., *Phys. Rev. Lett.*, 1987, **59**, 2903.
39. O'Shaughnessy, B., *J. Chem. Phys.*, 1991, **94**, 4042–7.
40. Friedman, B. and O'Shaughnessy, B., *Macromolecules*, 1993, **26**, 4888–98.
41. de Gennes, P.-G., *Scaling Concepts in Polymer Physics*, Cornell University Press, Ithaca, N.Y., 1985.

42. Zimm, B. H. and Stockmayer, W. H., *J. Chem. Phys.*, 1949, **17**, 1301–20.
43. Romantsova, I. I., PhD Thesis, Moscow, 1978.
44. Yerukhimovich, I.Ya, *Vysokomol. Soed A*, 1978, **20**, 114–18.
45. Edwards, S. F. and Freed, K. F., *J. Phys.*, 1970, **C3**, 739–49, 750–9, 760–70.
46. Allen, G., Burgess, J., Edwards, S. F. and Walsh, D. J., *Proc. Roy. Soc. Lond.*, 1973, **A334**, 453–91.
47. Gordon, M., Torkington, J. A. and Ross-Murphy, S. B., *Macromolecules*, 1977, **10**, 1090–100.
48. Ross-Murphy, S. B., *Polymer*, 1978, **19**, 497–502.
49. Yerukhimovich, I. Ya., *Vysokomol. Soed B*, 1978, **20**, 10–13.
50. Bonetskaya, N. K., Irzhak, V. I., Yel'yashevich, A. M. and Yenikolopyan, N. S., *Dokl. Akad. Nauk. SSSR*, 1975, **222**, 140–2.
51. Romantsova, I.I., Noah, O. V., Taran, Yu. A. and Platé, N. A., *Dokl. Akad. Nauk. SSSR*, 1977, **234**, 109–12.
52. Romantsova, I. I., Noah, O. V., Taran, Yu. A., Yel'yashevich, A. M., Gotlib, Yu. Ya. and Platé, N. A., *Vysokomol. Soed A*, 1977, **19**, 2800–7.
53. Khvan, A. M., Chupov, V. V., Noah, O. V. and Platé, N. A., *Vysokomol. Soed A*, 1985, **27**, 1243–8.
54. Yel'yashevich, A. M., *Vysokomol. Soed A*, 1978, **20**, 951–5.
55. Romantsova, I. I., Taran, Yu. A., Noah, O. V. and Platé, N. A., *Vysokomol. Soed A*, 1979, **21**, 1176–80.
56. Khvan, A. M., Noah, O. V., Zenkov, I. D., Shablygin, M. V. and Platé, N. A., *Vysokomol. Soed A*, 1986, **28**, 2251–3.
57. Khvan, A. M., PhD Thesis, Moscow, 1987.
58. Noah, O. V. and Platé, N. A., *Comput. Polymer Sci.*, 1992, **2**, 173–6.
59. Litmanovich, A. D., *Europ. Polymer J.*, 1980, **16**, 269–75.
60. Yashin, V. V., Strikitsa, M. N., Noah, O. V., Litmanovich, A. D. and Platé, N. A., *Vysokomol. Soed.*, 1993, **35**, 559–64.
61. Doi, M., *J. Phys. Ser. A*, 1976, **9**, 1465, 1479.
62. Zel'dovich, Ya. B. and Ovchinnikov, A. A., *Zh. ETF*, 1978, **74**, 1588.
63. Grassberger, P. and Scheunert, M., *Fortschritte der Phys.*, 1980, **28**, 547.

CHAPTER 6

Interchain Effects

6.1 THEORY

6.1.1 GENERAL REMARKS

In Chapters 3 to 5 the reactions of quasi-isolated macromolecules have been considered but interchain interactions have been neglected. Nevertheless, the interaction between macromolecules may affect the kinetics of the process and characteristics of the products obtained, even for reactions in diluted solutions. Shibaev *et al.* [1] observed that the chlorination of polyethylene (0.1% solution in chlorobenzene) with chlorine below 130 °C led to extremely heterogeneous in composition products, whereas above 130 °C the chlorinated polymer was relatively homogeneous. To interpret these results, the authors of reference [1] supposed that associates of macromolecules exist in the solution below 130 °C, the time of their life and rearrangements being commensurate with the duration of the reaction. As a result, highly chlorinated macromolecules are formed on the surfaces of the associates, whereas inside only a low degree of chlorination may be reached. In this case interaction between macromolecules affects in essence macrokinetics of the process (chlorine diffusion into associates and shielding of chains settled in the inner regions of associates). Much more interesting is the direct influence of interchain interactions on the reactivity of polymer functional groups.

The influence of the medium on reactivity is one of the most important features of chemical reactions in liquids [2, 3]. For macromolecular reactions in dilute solutions, it is more accurate to speak about the dependence of reactivity of the polymer functional group upon its microenvironment, i.e. the molecules or their fragments in direct contact with this group. Indeed, due to a high local concentration of polymer units and selective solvation and sorption effects, the average composition in the immediate proximity of a macromolecular coil in a dilute solution differs from the composition of the solution as a whole.

The local concentrations of unreacted and reacted units and also the affinity between chains and low-molecular components of the solution (solvent, reactants) change during a macromolecular reaction. Therefore, the microenvironment of polymer functional groups changes with conversion. This conclusion may also be applied to reactions in concentrated solutions and in polymer melts where the medium consists largely of macromolecules. The transformation of macromolecules leads to a substantial change in the medium and therefore in the microenvironment of functional groups.

Taking into account the influence of the microenvironment on reactivity, it may be concluded that the reactivity of polymer functional groups changes during a macromolecular reaction. This principle has been taken by Litmanovich [4,5] as the basis of the quantitative description of the kinetics of macromolecular reactions.

6.1.2 REACTION MODEL

Consider the irreversible transformation of A units into B units during a macromolecular reaction in polymer solution or melt. Let z be the coordination number of the quasi-lattice of solution or melt, each cell containing either one unit (A or B) or one molecule of a low-molecular component. Then, in the coordination sphere of each A group (infinite chains are considered, the end groups being neglected) two cells are occupied by the nearest units of the same chain (inner neighbors) and $(z-2)$ cells by units belonging to other chains or by low-molecular components (external neighbors). Those neighbors occupying the coordination sphere of an A group are supposed to influence its reactivity. If any pronounced influence of remote groups (such as effects in polyconjugated systems) is excluded, this assumption appears to be valid for a wide range of macromolecular reactions. Note that the $(z-2)$ external neighbors may include remote units of the same chain, but their influence on the reactivity of a given A group does not differ from that of other chain units (for the reaction model under consideration).

A groups with 0, 1 and 2 inner neighbors of B type have different reactivities because of the different nature of their inner neighbors and also because they might have different external microenvironments. However, it may be supposed that at any particular time the composition of the coordination sphere of a given A group is determined by the nature of the inner neighbors and does not depend upon the position of respective triads (AAA, AAB$^+$ and BAB) in a polymer chain. In other words, it is assumed that the external microenvironment of an A group with inner neighbors fixed and at a fixed time does not depend upon the distribution of units along the chain. For this case, it is sufficient to distinguish three types of A groups mentioned (according to the number of B-type inner neighbors), their instantaneous reactivities being characterized by rate coefficients k_0, k_1 and k_2 respectively. These coefficients are time dependent because the nature of the external neighbors for A groups changes with conversion.

6.1.3 STATISTICAL RELATIONSHIPS

To describe such a process, it is necessary to consider statistical properties of any macromolecule chosen at random which transforms in accordance with a reaction model taken above. First note that all relations previously derived [6,7] for a statistically stationary chain are valid in our case. Moreover, the reaction considered should belong to Markovian processes with locally interacting components [8]. Regularities of such a process for the special case of a macromolecular reaction with time-independent rate coefficients were investigated by Mityushin [9]. We present below the principal Mityusin results in a form convenient for the problem under consideration and use them to derive some important equations characterizing the statistical properties of a reacting chain.

Let Z, Y, U and V be sequences of A and B units, $|Z|$ and Z^i the length and ith unit of Z. We select such a small time interval Δt that A $\rightarrow$ B transfers at any place in the chain during this interval may be considered as mutually independent. Hence, the probability of a Z-sequence at a moment $(t + \Delta t)$ is determined as follows:

$$P_{t+\Delta t}(Z) = \sum_{|Y| = |Z| + 2} P_t(Y)P(Z/Y) \tag{6.1}$$

where

$$P(Z/Y) = \prod_{i=1}^{|Z|} P(Z^i/Y^i Y^{i+1} Y^{i+2}) \tag{6.2}$$

In Eqs. (6.1) and (6.2), $P_t(Y)$ is the probability of a Y-sequence at time t, $P(Z^i/Y^i Y^{i+1} Y^{i+2})$ the probability of a transfer $Y^{i+1} \rightarrow Z^i$ during an interval from t to $t + \Delta t$, $P(Z/Y)$ the probability of a Z-sequence formation as a result of a reaction of respective Y-sequence units during the same interval.

In accordance with the reaction model taken,

$$P(B^i/A^i A^{i+1} A^{i+2}) = k_0(t)\Delta t$$
$$P(B^i/A^i A^{i+1} B^{i+2}) = P(B^i/B^i A^{i+1} A^{i+2}) = k_1(t)\Delta t \tag{6.3}$$
$$P(B^i/A^i A^{i+1} B^{i+2}) = k_2(t)\Delta t$$

Here $k_0(t)$, $k_1(t)$ and $k_2(t)$ are mean time-dependent transient probabilities related to the unit of time. For convenience the same symbols are kept for transient probabilities and rate coefficients because both a probability of any sequence and its concentration are calculated from identical equations.

According to Mityushin's theorem [9] (see its description in reference [5]),

$$P_t(ZAAY) = \phi_t(Z)\psi_t(Y) \tag{6.4}$$

where $\phi_t(Z)$ and $\psi_t(Y)$ are functions of Z and Y respectively. Using (6.4) we can write

$$P_t(ZAA) = P_t(ZAAA) + P_t(ZAAB) = \phi_t(Z)[\psi_t(A) + \psi_t(B)] \tag{6.5}$$

Similarly,

$$P_t(AAY) = \psi_t(Y)[\phi_t(A) + \phi_t(B)] \tag{6.6}$$

$$P_t(AA) = [\phi_t(A) + \phi_t(B)][\psi_t(A) + \psi_t(B)] \tag{6.7}$$

It follows from Eqs. (6.4) to (6.7) that

$$P_t(ZAAY) = \frac{P_t(ZAA)P_t(AAY)}{P_t(AA)} \tag{6.8}$$

For a special case

$$2k_1(t) = k_0(t) + k_2(t) \tag{6.9}$$

i.e. the transient probabilities form an arithmetical progression, the following relation is valid [9]:

$$P_t(ZAY) = \phi_t(Z)\psi_t(Y) \tag{6.10}$$

and therefore

$$P_t(ZAY) = \frac{P_t(ZA)P_t(AY)}{P_t(A)} \tag{6.11}$$

Equations (6.8) and (6.11) are correct both for time-independent (cf. Chapter 3) and for time-dependent transient probabilities.

Now we shall derive the equations connecting probabilities of the sequences separated by sequences of A units of various length. We need probabilities of keeping an A unit unchanged during the interval from t to $t + \Delta t$. It follows from Eq. (6.3) that

$$P(A^i/A^iA^{i+1}A^{i+2}) = 1 - k_0(t)\Delta t = \exp[-k_0(t)\Delta t]$$
$$P(A^i/A^iA^{i+1}B^{i+2}) = P(A^i/B^iA^{i+1}A^{i+2}) = \exp[-k_1(t)\Delta t] \tag{6.12}$$
$$P(A^i/B^iA^{i+1}B^{i+2}) = \exp[-k_2(t)\Delta t]$$

We find a relation between the probabilities of sequences ZAAAY and ZAAY at a time t. Divide interval $(0, t)$ into small and equal intervals Δt, so that $t/\Delta t = m$ (m is an integer). Note that at $t = 0$ when a chain consists of A units only, the probability of sequence UA_nV is equivalent to that of a sequence of A units having length $|U| + n + |V|$ and is equal to unity. Taking this point into account, noting Eqs. (6.12) and also the independence of $A \rightarrow B$ transfers during Δt and changing for convenience $\displaystyle\sum_{\substack{|U|=|Z|+1 \\ |V|=|Y|+1}}$ through $\displaystyle\sum_{u,v}$, we may express the probabilities of sequences ZAAAY and ZAAY at a moment $0 + \Delta t$ as

$$P_{0+\Delta t}(\text{ZAAAY}) = \sum_{u,v} P_0(\text{UAAAV})P(\text{ZAAAY/UAAAV})$$

$$= \sum_{u,v} P(\text{ZA/UAA})P(\text{A/AAA})/P(\text{AY/AAV})$$

$$= \exp[-k_0(0)\Delta t] \sum_{u,v} P(\text{ZA/UAA})P(\text{AY/AAV}) \qquad (6.13)$$

$$P_{0+\Delta t}(\text{ZAAY}) = \sum_{u,v} P_0(\text{UAAV})P(\text{ZAAY/UAAV})$$

$$= \sum_{u,v} P(\text{ZA/UAA})P(\text{AY/AAV}) \qquad (6.14)$$

Comparing Eqs. (6.13) and (6.14), we get

$$P_{0+\Delta t}(\text{ZAAAY}) = P_{0+\Delta t}(\text{ZAAY})\exp[-k_0(0)\Delta t] \qquad (6.15)$$

Using Eq. (6.15) we find, for time $0 + 2\Delta t$,

$$P_{0+2\Delta t}(\text{ZAAAY}) = P_{0+2\Delta t}(\text{ZAAY})\exp\{-[k_0(0) + k_0(0+\Delta t)]\Delta t\} \qquad (6.16)$$

Continuation of the procedure gives

$$P_t(\text{ZAAAY}) = P_t(\text{ZAAY})\exp\left[-\sum_{i=0}^{m-1} k_0(0+i\Delta t)\Delta t\right]$$

and finally for $\Delta t \to 0$,

$$P_t(\text{ZAAAY}) = P_t(\text{ZAAY})\exp\left[\int_0^t -k_0(t)\,dt\right] \qquad (6.17)$$

Equation (6.17) characterizes the process with time-dependent transient probabilities. If k_0 is constant, Eq. (6.17) transforms into the equation

$$P_t(\text{ZAAAY}) = P_t(\text{ZAAY})\exp(-k_0 t)$$

derived in reference [9]. If $2k_1(t) = k_0(t) + k_2(t)$, the following relation is true [5]:

$$P_t(\text{ZAAY}) = P_t(\text{ZAY})\exp\left[\int_0^t -k_0(t)\,dt\right] \qquad (6.18)$$

It follows from Eq. (6.17) that

$$P_t(\text{ZA}_{n+1}\text{Y}) = P_t(\text{ZA}_n\text{Y})\exp\left[\int_0^t -k_0(t)\,dt\right] \qquad (n \geq 2) \qquad (6.19)$$

$$P_t(\text{ZA}_n\text{Y}) = P_t(\text{ZAAY})\exp\left[-(n-2)\int_0^t k_0(t)\,dt\right] \qquad (n \geq 2) \qquad (6.20)$$

When Eq. (6.9) is valid, Eq. (6.19) is correct for $n \geq 1$ and Eq. (6.20) transfers into

$$P_t(\mathrm{ZA}_n\mathrm{Y}) = P_t(\mathrm{ZAY}) \exp\left[-(n-1) \int_0^t k_0(t)\,\mathrm{d}t \right] \qquad (n \geq 1) \qquad (6.21)$$

Eq. (6.19) leads to the equations

$$P_t(\mathrm{ZA}_{n+1}) = P_t(\mathrm{ZA}_n) \exp\left[-\int_0^t k_0(t)\,\mathrm{d}t \right] \qquad (6.22)$$

and

$$P_t(\mathrm{A}_{n+1}) = P_t(\mathrm{A}_n) \exp\left[-\int_0^t k_0(t)\,\mathrm{d}t \right] \qquad (6.23)$$

which are correct for $(n \geq 2)$ in general and for $(n \geq 1)$ providing Eq. (6.9) is valid. Relations similar to Eqs. (6.22) and (6.23) also exist for the sequences ZA_n and A_n.

6.1.4 KINETIC DESCRIPTION

Using relations derived in the previous section, it is possible to describe a kinetic curve and also the units distribution and composition heterogeneity of the reaction products. The methods described in Chapter 3 are suitable for this purpose, the equation containing time-dependent transient probabilities (or rate coefficients).

It is easy to transform the McQuarrie equations [10]:

$$\frac{\mathrm{d}P(\mathrm{A})}{\mathrm{d}t} = -k_0 P(\mathrm{A}_3) - 2k_1[P(\mathrm{A}_2) - P(\mathrm{A}_3)] - k_2[P(\mathrm{A}) - 2P(\mathrm{A}_2) + P(\mathrm{A}_3)]$$

and

$$\frac{\mathrm{d}P(\mathrm{A}_n)}{\mathrm{d}t} = -2k_1[P(\mathrm{A}_n) - P(\mathrm{A}_{n+1})] - k_0[(n-2)P(\mathrm{A}_n) - 2P(\mathrm{A}_{n+1})] \qquad (n \geq 2)$$

into Eqs. (6.24) and (6.25), taking into account Eq. (6.23) and a dependence of k_0, k_1 and k_2 upon time:

$$\frac{\mathrm{d}P(\mathrm{A})}{\mathrm{d}t} = -k_2(t)P(\mathrm{A}) + \Big([k_2(t) - k_0(t)] + [k_0(t) - 2k_1(t) + k_2(t)]$$

$$\times \left\{ 1 - \exp\left[-\int_0^t k(t)\,\mathrm{d}t \right] \right\} \Big) P(\mathrm{A}_2) \qquad (6.24)$$

$$\frac{\mathrm{d}P(\mathrm{A}_n)}{\mathrm{d}t} = -\left(nk_0(t) - 2[k_1(t) - k_0(t)] \left\{ 1 - \exp\left[-\int_0^t k(t)\,\mathrm{d}t \right] \right\} \right) P(\mathrm{A}_n) \qquad (n \geq 2)$$

$$(6.25)$$

Equation (6.24) which has to be solved together with Eq. (6.25) at $n = 2$ describes the time dependence of the fraction of A units, i.e. the kinetic curve. So far as [7]

$$P(BA_nB) = P(A_n) - 2P(A_{n+1}) + P(A_{n+2}),$$

it is possible to calculate the probability of any sequence of A units, i.e. to describe the sequence distribution of unreacted units using Eqs. (6.24) and (6.25).

Sequence distribution of reacted units may be calculated, for example, by means of the so-called B-approximation (see Chapter 3):

$$\frac{dP(AB_jA)}{dt} = \delta_{j,1} k_0(t) P(A_3) + \frac{2k_1(t)(1 - \delta_{j,1}) P(AB_{j-1}A)}{P(BA)} - 2P(AB_jA)$$

$$\times \frac{k_1(t)P(BA_2) + k_2(t)P(BAB)}{P(BA)} + \frac{k_2(t)P(BAB)}{[P(BA)]^2}$$

$$\times \sum_{m=1}^{j-2} P(AB_mA)P(AB_{j-m-1}A) \tag{6.26}$$

where

$$\delta_{j,1} = \begin{pmatrix} 1 & \text{at } j = 1 \\ 0 & \text{at } j \neq 1 \end{pmatrix}$$

It is possible to solve Eq. (6.26) together with Eqs. (6.24) and (6.25) because $P(BA) = P(A) - P(A_2)$ and $P(BA_2) = P(A_2) - P(A_3)$. Using Eqs. (6.24) to (6.26), it is possible to calculate the probability of any sequence of A and B units [11] and, therefore, to describe completely the distribution of units.

Composition heterogeneity of the reaction products may be calculated, for instance, with the help of the modified first-order Markovian approximation (see Chapter 3). According to this method, the composition heterogeneity is expressed by means of the Gaussian distribution

$$w(y) = (2\pi)^{-1/2} \sigma^{-1} \exp\left[\frac{-(y - \bar{y})^2}{2\sigma}\right] \tag{6.27}$$

where $y = 1 - P(A)$ is the degree of conversion of a macromolecule, $\bar{y}$ is a mean value of y for the whole sample and σ is a standard deviation. Dispersion of the distribution is determined as follows:

$$\lim_{n \to \infty} \left(\frac{\sigma^2}{n}\right) = \frac{(1 - P_{aa})P_{ba}(1 - P_{ba} + P_{aa})}{(1 - P_{aa} + P_{ba})^3} \tag{6.28}$$

where n is the length (the number of units) of the chain fragment [12], P_{aa} and P_{ba} are the conditional probabilities of finding A to the right of A and to the right of B respectively. So far as $P_{aa} = P(A_2)/P(A)$ and $P_{ba} = P(BA)/P(B) = [P(A) - P(A_2)]/[1 - P(A)]$, the solution of Eqs. (6.24) and (6.25) makes it possible to calculate composition heterogeneity according to Eqs. (6.27) and (6.28).

To solve Eqs. (6.24) to (6.28), it is necessary to know the dependence of the rate coefficients upon the time (or on the conversion). Such a dependence is determined by the reaction mechanism and conditions and for particular reaction systems may be found experimentally or may be deduced from the features of the reaction or from general physicochemical considerations. In the last case the adequacy of the assumed dependence remains uncertain until there is experimental proof. The examples of application of the theory will be described in the next sections.

Thus, the approach suggested (to follow the fate of a single macromolecule accounting for a change in the functional group microenvironment during the reaction) employs the mathematics of the intrachain neighbor effect to describe a contribution of interchain interactions into the process kinetics and the composition and structure of the reaction products.

6.2 REACTIONS IN MELT

6.2.1 GENERAL REMARKS

Macromolecular reactions in melts need to attract the attention of theoreticians, at least in view of the applied significance of polymer modification during processing [13, 14]. Naturally, it should begin with studying relatively simple reaction models including interchain effects.

It has been pointed out in references [4] and [5] that the reactions proceed in such a way that a fragment of the A group splits off and leaves the melt are of special interest for studying the interchain effects in a polymer melt. In that case the reaction medium consists of macromolecules exclusively containing unreacted and reacted units, therefore making the mathematical description of the process a little easier. Besides, due to high concentration of functional groups of the polymer in a bulk state, a contribution of interchain interactions into the reaction kinetics might be especially significant (certainly if the interaction between unreacted and reacted groups affects the polymer reactivity). Thermal decomposition of *tert*-butyl esters of polymeric acids may be related to such reactions.

tert-Butyl esters of low-molecular carboxylic acids decompose by heating to respective acid and isobutylene. The decomposition of *tert*-butyl acetate [15] and *tert*-butyl propionate [16] in a gas phase (at 243–303 °C and 5–275 torr) proceeds up to 90% conversion as a first-order monomolecular reaction, activation energy being 40.5 and 39.1 kcal/mol respectively.

However, for thermal decomposition of poly(*tert*-butyl acrylate) (PTBA) in a melt at 161–202 °C, Schaefgen and Sarasohn [17] observed a sharp acceleration beginning from conversion at about 10%. Having considered various possible mechanisms of the acceleration, the authors concluded that the most probable is an intrachain acceleration by a relatively long sequence of carboxylic groups formed in the direct neighborhood of the reacting ester group. No proper equation for describing the reaction kinetics has been derived in reference [17].

Moreover, the hypothesis itself seems incorrect: a formation of long sequences of carboxylic groups at low conversion (less than 10%) is hardly possible provided an isolated carboxylic groups does not accelerate the decomposition of a neighboring ester group.

Two years previously, Grant and Grassie [18] studied the influence of low-molecular carboxylic acids on the thermal decomposition of poly(*tert*-butyl methacrylate). Benzoic and naphtoic acids did not produce any significant effect but 3,5-dinitrobenzoic acid substantially enhanced the rate of isobutylene elimination. Surprisingly, Grant and Grassie perceived in this result neither more nor less than the ability of COOH groups to accelerate the decomposition of the inner neighbor and concluded that the reaction propagates autocatalytically along the macromolecular chain.

Thus, the authors of references [17] and [18] interpreted the acceleration of the reaction in terms of intramolecular interactions exclusively. It is worth noting that neither Grant and Grassie nor Schaefgen and Sarasohn succeeded in describing the reaction kinetics quantitatively.

Litmanovich and Cherkezyan [19, 20] supposed interchain interaction to contribute substantially to isobutylene elimination. PTBA has been chosen for an experimental test because in the case of poly(*tert*-butyl methacrylate), parallel with the isobutylene elimination, depolymerization also occurs [18].

6.2.2   DECOMPOSITION OF POLY(*tert*-BUTYL ACRYLATE)

Evidence of the Interchain Effects

During PTBA decomposition, carboxylic groups formed as a result of isobutylene elimination undergo subsequent dehydration and to a less extent decarboxylation. However, in up to 80% conversion of ester groups, no more than 6–8% of water is evolved, when referred to the whole quantity of the water evolved at a given temperature (see Fig. 6.1). Therefore, in up to 75% conversion of ester groups, Litmanovich and Cherkezyan [19, 20] considered the PTBA decomposition as a reaction of isobutylene elimination only, neglecting the subsequent transformation of carboxylic groups. Within this conversion range, isobutylene elimination is subjected to a quantitative kinetic description.

To test whether an acceleration of the reaction is caused by interchain interaction, the PTBA decomposition was carried out in a blend with poly(acrylic acid) (PAA) prepared by means of freeze-drying of 2% solution of these polymers in a *tert*-butyl alcohol–water mixture (100:14 by volume). Such a method apparently leads to blends with a well-developed interface [21–23], so that conditions favorable for observing the possible interchain interactions are secured. Kinetics of PTBA decomposition at 180 C both in pure state and in a mixture with PAA are shown in Fig. 6.2. It can be noted that the initial rate of PTBA decomposition increases substantially in the PAA presence. Since at the

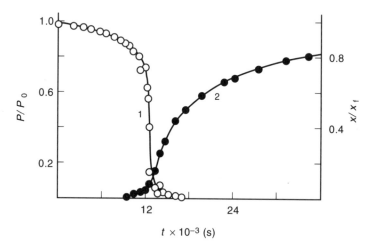

Fig. 6.1 Decomposition of PTBA at 180 °C: 1, ester groups diminution; 2, water evolution, P and P_0 are current and initial ester groups concentrations; x and x_f are current and final quantities of the water evolved. (Reproduced by permission of Nauka, Moscow, Russia, from *Vysokomol. Soed. A*, 1985, **27**, 1865)

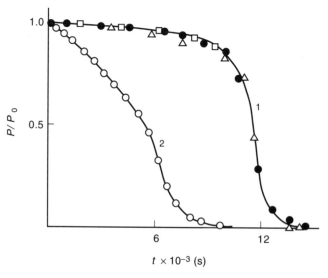

Fig. 6.2 Decomposition of PTBA at 180 °C: 1, in pure state; 2, in a PTBA–PAA mixture (2:1, unit/unit). (Reproduced by permission of Nauka, Moscow, Russia, from *Vysokomol. Soed. A*, 1985, **27**, 1865)

beginning of the reaction there are no carboxylic groups besides those belonging to PAA, the increase in the initial rate undoubtedly proves a significant contribution of interchain interaction of acrylic acid units and TBA ester groups into the acceleration of PTBA decomposition.

Supposed Mechanism

Based on the shape of kinetic curves and preliminary calculations, the acceleration has been supposed [19, 20] to be caused by interaction of one ester and two carboxylic groups. Using spherical atomic models, two structures have been revealed to be favorable for acceleration: one of the two carboxylic groups is an inner neighbor, the other being external, or both of them are external neighbors. Therefore interchain interaction is supposed to be an inalienable component of the acceleration. Thus PTBA decomposition is supposed to include thermal and autocatalytic stages (see Scheme 6.1, where α, β and γ are true rate constants of the respective stages):

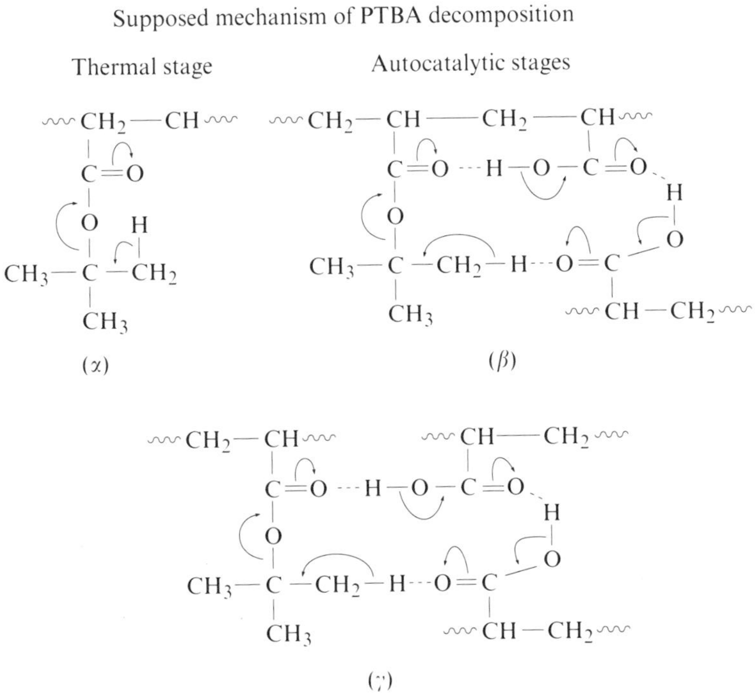

Supposed mechanism of PTBA decomposition

Thermal stage   Autocatalytic stages

Scheme 6.1

In the thermal stage, isobutylene elimination is accompanied by transfer of the methyl proton to the carbonyl oxygen. In the autocatalytic stages, mobile carboxyl proton transfers to the TBA carbonyl oxygen. Mobility of this proton apparently increases due to the association of two carboxylic groups. Electron density redistribution indicated in the scheme facilitates the methyl proton transfer to the other carboxylic group (more precisely to the carboxylate anion formed). Using the IR technique, the existence of H bonds has been observed in references [19] and [20] under experimental conditions, which supports the supposed mechanism of acceleration because such H bonds are necessary to stabilize an intermediate reaction complex.

These ideas concerning the mechanism of both isobutylene elimination and catalytic action of carboxylic acids were also confirmed by means of quantum chemical calculations [24].

Reaction Kinetics

It is possible [4, 5] (see Section 6.1) to use any equation derived for an ordinary (i.e. intrachain) neighbor effect to describe the reaction with the interchain effect provided the rate constants are changed for time-dependent rate coefficients. The simple and convenient Keller [25] equations were used to describe the PTBA decomposition kinetics:

$$\frac{dP(AAA)}{dt} = -(k_{aa} + 2\phi)P(AAA)$$

$$\frac{dP(AAB^+)}{dt} = 2\phi P(AAA) - (k_{ab} + \phi)P(AAB^+) \qquad (6.29)$$

$$\frac{dP(BAB)}{dt} = \phi P(AAB^+) - k_{bb}P(BAB)$$

where

$$\phi = \frac{k_{aa}P(AAA) + k_{ab}P(AAB^+)/2}{P(AAA) + P(AAB^+)/2} \qquad (6.30)$$

Here and further, for the sake of convenience and brevity, the symbols k are used instead of $k(t)$, while subindexes show the inner neighbors.

To solve Eqs. (6.29) the dependence of rate coefficients upon conversion should be known. The dependence is determined by the reaction mechanism and also by the distribution of reacted and unreacted units through the volume. Let us imagine a polymer melt as a totality of small elementary reaction volumes, i.e. microreactors. Taking a reptation mechanism for translational movements of macromolecules in a melt [26], the magnitude of the translational displacement of a macromolecule during a kinetic experiment ($\sim 10^4$ s) has been estimated as being equal to $\sim 10^{-4}$ cm [19]. Thus diffusional transfer of macromolecules throughout the reaction system is negligible. However, within any microreactor the mixing of units due to reptation is supposed to be perfect, so that for a given A unit the probability of encountering an external neighbor of the type of A or B is proportional to the fraction of the latter in the microreactor. Supposing, as a first approximation, that microreactors do not differ one from another in an A/B ratio and taking into account Scheme 6.1, the dependence of rate coefficients upon conversion may be expressed as follows:

$$k_{aa} = \alpha + \gamma[1 - P(A)]^2$$
$$k_{ab} = \alpha + \beta[1 - P(A)] + \gamma[1 - P(A)]^2 \qquad (6.31)$$
$$k_{bb} = \alpha + 2\beta[1 - P(A)] + \gamma[1 - P(A)]^2$$

Using Eqs. (6.29) to (6.31), the α, β and γ values have been calculated for PTBA decomposition at 165–200 °C (see Table 6.1). To test the validity of these values, for a given set of experimental data substantially different sets of initial values of the rate constants have been tested in the process of minimizing discrepancies between the calculated kinetic curve and experimental data. In all cases the optimization process led to a unique set of α, β and γ values. This indicates the absence of the local minima on the used region of the search surface [27]. Approximate estimations of the influence of experimental errors on the accuracy of the rate constants calculations have been made [91] in the following way. Taking 5% normally distributed relative errors of the kinetic experiment the simulated sets of the 'experimental points' were varied. For each of the sets the rate constants have been calculated. As a result the values of α appeared to vary within the limits of $\sim 15\%$ deviations and those of β and γ within $\sim 50\%$ deviations. Experimental data on PTBA decomposition at various temperatures are described quantitatively using the chosen reaction model and optimizing rate constants collected in Table 6.1 (see Fig. 6.3).

All three rate constants obey the Arrhenius law (see Fig. 6.4). The activation energy of the thermal stage (40 kcal/mol) is close to that for low-molecular

Table 6.1. Rate constants[a] of PTBA decomposition at various temperatures

$T(°C)$	$\alpha \times 10^6$	$\beta \times 10^4$	$\gamma \times 10^4$
165	1.0	7.0	3.0
170	1.7	10.8	4.2
175	2.5	15.8	6.7
180	5.0	22.5	9.7
185	8.3	31.7	13.3
190	15.0	43.3	18.3
195	20.0	51.7	25.0
200	36.7	73.3	33.3

[a]All rate constants have a dimension s^{-1}, in accordance with the description of the reaction kinetics in terms of probabilities. (Reproduced by permission of Nauka, Moscow, Russia, from *Vysokomol Soed. A*, 1985, **27**, 1865).

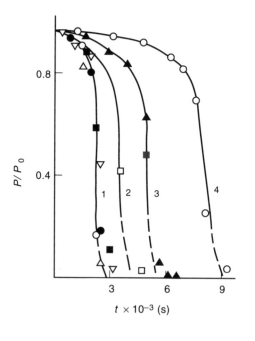

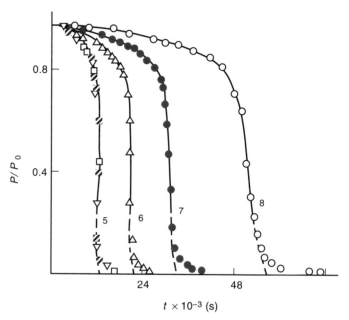

Fig. 6.3 Kinetics of PTBA decomposition at various temperatures (°C): 200(1), 195(2), 190(3), 185(4), 180(5), 175(6), 170(7) and 165(8). Points, experimental data; curves, calculations (for rate constants values given in Table 6.1). (Reproduced by permission of Nauka, Moscow, Russia, from *Vysokomol. Soed. A*, 1985, **27**, 1865)

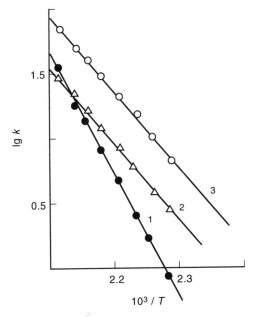

Fig. 6.4 Arrhenius plot for rate constants of PTBA decomposition: $\alpha \times 10^6(1)$, $\beta \times 10^4(3)$, $\gamma \times 10^4\,\mathrm{s}^{-1}(2)$. (Reproduced by permission of Nauka, Moscow, Russia, from *Vysokomol. Soed. A*, 1985, **27**, 1865)

analogs, viz. *tert*-butyl acetate [15] and *tert*-butyl propionate [16]. Activation energies of autocatalytic stages (26 kcal/mol) are substantially less and are equal to each other. Meanwhile the pre-exponent of β is greater than that of γ, i.e. acceleration by one inner and one external carboxylic group is more favorable (due to entropy and steric factors) than that produced by two external COOH groups. These results are in accordance with the reaction mechanism postulated.

Up to this point all microreactors have been considered to be identical in composition. However, remember that PTBA decomposition begins from the slow thermal stage; then, at about several per cent conversion, a sharp acceleration occurs. Therefore, as the 'launching' of the fast stage is not realized at the same time in various microreactors, these differ in composition one from another. Thus the experimental kinetic curve is actually a result of superposition of the processes proceeding in various microreactors differing in the degree of conversion. The question arises as to whether the mathematical model given above (neglecting nonsimultaneity of the reaction launching) permits true values of the rate constants to be estimated.

The problem has been considered by Cherkezyan and Litmanovich [28]. The equation was derived to connect the kinetic curve for a sole microreactor $p(t)$ and for a sample as a whole $P(t)$ (just the latter is measured in experiment):

$$p(t) = P(t) + \frac{\mu^{-1} dP(t)}{dt} \tag{6.32}$$

where μ^{-1} is a mean delay time of the reaction beginning in a microreactor. By determining $P(t)$ experimentally it is possible to use the function $p(t)$ calculated from Eq. (6.32) to find the true values of the rate constants α, β and γ. It was shown in reference [28] that even for very small microreactors containing 10^3 units only (when $\mu^{-1} \sim 100\,\text{s}$) the values of the rate constants calculated from the experimental curve $P(t)$ within the limits of errors coincide with ones calculated from $p(t)$. Thus a treatment of PTBA decomposition kinetics performed in references [19] and [20] appears to be correct.

The influence of nonsimultaneity of the reaction launching on the calculations of a composition distribution of partially decomposed PTBA has also been considered in reference [28]. Equations (6.26) to (6.28) were used to calculate differential composition distributions for a microreactor $w(t, p)$ and for a sample as a whole $W(t, P)$ which are connected as follows:

$$\frac{dW(t, P)}{dt} = \mu[w(t, p) - W(t, p)] \tag{6.33}$$

Using Eqs. (6.26) to (6.28) and (6.33) and taking $\mu^{-1} = 100\,\text{s}$, the $W(t, P)$ values for various P were calculated and then distribution functions for the sample as a

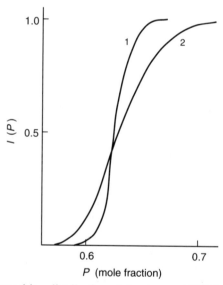

Fig. 6.5 Integral composition distribution of decomposed PTBA calculated neglecting a delay of the reaction launching (1) and with account of that (2) for $P = 0.62$ and $\mu^{-1} = 100\,\text{s}$. (Reproduced by permission of Nauka, Moscow, Russia, from *Vysokomol. Soed. B*, 1986, **28**, 820)

whole might be constructed at any conversion degree [28]. In this case for $P = 0.62$, the delay of the reaction launching leads to an increase in the composition heterogeneity which could be found out experimentally (see Fig. 6.5).

6.2.3 DECOMPOSITION OF *tert*-BUTYL ACRYLATE–STYRENE COPOLYMERS

When intermolecular effects are studied, valuable information can be obtained by changing the concentration of the interacting groups using an appropriate diluent. It is difficult to select such a diluent for reaction in polymer melt at relatively high temperatures. However, the use of some copolymers appears to be suitable in this case; the copolymers should contain reacting pendant groups side by side with pendant groups that are neutral, i.e. remain unchanged themselves and do not affect the reaction being studied.

Grant and Grassie [18] used copolymers of *tert*-butyl methacrylate and styrene (STY) to elucidate the mechanism of acceleration in the decomposition of poly(*tert*-butyl methacrylate). They found that an increase in STY content in copolymer led to a decrease in the rate of decomposition of ester groups. Grant and Grassie considered STY units exclusively as an intrachain diluent separating ester groups from carboxylic groups formed in the same chain (as a result of isobutylene elimination) and caused the acceleration of the reaction. However, STY units appear to serve also as an interchain diluent, diminishing the probability of encounter between an ester group and carboxylic groups belonging to other chains.

Cherkezyan and Litmanovich [29, 30] studied the decomposition of *tert*-butyl acrylate (TBA) and STY copolymers with the purpose of obtaining some additional information related to a contribution of interchain interactions into the kinetics of PTBA decomposition.

The kinetics of degradation of TBA–STY copolymers of different composition has been studied at 180 °C. Under experimental conditions, decomposition of TBA units proceeds (STY units being unchanged) and up to 75% conversion of ester groups the transforming polymer contains TBA, acrylic acid and STY units designated by A, B and C respectively [29]. As shown in Fig. 6.6, the greater the STY content in copolymer the slower is the decomposition of TBA units. Nevertheless, for all samples, the reaction proceeds with a pronounced acceleration. These data permit some essential features of the reaction mechanism to be elucidated.

Consider once more the Schaefgen and Sarasohn [17] hypothesis according to which the acceleration of PTBA decomposition is caused by a long sequence of carboxylic groups formed in the direct neighborhood of the reacting ester group in the same chain. The fraction of A units $Q(A_n)$ belonging to sequences CA_iC of length $i \geq n$ may be written as follows [31]:

$$Q(A_n) = nP_{AA}^{n-1} - (n-1)P_{AA}^n \qquad (6.34)$$

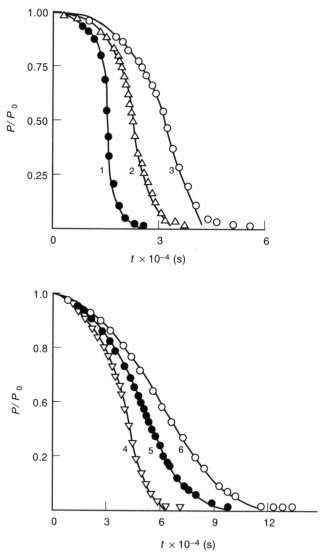

Fig. 6.6 Kinetics of TBA–STY copolymers decomposition at 180 °C. Points, experimental data; curves, calculation (for rate constant values given in Table 6.2). Composition of the samples μ_A (mole fraction of TBA): 0.89(1), 0.69(2), 0.53(3), 0.46(4), 0.38(5), 0.32(6). P and P_0 are current and initial mole fractions of unreacted TBA units. (Reprinted from *European Polymer J.*, **21**, 623, 1985, with kind permission from Elsevier Science Ltd, The Boulevard, Langford Lane, Kidlington OX5 1GB, UK)

where

$$P_{AA} = \frac{(1 - 2r_A r_C - q) + [(1 - 2r_A r_C - q)^2 + 4r_A r_C(1 - r_A r_C)]^{1/2}}{2(1 - r_A r_C)}$$

$q = (1 - m_A)/m_A$, m_A is a mole fraction of TBA in copolymer $r_A = 0.22$ and $r_C = 1.25$ are reactivity ratios [32]. For a sample of composition $m_A = 0.32$, Eq. (6.34) gives, for $n = 3$, $Q(A_3) = 0.06$. Therefore about 6% of TBA units form sequences containing three or more A units.

Only a small part of such TBA units might find themselves in the neighborhood of a sequence of two or more carboxylic groups (formed during the reaction) and therefore undergo an autocatalytic decomposition. In other words, the accelerating effect for this sample would be very small. However, actually an overwhelming part of the copolymer of composition $m_A = 0.32$ decomposes autocatalytically (see Fig. 6.6). Hence the hypothesis of Schaefgen and Sarasohn is not correct. Cherkezyan and Litmanovich [29, 30] tried to describe quantitatively the kinetics of copolymers TBA–STY decomposition in terms of interchain interactions.

In the chain of reacting TBA–STY, there are six types of A-centered triads indicated in Scheme 6.2, together with corresponding rate coefficients:

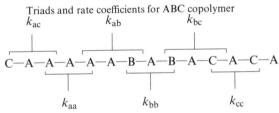

Triads and rate coefficients for ABC copolymer

k_{ac} k_{ab} k_{bc}

C—A—A—A—A—A—B—A—B—A—C—A—C—A

k_{aa} k_{bb} k_{cc}

Scheme 6.2

Using Keller's method [25] the following set of equations has been derived:

$$\frac{dP_{aa}}{dt} = -(k_{aa} + 2\phi)P_{aa}$$

$$\frac{dP_{ab}^*}{dt} = -(k_{ab} + \phi)P_{ab}^* + 2\phi P_{aa}$$

$$\frac{dP_{bb}}{dt} = -k_{bb}P_{bb} + \phi P_{ab}^* \tag{6.35}$$

$$\frac{dP_{ac}^*}{dt} = -(k_{ac} + \phi)P_{ac}^*$$

$$\frac{dP_{bc}^*}{dt} = -k_{bc}P_{bc}^* + \phi P_{ac}^*$$

$$\frac{dP_{cc}}{dt} = -k_{cc}P_{cc}$$

where

$$\phi = \frac{k_{aa}P_{aa} + k_{ab}P^*_{ab}/2 + k_{ac}P^*_{ac}/2}{P_{aa} + P^*_{ab}/2 + P^*_{ac}/2} \qquad (6.36)$$

and $P_{aa} = P(AAA)$, $P^*_{ab} = P(AAB^+)$ and so on. Equations (6.35) describe the time dependence of the probabilities of triads and their sum gives the kinetic curve because

$$P_{aa} + P^*_{ab} + P_{bb} + P^*_{ac} + P^*_{bc} + P_{cc} = P(A)$$

The same mechanism for the reaction of copolymers as for PTBA decomposition has been taken in reference [29], as well as the assumption about a uniform distribution of A, B and C units in the microreactor so that, for any reacting A unit the probability of encountering A, B or C is proportional to the concentration of respective units in the microreactor.

An additional hypothesis has also been proposed. STY units are supposed to change the reaction rate but not only by the effect of a dilution. Like any other 'solvent', STY units might also influence the ability of the ester and carboxylic groups to form an appropriate intermediate complex. STY units might therefore change the values of the autocatalytic rate constants. The hypothesis refers this possible effect mainly to an interaction of the ester group with the external carboxylic groups. As for the interaction between an ester group and its inner carboxylic neighbor, this interaction is supposed to be insensitive to the nature of the second inner neighbor of the ester group. This assumption is not in contradiction with the mechanism of PTBA decomposition, according to which the interaction between an ester group and its inner carboxylic neighbor is also independent of the nature of the second inner neighbor—regardless of whether it is A or B (see Section 6.2.2).

Noting the hypotheses considered above, it is possible to write the following expressions for the rate coefficients

$$k_{aa} = k_{ac} = k_{cc} = \alpha + \gamma[P_0(A) - P(A)]^2$$
$$k_{ab} = k_{bc} = \alpha + \beta[P_0(A) - P(A)] + \gamma[P_0(A) - P(A)]^2 \qquad (6.37)$$
$$k_{bb} = \alpha + 2\beta[P_0(A) - P(A)] + \gamma[P_0(A) - P(A)]^2$$

Here $P_0(A)$ is the value of $P(A)$ in the initial copolymer, so that $P_0(A) - P(A)$ is the probability of an acrylic acid unit in the transforming polymer chain.

Triads probabilities of initial copolymer samples necessary for solution of Eqs. (6.35) to (6.37) were calculated by methods described in Chapter 4 using reactivity ratios for TBA and STY copolymerization found in reference [32].

Using experimental data and Eqs. (6.35) to (6.37), the rate constants α, β and γ for six copolymer samples of various compositions were found. The constants values are collected in Table 6.2. The kinetic curves corresponding to these values describe well the experimental data (see Fig. 6.6). From the table data it can be

Table 6.2. Rate constants for decomposition of TBA–STY copolymers at 180 °C

Copolymer composition $m_A{}^a$	Rate constants (s^{-1})		
	$\alpha \times 10^6$	$\beta \times 10^4$	$\gamma \times 10^4$
1.00	5.0	22.5	9.7
0.89	5.0	33.3	2.0
0.69	5.0	21.0	4.6
0.53	5.0	19.8	7.5
0.46	5.0	20.8	7.1
0.38	5.0	46.3	6.6
0.32	5.0	42.0	6.6

$^a m_A$ is mole fraction of TBA.
(Reprinted from *European Polymer J.*, **21**, 623, 1985, with permission from Elsevier Science Ltd. The Boulevard, Langford Lane, Kidlington OX5 1GB, UK).

seen that there is no obvious dependence of the rate constants β and γ on the copolymer composition, the values of these constants not being varied within the limits of error for their estimation [29]. The rate constant α is also not changed for the samples studied. Hence STY units appear to be true inert diluents in the system considered.

As a whole, the results obtained for TBA–STY copolymers confirm the conclusion about determining contribution of interchain interaction in the PTBA decomposition formulated in Section 6.2.2.

6.3 REACTIONS IN POLYMER BLENDS

6.3.1 GENERAL REMARKS

A chemical reaction in polymer blend is an effective route for modification of its properties, in particular for blend homogenization. As examples, a series of publications dedicated to so-called reactive compatibilization [33–36] and also to blends reinforcement by reactive compounding [13] may be pointed out.

A very interesting investigation of competition between phase separation and chemical reaction occurring in a mixture of polycarbonate and poly(ethylene terephtalate) or polycarbonate and polyarylate has been carried out by Tanaka *et al.* [37]. However, transesterification and similar processes are in essence the changes of the polymer backbone and will not be considered here.

As to macromolecular reactions, it is expedient to perform a partial transformation of polymers which would lead to an increase in attractive interactions between the blend components. (In some cases the compatibility may be realized by entropy contributions exclusively, in spite of the unfavorable character of pair interactions of all different segments [38]; however, this is a rare situation.) For example, poly(vinyl chloride) (PVC) displays a tendency to intermolecular inter-action with carbonyl-containing polymers [39], in particular with poly(methyl methacrylate) (PMMA) [40]. Nevertheless, PVC and PMMA are incompatible. However, when Jayabalan [41] foamed a mixture of PVC and PMMA at 200 °C, an impact-resistant foam was formed and only one glass temperature was observed for the material obtained. An explanation is [41] that a partial dehydrochlorination of PVC occurs and the mobile allylic H atoms generated form hydrogen bonds with carbonyl groups of PMMA, thus improving inter-molecular interactions favorable for blend homogenization.

Numerous experiments show that the kinetics of a chemical transformation in a blend differs from that of the same polymer in a pure state [42–44]. For example, thermal decomposition of PVC and poly(vinyl acetate) (PVAc) in the blend proceeds faster for both polymers as compared with the reaction of pure components. It is assumed that HCl formed in the course of PVC dehydro-chlorination diffuses in PVAc and acts as a catalyst of acetic acid elimination. In turn, the acetic acid diffuses in PVC and catalyzes HCl elimination. Afterwards the processes initiated in such a way propagate along the chains due to allylic activation [43]. In a blend of PVC and PMMA, chlorine atoms (intermediate particles of PVC dehydrochlorination) initiate PMMA depolymerization at the temperature of PVC decomposition, whereas HCl reacts with PMMA pendant groups and anhydride formed inhibits the splitting of the PMMA chain at higher temperatures [43].

In both cases mentioned these were low-molecular reaction products that diffused into the phase of the other component of a blend and affected the reaction kinetics of the latter. Below we shall discuss the problems of those macromolecular reactions in blends that proceed with the interchain effect. First of all, it means a direct influence of the interchain interactions on the reactivity of polymer transforming groups.

6.3.2   DECOMPOSITION OF PTBA IN BLENDS

The influence of different factors on the reaction kinetics has been revealed during a study of thermal decomposition of PTBA pendant groups in blends with other polymers. In all cases the blends were prepared from a solution in a proper solvent by freeze-drying.

Cherkezyan and co-workers [45, 46] studied the decomposition of PTBA in blends with polyethyleneimine (PEI). PEI does not affect the initial rate of the reaction, but the acceleration decreases substantially: the ratios of maximum and initial effective rate constants are 200, 83, 15 and 3 for pure PTBA and blends of a composition PTBA/PEI 3:1, 1:1 and 1:3 (unit per unit) respectively (see Fig. 6.7).

More details have been obtained for the blend of composition 1:1 [46]. The initial blend had two T_g corresponding to T_g of pure components. At conversions of 48% and more the blend had only one T_g, higher than T_g of PTBA decom-

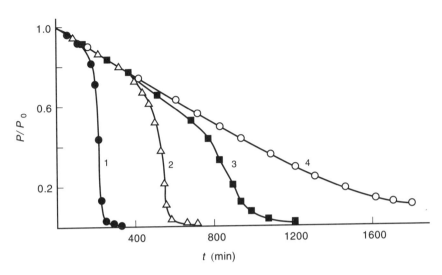

Fig. 6.7 Kinetics of PTBA decomposition in pure state (1) and in PTBA–PEI blends of composition 3:1(2), 1:1(3), 1:3(4). P and P_0 are current and initial mole fractions of unreacted TBA units. (Reproduced by permission of Nauka, Moscow, Russia, from *Vysokomol. Soed. B*, 1981, **23**, 645)

posed to the same degree of conversion. At the conversion of 100% a glass transition was not observed at all and the final blend became insoluble. According to IR spectra amide bonds are formed during the process.

As the PTBA decomposition proceeds to form carboxylic groups, the interaction between PEI and transforming PTBA increases and heterogeneous (though highly dispersed) initial blend becomes homogeneous. Therefore PEI acts as a diluent decreasing the interchain interaction of carboxylic and ester groups. The formation of interchain amide links increases a rigidity of the system and consumes irreversibly the carboxylic groups accelerating the PTBA decomposition. The dotted line on Fig. 6.8 is the calculated kinetic curve corresponding to the maximum possible action of PEI as an inert diluent. It can be seen that it is not a dilution but an interchain amide links formation that contributes mainly in the decrease of the acceleration.

It has already been mentioned in Section 6.2 that the rate of PTBA decomposition increased in a blend with poly(acrylic acid) (PAA) due to interchain interaction of carboxylic and ester groups. Cherkezyan *et al.* [47] showed that a dependency of the initial rate of PTBA decomposition on the composition of PTBA–PAA blend was described by two straight lines of different inclines (see Fig. 6.9). The authors assumed that an initial acceleration occurs in an interface and took into account their results concerning a linear depen-dence of the initial rate on the composition for blends of PTBA with crystalline bicyclo-(4,2,0)-

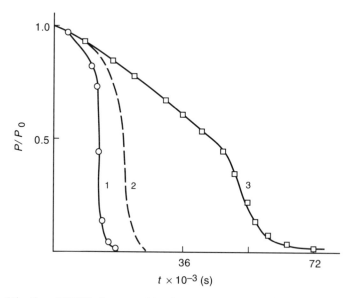

Fig. 6.8 Kinetics of PTBA decomposition in pure state (1) and in the PTBA–PEI blend 1:1(3); curve 2, calculation with account of an effect of a dilution. (Reproduced by permission of Nauka, Moscow, Russia, from *Vysokomol. Soed. B*, 1984, **26**, 112)

1(6)-octene-3,4-dicarboxylic acid [47]. Thus they concluded that the size of particles of the dispersed phase does not change within the linearity region whereas the intersection point apparently meets a phase inversion.

An influence of the PTBA–PAA blend initial structure on the reaction kinetics has been studied in more detail by Ermakov and co-workers [48, 49]. The blends of the same composition PTBA/PAA = 1:10 were prepared by one and the same procedure (freeze drying), but from different solvents: *tert*-butanol–water (7:1) and dioxane–water (50:1). Very different kinetic curves for the PTBA decomposition in these blends have been observed [48] (see Fig. 6.10). It is natural to connect the differences in the reaction kinetics with PTBA particles size and the fraction of PTBA chains situated on the interface, that is with a degree of interpenetration of the blend components. Note that the relaxation processes (by means of which both blends could transfer to one and the same equilibrium state at 180 °C) proceed apparently slowly enough. Therefore the initial structure of the blend has a significant effect not only on the initial reaction rate but also on the character of the whole kinetic curve.

According to FTIR spectroscopy data [49], in the IR spectrum of the 1:10 blend prepared from *tert*-butanol–water, some bands characteristic for PTBA are absent. Ermakov *et al.* [49] conclude that PTBA does not form compact dispersed phase particles in this blend. The spectrum of the blend obtained from

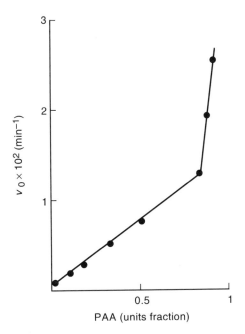

Fig. 6.9 Dependence of the initial rate of PTBA decomposition on the composition of PTBA–PAA blends. (Reproduced by permission of Nauka, Moscow, Russia, from *Vysokomol. Soed. B*, 1985, **27**, 225)

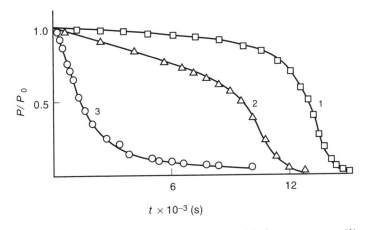

Fig. 6.10 Kinetics of PTBA decomposition at 180 °C in pure state (1) and in PTBA–PAA (1:10) blends prepared from solvent: dioxane–water (2) and *tert*- butanol–water (3). P and P_0 are current and initial mole fractions of TBA units in transforming PTBA. (Reproduced by permission of Nauka, Moscow, Russia, from *Vysokomol. Soed. A*, 1988, **30**, 2595)

dioxane–water comprises all bands characteristic of PTBA. Thus the structure of the first blend appears to be more favorable for interchain interaction.

It is interesting that for both of these blends DSC measurements fixed only one glass transition corresponding to T_g of PAA [48]. This means that the PTBA particles, if they are present in the blends, have too small a size to be fixed by the DSC method.

The extent of component interpenetration in PTBA–PAA blends has also been estimated [49] using the nonradiative energy transfer between the fluorescent labels introduced in the polymers [50]. Fluorescent-labeled PTBA and PAA were prepared by radical copolymerization of corresponding monomers with small amounts of α-naphthylacrylate and methyl-1-pyrenylacrylate respectively. The contents of labels were 1 per 195 units and 1 per 1100 units for PTBA and PAA respectively. PTBA–PAA blends of compositions 1:10 and 1:1 containing labeled macromolecules were obtained from different solvents. The results of the reflected fluorescence measurements for films of the blends (prepared by molding under a pressure of 10–15 kbar) as well as the data on the initial PTBA decomposition rates in these blends are collected in Table 6.3.

It is seen that emission intensity of the acceptor to that of the donor ratio I_{398}/I_{337}, characterizing the efficiency of the donor-to-acceptor nonradiative energy transfer and being proportional to the extent of interpenetration of the PTBA and PAA macromolecules, depends on the blend composition for the same solvent. For blends of the same composition the I_{398}/I_{337} ratios and, consequently, the interaction between components are greater in blends obtained from aqueous *tert*-butanol.

It can be seen that both the maximal I_{398}/I_{337} ratio and maximal initial reaction rate are observed for the 1:10 blend obtained from aqueous *tert*-butanol. Note that the I_{398}/I_{337} ratios are approximately the same for the 1:1 and 1:10 PTBA–PAA blends obtained from aqueous *tert*-butanol and aqueous dioxane respectively. At the same time, the initial decomposition rates in these blends are very close to one another. The minimal values of both the I_{398}/I_{337} ratio and initial reaction rate are observed for the 1:1 blend obtained from aqueous dioxane.

Table 6.3. Acceptor-to-donor emission intensity ratio (I_{398}/I_{337}) and initial decomposition rates (v_0) of TBA groups in PTBA–PAA blends

Composition of blend (unit/unit)	Solvent	I_{398}/I_{337}	$v_0 \times 10^4 (s^{-1})$
1:1	Dioxane–water	4.3	0.2
1:1	*tert*-Butanol–water	9.8	1.0
1:10	Dioxane–water	9.6	0.8
1:10	*tert*-Butanol–water	18.0	1.6

(Reproduced by permission of Nauka, Moscow, Russia, from *Polymer Sci.*, 1992, **34**, 513).

Thus, the results of investigating the structure of PTBA–PAA blends confirm the assumption [48] that the differences in reaction kinetics for blends prepared from various solvents are attributed to different structures of the blends. It is worth underlining the high sensitivity of the kinetic characteristics to the structural factors (see Fig. 6.10) that make it possible to consider the relatively readily available kinetic data as a valuable source of information on the structure of polymer blends. In this connection it would be expedient to work out mathematical models accounting for a dependence of the reaction kinetics on the blend structure.

6.3.3 THEORETICAL CONSIDERATIONS

Ermakov and Litmanovich [48] suggested relatively simple models for the PTBA decomposition in 1:10 blends with PAA. In the matrix of PAA there are spheric PTBA particles of radius r_0 surrounded by an interface layer of thickness λ (see Fig. 6.11). For the sake of simplicity a linear gradient of reacting PTBA concentration in the interface layer is supposed. It is also assumed in reference [48] that the interdiffusion of PTBA and PAA does not occur, so that r_0 and λ remain unchanged during the reaction. However, a diffusional movement of reacting PTBA macromolecules between a sphere and a layer is possible.

In an interface layer, due to interaction with PAA, macromolecules of PTBA decompose more quickly than in a sphere. Therefore an initial reaction rate is determined by the ratio of PTBA fractions situated in a sphere and in a layer. The subsequent course of the reaction (within the framework of the hypothesis related to an interdiffusion) depends on a diffusional exchange of decomposing PTBA between the sphere and the layer. Two models are considered corresponding to the following extreme cases. The model of a 'perfect exchange' corresponds to an instantaneous mixing of reacting PTBA chains. The model of 'chemical relay race' implies that translational diffusion is forbidden both for PAA and for PTBA, so that the accelerating action of PAA, as well as of carboxyls formed from PTBA, transfers from the layer to the center of the sphere as a result of the reaction itself. A change of the profile of ester group concentration during the reaction for both models is shown schematically in Fig. 6.11.

For each model the kinetic equations have been derived, the mechanism of PTBA decomposition described in Section 6.2 being used. The equations contain the rate constants α, β, γ found in reference [19]. As a partial PAA dehydration proceeds in the experimental conditions (180 °C), the corresponding rate constant [51] is also included. Then the kinetic curves were calculated for both models, r_0 and λ being varied to estimate the influence of structural parameters on the reaction kinetics. Calculated curves are shown in Fig. 6.12, together with experimental data for blends prepared from different solvents (cf. Fig. 6.10).

For both blends the model of 'chemical relay race' describes kinetic data better than the model of 'perfect exchange'. However, at high degrees of conversion

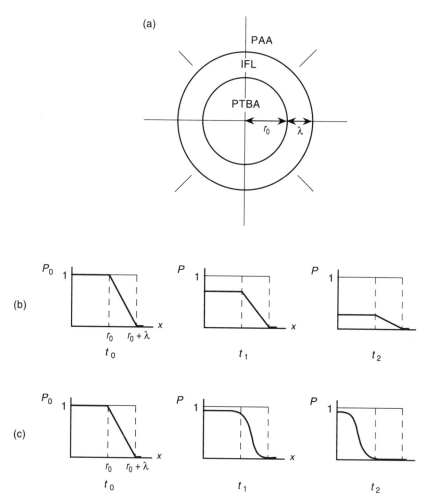

Fig. 6.11 Supposed structure of (a) PTBA–PAA blend and schematic change of TBA units concentration with conversion for the models (b) 'perfect exchange' and (c) 'chemical relay race'; IFL; interface layer. (Reproduced by permission of Nauka, Moscow, Russia, from *Vysokomol. Soed. A*, 1988, **30**, 2595)

calculated curves deviate from experiment. Though none of the models describe the kinetics completely their analysis together with experimental data leads to useful preliminary conclusions about the possible structure of the blend. For the blend prepared from *tert*-butanol–water, both models describe the initial rate provided as high a fraction of PTBA as 0.9 (calculated on the whole PTBA quantity) is situated in an interface layer. A possible pattern therefore corre-

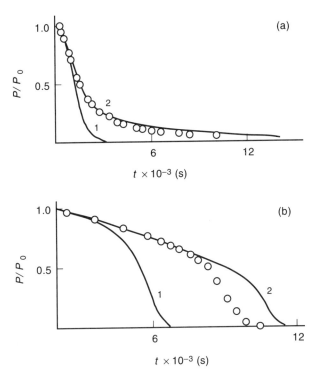

Fig. 6.12 Decomposition of PTBA–PAA (1:10) blends prepared from solvent: (a) *tert*-butanol–water and (b) dioxane–water. Points, experimental data; curves, calculation for models 'perfect exchange' (1) and 'chemical relay race' (2). (Reproduced by permission of Nauka, Moscow, Russia, from *Vysokomol. Soed. A*, 1988, **30**, 2595)

sponds to extreme segmental mixing, i.e. a friable structure of PTBA particles and very small spheres. On the other hand, for the blend prepared from dioxane–water, the fraction of PTBA in an interface layer is less than 0.1. In this case, calculations correspond to the existence of compact PTBA particles surrounded by a relatively small interface layer. It is worth noting that these preliminary predictions are in qualitative accordance with above-mentioned experimental data related to the blends structure [49], obtained later.

The first mathematical models described an influence of the initial blend structure on the reaction kinetics. It is evident, however, that not only initial structure but also its change during the reaction affects the process. Thus the problem arises: to include the components interdiffusion in the theory of macromolecular reactions in polymer blends.

6.3.4   REACTION AND INTERDIFFUSION IN POLYMER BLEND

As to interdiffusion in polymer blends as such (without a reaction), the problem is investigated very intensively by both theoretical and experimental methods [52, 53]. A blend of compatible flexible entangled polymers A and B is mainly considered. Two simple relations are derived that connect an interdiffusion coefficient D_m with self-diffusion coefficients D_s^A and D_s^B, namely fast-mode diffusion [54, 55]

$$D_m^F = [(1 - \phi)N_A D_s^A + \phi N_B D_s^B]\left[\frac{1 - \phi}{N_A} + \frac{\phi}{N_B} - 2\chi\phi(1 - \phi)\right] \quad (6.38)$$

and slow-mode diffusion [56, 57]

$$D_m^S = \left(\frac{1 - \phi}{N_A D_s^A} + \frac{\phi}{N_B D_s^B}\right)^{-1}\left[\frac{1 - \phi}{N_A} + \frac{\phi}{N_B} - 2\chi\phi(1 - \phi)\right] \quad (6.39)$$

where ϕ is a volume fraction of polymer A, N_A and N_B are degrees of polymerization and χ is the Flory–Huggins interaction parameter.

According to some recent publications [58, 59], Eqs. (6.38) and (6.39) are approximate relations. Some experimental data are described by slow-mode diffusion [60, 61], whereas the fast-mode relations fit the other sets of experiment very well [62, 63]. Nevertheless, the approaches of both a slow-mode and a fast-mode diffusion apparently may be used for analysis of various processes in polymer blends in order to estimate a contribution of the interdiffusion.

The diffusion in a mixture of A and B polymers (without a reaction) is characterized by the time-dependent profile of A chains concentration (ϕ). When a macromolecular reaction proceeds at least two new factors are contributed into the problem. First of all, unlike in low-molecular systems, in polymer ones these are chains that diffuse but these are units that react. Therefore the concentration profile of the reacting chains and that of the unreacted units, in the general case, do not coincide and such a divergence may affect both the driving forces of an interdiffusion and the reaction kinetics. Second, the products of the macro-molecular reaction are heterogeneous in composition, so that a binary initial blend becomes essentially multicomponent during the process. Thus the problem is significantly complicated.

Yashin *et al.* [64], having considered a simple model of the reaction in blend, tried to describe a mutual influence of the chemical reaction and interdiffusion on an evolution of the reacting system. Let the irreversible macromolecular reaction occur in an incompressible blend of polymers A and B: A units transform into B units, A → B, provided B units may affect the reactivity of A units (the interchain effect by direct contact of A and B units belonging to different chains), A + B → 2B.

At any time, a structure of the blend is characterized by both a spatial distribution of reacting chains with a length N_A (and unchangeable though affecting the reaction chains with a length N_B) and that of the A and B units. The corresponding volume fractions are denoted $\phi(\mathbf{r}, t)$, $\psi(\mathbf{r}, t)$, $\rho_A(\mathbf{r}, t)$ and $\rho_B(\mathbf{r}, t)$. Due to incompressibility,

$$\phi(\mathbf{r}, t) + \psi(\mathbf{r}, t) = 1$$

$$\rho_A(\mathbf{r}, t) + \rho_B(\mathbf{r}, t) = 1$$

Thus the two variables, namely $\phi(\mathbf{r}, t)$ and $\rho_A(\mathbf{r}, t)$, are sufficient to describe the blend structure and so the further subscript A will be omitted.

A change in the blend structure is described by general balance equations:

$$\frac{\partial}{\partial t} \phi(\mathbf{r}, t) = -\operatorname{div} \mathbf{J}_\phi(\mathbf{r}, t) \tag{6.40}$$

$$\frac{\partial}{\partial t} \rho(\mathbf{r}, t) = -\operatorname{div} \mathbf{J}_\rho(\mathbf{r}, t) + f[\rho(\mathbf{r}, t)] \tag{6.41}$$

because a volume fraction of reacting chains ϕ in any closed region of the system changes by the flow of these chains across the region boundary $\mathbf{J}_\phi$ (Eq. (6.40)), whereas a volume fraction of A units changes both by the flow across the region boundary $\mathbf{J}_\rho$ and by their chemical transformation inside the region (Eq. (6.41)), where $f(\rho)$ describes the reaction kinetics.

At certain conditions formulated in reference [64], the reacting blend may be considered as a local-equilibrium one. This means that a state of regions containing many chains but having a small volume as compared to the inhomo-geneity scale is determined at any time by quasi-equilibrium density of a free energy $F(\phi, \rho)$, which in turn depends on the local meanings of $\phi(\mathbf{r}, t)$ and $\rho(\mathbf{r}, t)$. The flows $\mathbf{J}_\phi$ and $\mathbf{J}_\rho$ depend on these local meanings and their gradients only. These dependencies are expressed by relations of linear nonequilibrium thermodynamics [65, 66]:

$$\mathbf{J}_\phi = -\Lambda_{\phi\phi} \operatorname{grad} \mu_\phi - \Lambda_{\phi\rho} \operatorname{grad} \mu_\rho \tag{6.42}$$

$$\mathbf{J}_\rho = -\Lambda_{\rho\phi} \operatorname{grad} \mu_{\rho\phi} - \Lambda_{\rho\rho} \operatorname{grad} \mu_\rho \tag{6.43}$$

where

$$\mu_\phi = \frac{\partial}{\partial \phi} F(\phi, \rho) \qquad \mu_\rho = \frac{\partial}{\partial \rho} F(\phi, \rho) \tag{6.44}$$

are chemical potentials and Λ are kinetic coefficients. According to Onsager's theorem,

$$\Lambda_{\phi\rho} = \Lambda_{\rho\phi} \tag{6.45}$$

Thus the problem comes to the solution of two tasks. First, it requires an expression for the free energy density $F(\phi, \rho)$ of the local equilibrium blend of a homopolymer B and a random copolymer AB with ϕ and ρ fixed. Second, it requires the kinetic coefficients Λ determining an interdiffusion in the blend.

For such a blend $F(\phi, \rho)$ (for a lattice model, the free energy per one cell) in the region with coordinates $\mathbf{r}$ depends on the local meanings of $\phi(\mathbf{r}, t)$ and $\rho(\mathbf{r}, t)$ but not on their derivatives on $\mathbf{r}$. Therefore the free energy density F as a function of ϕ and ρ is just like that for a spatially homogeneous system.

It is necessary to note that Yashin *et al.* [64] consider the most simple type of macromolecular reactions when a reaction rate depends on the local concentration of A units and—in the case of the interchain effect—on that of B units. The copolymer AB forming in such a process has the Bernoullian distribution in composition and structure. For that particular case, the final expression of free energy may be written as follows [64]:

$$\frac{F(\phi, \rho)}{kT} = \frac{\phi}{N_A} \ln \phi + \frac{1-\phi}{N_B} \ln(1-\phi) + \chi\rho(1-\rho) + \rho \ln \frac{\rho}{\phi} + (\phi-\rho)\ln \frac{\phi-\rho}{\phi} \quad (6.46)$$

The first three terms on the right-hand side of Eq. (6.46) are entropy and enthalpy constituents of the Flory–Huggins free energy for a binary mixture of components having chain lengths N_A and N_B. The rest of the terms reflect and additional entropy contribution caused by the nonidentity of copolymer macromolecules.

We now come to kinetic coefficients. Two kinds of diffusion movements in the system under consideration should be distinguished. First, there is the relative movement of copolymer AB chains and homopolymer B chains. Second, there is the relative movement of copolymer chains having different compositions. Strictly speaking both types of movements should be taken into account. A simplified theory [64] includes the first of them only because that one is assumed to determine mainly an interdiffusion, the local composition heterogeneity of the copolymer being neglected.

Then the flows are connected with the relation

$$\mathbf{J}_\rho = \frac{\rho}{\phi} \mathbf{J}_\phi \quad (6.47)$$

and from Eqs. (6.42), (6.43), (6.45) and (6.47), it follows that

$$\Lambda_{\rho\phi} = \Lambda_{\phi\rho} = \frac{\rho}{\phi} \Lambda_{\phi\phi} \quad (6.48)$$

$$\Lambda_{\rho\rho} = \left(\frac{\rho}{\phi}\right)^2 \Lambda_{\phi\phi} \quad (6.49)$$

Therefore Eq. (6.42) may be written as follows:

$$\mathbf{J}_\phi = -\Lambda_{\phi\phi}\left(\operatorname{grad}\mu_\phi + \frac{\rho}{\phi}\operatorname{grad}\mu_\rho\right) \quad (6.50)$$

It is only necessary to calculate $\Lambda_{\phi\phi}$, which may be found within the frames of a fast-mode theory using the Brochard-Wyart approach [67].

For interdiffusion of homopolymer B and random copolymer,

$$\Lambda_{\phi\phi} = \frac{\phi(1-\phi)}{kT}[\phi N_B D_S^B + (1-\phi)N_A D_S^{cpl}] \tag{6.51}$$

where D_S^B and D_S^{cpl} are self-diffusion coefficients and

$$D_S^{cpl(B)} = \frac{kT}{Z_{cpl(B)}}\frac{N_e}{N_{A(B)}^2} \tag{6.52}$$

where N_e is a number of units between two entanglements, Z_{cpl} and Z_B are friction coefficients for movements of corresponding macromolecules along the tube, in the reptation model (per one unit):

$$Z_{cpl} = \frac{\rho}{\phi}[z_{AA}\rho + z_{AB}(1-\rho)] + \left(1 - \frac{\rho}{\phi}\right)[z_{BB}(1-\rho) + z_{AB}\rho] \tag{6.53}$$

$$Z_{BB} = z_{BB}(1-\rho) + z_{AB}\rho \tag{6.54}$$

where z_{AA}, z_{BB} and z_{AB} are microscopic friction coefficients between corresponding couples of units. Equations (6.51) to (6.54) give expressions for $\Lambda_{\phi\phi}$.

Finally, the balance equations (6.40) and (6.41) transform into

$$\frac{\partial\phi}{\partial t} = \text{div}[D_\phi(\phi,\rho)\,\text{grad}\,\phi] + \text{div}[L_\phi((\phi,\rho)\,\text{grad}\,\rho] \tag{6.55}$$

$$\frac{\partial\rho}{\partial t} = \text{div}[D_\rho(\phi,\rho)\,\text{grad}\,\rho] + \text{div}[L_\rho(\phi,\rho)\,\text{grad}\,\phi] + f(\rho) \tag{6.56}$$

where the diffusion coefficients are

$$D_\phi(\phi,\rho) = \left(\frac{1-\phi}{N_A} + \frac{\phi}{N_B}\right)[\phi N_B D_S^B + (1-\phi)N_A D_S^{cpl}] \tag{6.57}$$

$$L_\phi(\phi,\rho) = -2\chi\rho(1-\phi)[\phi N_B D_S^B + (1-\phi)N_A D_S^{cpl}] \tag{6.58}$$

$$D_\rho(\phi,\rho) = \frac{\rho}{\phi}L_\phi(\phi,\rho) \qquad L_\rho(\phi,\rho) = \frac{\rho}{\phi}D_\phi(\phi,\rho) \tag{6.59}$$

and the self-diffusion coefficients are calculated from Eq. (6.52). The function $f(\rho)$ in Eq. (6.56) describing a local chemical transformation itself has been used in reference [64] as follows:

$$f(\rho) = -\alpha\rho - \beta\rho(1-\rho) \tag{6.60}$$

where α and β are rate constants of spontaneous and catalytic (interchain effect) transformations of A units respectively.

The influence of various factors on a change in the blend structure (initially, the two joint films of A and B homopolymers) has been studied, microscopic friction coefficients being taken equal to one another in the calculations.

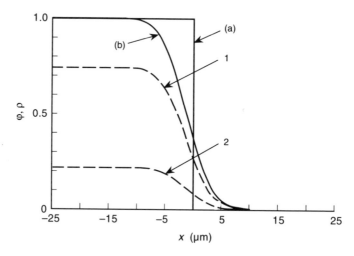

Fig. 6.13 Influence of the rate constant α on the blend structure without the interchain effect ($\beta = 0$). Calculations for $N_A = N_B = 1000$; $\chi = 0$; $\alpha = 2 \times 10^{-6}$ (1) and $1 \times 10^{-5}\,\mathrm{s}^{-1}$ (2). Dotted lines, profile ρ at $t = 1.5 \times 10^5\,\mathrm{s}$; solid lines, profile ϕ at (a) $t = 0$ and (b) $t = 1.5 \times 10^5\,\mathrm{s}$. (Reproduced by permission of Nauka, Moscow, Russia, from *Vysokomol. Soed. A*, 1994, **36**, 955)

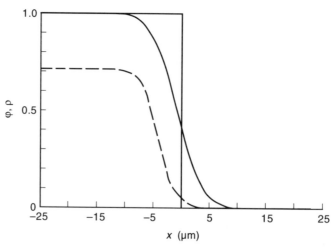

Fig. 6.14 Influence of interchain effect on the blend structure. Calculation for $\alpha = 4 \times 10^{-7}$, $\beta = 2 \times 10^{-5}\,\mathrm{s}^{-1}$; the other parameters as in Fig. 6.13. (Reproduced by permission of Nauka, Moscow, Russia, from *Vysokomol. Soed. A*, 1994, **36**, 955)

Even for $\beta = 0$ (no interchain effect) profiles of ϕ and ρ are separated; the greater the rate constant α the greater are the differences between profiles (see Fig. 6.13). However, the ratio ρ/ϕ does not depend on coordinate x; i.e. the mean composition of the reaction product is the same for any small region of the blend.

When an accelerating interchain effect is also manifested the situation is more complicated. In regions enriched with polymer B, the acceleration increases. The composition of a reacting polymer depends on coordinate x and not only do the ϕ and ρ profiles diverge but the intervals of their change do not coincide (see Fig. 6.14).

Yashin *et al.* [64] consider some peculiarities of the process caused by the interchain effect which manifest themselves especially clearly when the reaction occurs by contact of A and B units exclusively, that is $\alpha = 0$ and $f(\rho) = -\beta\rho$ $(1 - \rho)$. For components differing in chain length, the ϕ profile becomes asymmetric relative to the initial boundary. Nevertheless, the two blends presented in Fig. 6.15 could appear to be identical: the ϕ profiles would coincide after two turns. However, ρ profiles differ sharply. The reason is that in a region occupied by polymer A the reaction does not begin before B chains diffuse into that region. For a fixed time interval the short B chains ($N_A/N_B = 4$) penetrate more deeply into the polymer A region and stimulate the reaction of a significant fraction of A chains. For the same time interval the long B chains ($N_A/N_B = \frac{1}{4}$; note that the product $N_A N_B$ is taken to be constant) initiate a lesser fraction of A chains to

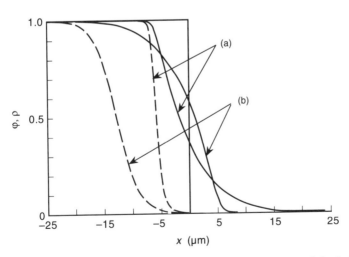

Fig. 6.15 Influence of the N_A/N_B ratio on the blend structure. Calculations for $N_A/N_B = 1/4$ (a) and 4(b); $N_A N_B = 10^6$; $\alpha = 0$, $\beta = 5 \times 10^{-5}$ s^{-1}; the other parameters as in Fig. 6.13. (Reproduced by permission of Nauka, Moscow, Russia, from *Vysokomol. Soed. A*, 1994, **36**, 955)

react. Therefore a ratio of chain lengths substantially affect both the formation of the blend structure and the reaction kinetics.

For the case of a strong interchain effect the blend structure pattern shown in the figures may to some extent be distorted if the relative movements of copolymer chains of different compositions are neglected. Therefore the task arises of including a description of that type of movement into the theory.

When blend processing includes a chemical reaction the intimate structure of the blend is characterized not only by spatial distribution of component chains but also by that of units because these distributions do not coincide. Meanwhile that intimate structure determines the properties of the blend, in the same way as the primary structure of a macromolecule (the units distribution, a microtacticity) determines the properties of a corresponding material. Therefore the advance in a working out of the theory of macromolecular reactions in polymer blends is an actual and important problem.

6.4   REACTION IN A GLASSY STATE

6.4.1   GENERAL REMARKS

Chemical transformations of polymers in a glassy state usually proceed much more slowly as compared with the reactions of melted or elastic samples. It refers, for example, to the cyclization of polyhydazides [68, 69] or to the intramolecular cyclization of polyamido acids [70, 71]. A decrease in the reaction rate is assumed to be caused by low mobility of macromolecules and by kinetic nonequivalency of transforming units [72], so that a transfer of unfavorable conformations for the reaction into favorable ones is hampered.

Meanwhile, Kamogava *et al.* [73] observed a higher rate in a glassy state than in solution for *cis–trans* isomerization of aromatic azo groups in copolymers of styrene with vinylaminoazobenzenes. Similar results were obtained by Paik and Morawetz [74] for isomerization of azoaromatic groups in copolymers of methyl methacrylate with small amounts of 4-*N*-methacrylaminoazobenzene or 4-*N*-methacrylamino-1,1-azonaphtalene: about 15% of azo groups reacted very quickly. Such a phenomenon has been explained in terms of nonuniform distribution of a free volume in a glassy sample [74, 75].

Our task, however, is to look at the contribution of the interchain effects (a direct action of reacted groups on the reactivity of the yet unreacted groups) in chemical transformation of a glassy polymer. It is natural to consider the problem using a reaction proceeding deliberately with an interchain effect, namely a decomposition of pendant *tert*-butylcarboxylate (TBC) groups.

Monahan [76, 77] studied PTBA photodecomposition below the polymer T_g. The process proceeded mainly as the very slow first-order reaction; only at a high degree of conversion has a slight acceleration been observed. Owing to these features PTBA appears to be inconvenient to study for the purpose formulated above (besides, it would be necessary to take into account the influence of irradiation on the decomposition of the TBC groups.

Otsu and co-workers [78, 79] synthesized poly(di-*tert*-butyl fumarate) (PDTBF) and, by heating it about 170 °C, transformed this polymer into the high-molecular poly(fumaric acid) (PFA). (Otsu notes that a high-molecular PFA can not be obtained by the polymerization of corresponding monomer, so this is a good example of the advantages of macromolecular reactions.) According to the Otsu data, PDTBF decomposition (with isobutylene elimination) proceeds very fast below T_g of the polymer. However, these authors [78, 79] did not study the reaction kinetics.

6.4.2 DECOMPOSITION OF POLY(DI-*tert*-BUTYL FUMARATE)

The kinetics of PDTBF decomposition has been studied by Ermakov and co-workers [80, 81]. It is seen in Fig. 6.16 that the PDTBF decomposition, after a slow initial stage, proceeds with very sharp acceleration. Note that the acceleration for glassy PDTBF is much greater than for melted PTBA. For example, at 170 °C the ratio of reaction rates at 50% and extrapolated to 0% of

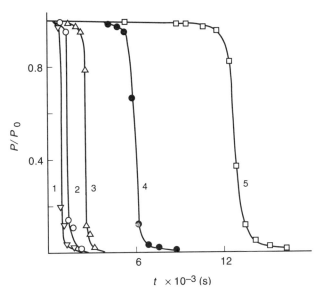

Fig. 6.16 Kinetic of PDTBF decomposition at various temperatures (°C): 180(1), 170(2), 160(3), 150(4) and 140(5). P_0 and P are initial and current concentrations of TBC groups. (Reproduced by permission of Nauka, Moscow, Russia, from *Vysokomol. Soed. A*, 1989, **31**, 793)

conversion v_{50}/v_0 is approximately 1500 for PDTBF and 90 for PTBA [80]. The question then arises as to what kind of mechanism of acceleration is realized in the case of PDTBF.

By analogy with PTBA it is natural to assume that the acceleration is caused by interaction of ester and carboxylic groups. Analysis of spherical atomic models showed [80] that the interaction of the TBC group with the nearest carboxyl neighbor in the chain of partly decomposed PDTBF is impossible due to steric hindrances (unlike PTBA, for which an interaction of the TBC group with the nearest carboxyl neighbor brings an essential contribution into acceleration; see Section 6.2.2). Obviously, for PDTBF the acceleration is realized by interchain interaction exclusively and a supermolecular structure of the polymer is especially favorable for interchain effects.

Ermakov *et al.* [80] also studied the decomposition of the TBC group in copolymers of di-*tert*-butyl fumarate (DTBF) and styrene (STY) of various compositions at 170 °C. STY units are considered to be inert diluents, hindering an interaction of TBC and carboxylic groups. Indeed, the more STY content is in

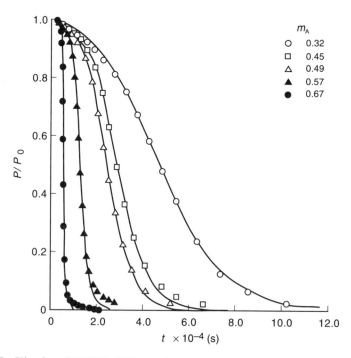

Fig. 6.17 Kinetics of DTBF–STY copolymer decompositions at 170 °C; m_A = DTBF fraction in copolymer; points, experimental data; curves, calculations according to Eq. (6.61). P_0 and P are initial and current TBC group fractions. (Reproduced by permission of John Wiley & Sons Ltd from *J. Polymer Sci. Chem.*, 1993, **31**, 395)

a copolymer, the less is the reaction rate. Nevertheless, for all copolymer samples, the pronounced acceleration has been observed (see Fig. 6.17).

It is necessary to note that steric hindrances preventing an interaction of TBC groups with the nearest (inner) carboxyl neighbors also exist for DTBF–STY copolymers. However, T_g of all copolymer samples are lower than the temperature of kinetic experiments [80]. Thus translational movements are not forbidden and interchain interactions can proceed by usual collisions of corresponding groups. Ermakov *et al.* [82] tried to describe the reaction kinetics for copolymers using the very simple equation

$$\dot{P} = -P[\alpha + \beta(P_0 - P)] \tag{6.61}$$

where P is a mole fraction of TBC groups in reacting copolymer, P_0 is its initial value, α is the rate constant of thermal decomposition and β is that of autocatalytic interaction of TBC and carboxylic groups. This simple model describes well the reaction kinetics for each copolymer sample (see Fig. 6.17). However, while the value of α does not depend on the sample composition, a completely different picture is observed for β: its value increases greatly with an increase in DTBF content in copolymer (see Fig. 6.18). The β increase is especially sharp for the samples containing large amount of DTBF units.

Ermakov *et al.* [82] tried to connect the change of β with some parameters of the units distribution in a copolymer chain. Using Ring's equations (6.34) and reactivity ratios for DTBF and STY copolymerization found in reference [80],

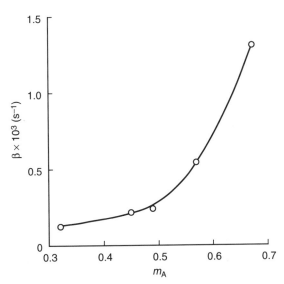

Fig. 6.18 Autocatalytic rate constant β versus m_A. (Reproduced by permission of John Wiley & Sons Ltd from *J. Polymer Sci. Chem.*, 1993, **31**, 395)

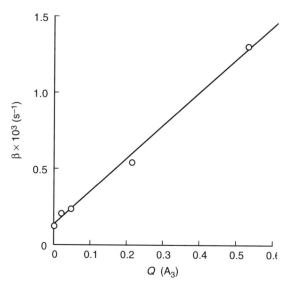

Fig. 6.19 Autocatalytic rate constant β versus $Q(A_3)$. (Reproduced by permission of John Wiley & Sons Ltd from *J. Polymer Sci. Chem.*, 1993, **31**, 395)

they calculated the values of $Q(A_n)$, i.e. the fraction of A units incorporated into sequences CA_iC containing $i \geq n$ A units (here A and C denote DTBF and STY respectively). For $n = 3$, β appears to depend linearly on $Q(A_3)$ (see Fig. 6.19). The linearity allows one to assume that even such short sequences of DTBF units reveal a tendency to form some supermolecular structures obviously favorable for interchain interaction. It is natural to conclude that such structures also exist in a sample of PDTBF under conditions of kinetic experiment.

Antipov *et al.* [83] studied the PDTBF structure and its change during a decomposition using X-ray measurements. The following results were obtained: the structure of PDTBF is more ordered than that of ordinary amorphous polymers; at intermediate degrees of conversion the X-ray scattering profiles contain maxima corresponding to both unreacted PDTBF and a final product, i.e. poly(fumaric acid) (PFA).

Ermakov *et al.* [82] decomposed PDTBF at 150 °C up to 70% of conversion and separated the product into three fractions by selective dissolution first in benzene and then in water. It is seen from Table 6.4 that partly decomposed PDTBF really contains both almost pure initial polymer—PDTBF—and practically pure final product of the reaction—PFA.

Figure 6.20 presents the kinetics of the decomposition of TBC groups in partly transformed polymer (before selective dissolution) and in fraction 1. One can see that the initial reaction rate in the fraction is considerably less than in the untreated sample (which may be considered to be a blend of PDTBF and PFA).

Table 6.4. Results of selective dissolution of partly decomposed PDTBF

Fraction number	Solvent	Weight portion	Composition (fraction of DTBF)
1	Benzene	0.18	0.99
2	Water	0.51	0.01
3	Unsolved	0.31	0.48

(Reproduced by permission of John Wiley & Sons Ltd from *J. Polymer Sci. Chem.*, 1993, **31**, 395).

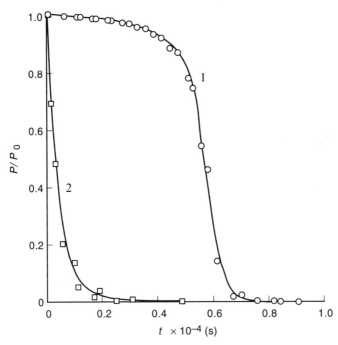

Fig. 6.20 Kinetics of TBC groups decomposition at 140 °C in fraction 1(1) (see Table 6.4) and untreated partly decomposed PDTBF (2). (Reproduced by permission of John Wiley & Sons Ltd from *J. Polymer Sci. Chem.*, 1993, **31**, 395)

Evidently the presence of PFA in the partly decomposed sample strongly increases the decomposition of almost pure PDTBF, so that the sharp acceleration observed for this sample can be explained by intimate mixing of the PDTBF and PFA in such a 'blend' and by interchain interaction of the components.

Taking into account all the data obtained, Ermakov *et al.* [82] assumed PDTBF decomposition in a glassy state to be a topochemical reaction. The

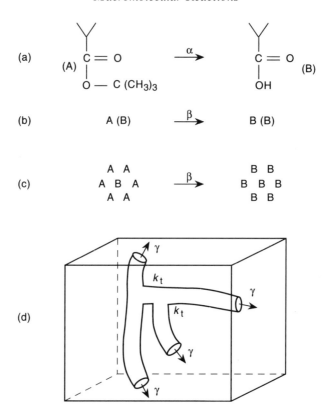

Fig. 6.21 A scheme of PDTBF decomposition in the glassy state: (a) thermal initiation, (b) autocatalytic action of carboxylic groups, (c) formation of cluster, (d) growth and branching of thread-like germ. (Reproduced by permission of John Wiley & Sons Ltd from *J. Polymer Sci. Chem.*, 1993, **31**, 395)

reaction mechanism suggested includes the interchain autocatalytic action of carboxylic groups and, typical for topochemical reactions [84], the processes of the formation and propagation of the new phase germs.

An initiation step proceeds as a thermal decomposition of TBC groups with a rate constant α (Fig. 6.21a). Then a spherical cluster grows with a rate constant β due to the interchain autocatalytic interaction of TBC and carboxylic groups (Fig. 6.21b). In such a way a primary germ of the new phase is formed consisting of a critical number (n_c) of joined carboxylic groups (Fig. 6.21c). After that a new factor contributes to germ propagation. As the volume of the carboxylic group is essentially less than that of the TBC group, the segments of PFA formed gain a tendency to collapse. Therefore mechanical stresses arise [85] which, obviously, facilitate the reaction. The propagation of germs is supposed to proceed along ordered regions (according X-ray data [83]) with a rate constant of the reaction front propagation γ, and germs have a thread-like shape (Fig. 6.21d). Secondary

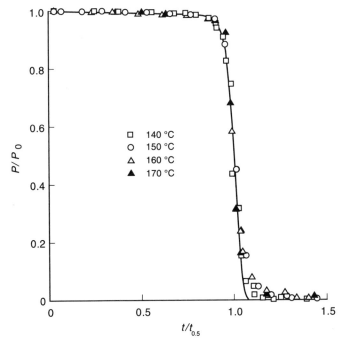

Fig. 6.22 Kinetics of PDTBF decomposition. Points, experimental data; curve, calculation. (Reproduced by permission of John Wiley & Sons Ltd from *J. Polymer Sci. Chem.*, 1993, **31**, 395)

germs also appear in the system as a result of branching of growing primary germs. Branching is characterized by k_+ quantity, which is the number of branches referred to the germ length equal to its diameter. Secondary germs also grow with the rate constant γ.

Based on this reaction scheme a corresponding mathematical model has been worked out in reference [82]. The calculated kinetic curve fits the experimental data closely in coordinates P/P_0 versus $t/t_{0.5}$, where $t_{0.5}$ is a half-conversion time (see Fig. 6.22). The coincidence of the kinetic curves for different temperatures means that the reaction mechanism is not changed within the temperature interval investigated (140–170 °C).

The rate constant α is of an order of magnitude usual for the thermal decomposition of TBC groups, say at 170 °C, $\alpha \times 10^6 = 4.1$ and $1.7\,\mathrm{s}^{-1}$ for PDTBF and PTBA respectively. The autocatalytic rate constant β exceeds α by two orders of magnitude. However, the rate constant γ (reflecting an influence of structural factors), which is five orders higher than α, gives the main contribution to acceleration. Thus for the decomposition of TBC groups in a glassy polymer the structural factors reinforce the 'pure' autocatalytic action of carboxylic groups and increase the acceleration of the reaction.

6.5 CONCLUSIONS

We have discussed above the reactions in a bulk state for macromolecules containing reactive groups in each unit. (The diffusion-controlled irreversible reaction in unentangled melts between two macromolecules, each bearing only one reactive group at the end or in the middle of the chain [86], has not been discussed.)

This chapter contains some new theoretical and experimental approaches to an investigation of the interchain effects for reactions in polymer melts, blends and glasses. It is shown, in particular, that the kinetic data may serve as a valuable source of information concerning some features of a structure of the reacting system. The results obtained, however, should be considered as the very beginning of a creation of the theory of macromolecular reactions in a polymer bulk state. The problem needs further fundamental development. It refers especially to the reactions in polymer blends, including flow systems and specific interactions between the components [87].

REFERENCES

1. Shibaev, V. P., Platé, N. A., Grushina, R. K. and Kargin, V. A., *Vysokomol. Soed*, 1964, **6**, 231–6.
2. Entelis, S. G. and Tiger, R. P., *Kinetika Reaktsij v Zhidkoj Phase* (*Reaction Kinetics in a Liquid Phase*), Khimiya, Moscow, 1973.
3. Moelwyn-Hughes, *The Chemical Statics and Kinetics of Solutions*, Academic Press, London, 1971.
4. Litmanovich, A. D., *Dokl. Akad. Nauk. SSSR*, 1978, **240**(1), 111–13.
5. Litmanovich, A. D., *Europ. Polymer J.*, 1980, **16**, 269–75.
6. Coleman, B. D. and Fox, T. G., *J. Polymer Sci. A*, 1963, **1**, 3183–96.
7. Ito, K. and Yamashita, Y., *J. Polymer Sci. A*, 1965, **3**, 2165–87.
8. Dobrushin, R. L., *Problemy Peredachi Informatsii*, 1971, **7**, 57–65.
9. Mityushin, L. G., *Problemy Peredachi Informatsii*, 1973, **9**, 81–5.
10. McQuarrie, D. A., *J. Appl. Prob.*, 1967, **4**, 413–78.
11. Platé, N. A., Litmanovich, A. D. and Noah, O. V., *Makromolekularnye Reaktsii* (*Macromolecular Reactions*), Khimiya, Moscow, 1977.
12. Noah, O. V., Litmanovich, A. D. and Platé, N. A., *J. Polymer Sci. Polymer Phys. Ed.*, 1974, **12**, 1711–25.
13. Rätzsh, M., *Makromol. Chem. Macromol. Symp.*, 1987, **12**, 165–202.
14. Xanthos, M. (Ed.), *Reactive Extrusion*, Hanser Publishers, Munich, 1992.
15. Rudy, C. and Fugassi, P., *J. Phys. Colloid. Chem.*, 1948, **52**, 357–63.
16. Warrick, E. and Fugassi, P., *J. Phys. Colloid. Chem.*, 1948, **52**, 1314–19.
17. Schaefgen, J. R. and Sarasohn, I. M., *J. Polymer Sci.*, 1962, **58**, 1049–61.
18. Grant, D. H. and Grassie, N., *Polymer*, 1960, **1**, 445–55.
19. Litmanovich, A. D. and Cherkezyan, V. O., *Europ. Polymer J.*, 1984, **20**, 1041–4.
20. Litmanovich, A. D. and Cherkezyan, V. O., *Vysokomol. Soed A*, 1985, **27**, 1865–70.
21. Shultz, A. R. and Young, A. L., *Macromolecules*, 1980, **13**, 663–8.

22. Litmanovich, A. D., Cherkezyan, V. O. and Khromova, T. N., *Vysokomol Soed B*, 1981, **23**, 645.
23. Jachowicz, J. and Morawetz, H., *Macromolecules*, 1982, **15**, 828–31.
24. Avakyan, V. G., Litmanovich, A. D. and Cherkezyan, V. O., *Izv. Akad. Nauk. SSSR Ser. Khim.*, 1984, 329–34.
25. Keller, J. B., *J. Chem. Phys.*, 1962, **37**, 2584–6.
26. De Gennes, P. G., *Scaling Concepts in Polymer Physics*, Cornell University Press, Ithaca, London, 1979.
27. Olonovsky, A. N., Stroganov, L. B., Noah, O. V. and Platé, N. A., *Vysokomol Soed A*, 1983, **25**, 882–8.
28. Cherkezyan, V. O. and Litmanovich, A. D., *Vysokomol Soed B*, 1986, **28**, 820–3.
29. Cherkezyan, V. O. and Litmanovich, A. D., *Europ. Polymer J.*, 1985, **21**, 623–6.
30. Cherkezyan, V. O. and Litmanovich, A. D., *Vysokomol Soed B*, 1985, **27**, 886–8.
31. Ring, W., *J. Polymer Sci. B*, 1963, **1**, 323–7.
32. Litmanovich, A. D., Cherkezyan, V. O., Khromova, T. N. and Koshevnik, A. Yu., *Vysokomol Soed B*, 1982, **24**, 237–8.
33. Braun, S. B., *Polymer Prepr.*, 1992, **33**(2), 598–9.
34. Akkareddi, M. K. and Van Buskirk, B., *Polymer Prepr*, 1992, **33**(2), 602–3.
35. Park, I. and Paul, D. R., *Polymer Prepr.*, 1992, **33**(2), 608–9.
36. Hobbs, S. Y., Stanley, J. J. and Phanstiel, O., *Polymer Prepr.*, 1992, **33**(2), 614–15.
37. Tanaka, H., Suzuki, T., Hayashi, T. and Nishi, T., *Macromolecules*, 1992, **25**, 4453–6.
38. Ellis, T. S., *J. Polymer Sci. Polymer Phys.*, 1993, **31**, 1109–25.
39. Benedetti, E., Posar, F., D'Alessio, A., et. al., *Br. Polymer J.*, 1985, **17**, 34–7.
40. Kögler, G. and Mirau, P. A., *Macromolecules*, 1992, **25**, 598–604.
41. Jayabalan, M., *J. Appl. Polymer Sci.*, 1982, **27**, 43–52.
42. Paul, D. R., Barlow, J. W. and Keskkula, H., in *Encyclopedia of Polymer Science and Engineering*, 2nd edn, Wiley, New York, 1988, vol. 12, pp. 399–461.
43. Grassie, N. and Scott, G., *Polymer Degradation and Stabilization*, Cambridge University Press, Cambridge, 1985, Chap. 2.
44. McNeil, I. C., in *Developments in Polymer Degradation*, Vol. 1, Ed. N. Grassie, Applied Science Publishers, London, 1977.
45. Litmanovich, A. D., Cherkezyan, V. O. and Khromova, T. N., *Vysokomol Soed B*, 1981, **23**, 645.
46. Cherkezyan, V. O., Litmanovich, A. D., Godovsky, Yu. K., Litmanovich, A. A. and Khromova, T. N., *Vysokomol Soed B*, 1984, **26**, 112–15.
47. Cherkezyan, V. O., Artamonova, S. D., Khromova, T. N., Litvinov, I. A. and Litmanovich, A. D., *Vysokomol Soed B*, 1985, **27**, 225–7.
48. Ermakov, I. V. and Litmanovich, A. D., *Vysokomol Soed A*, 1988, **30**, 2595–601.
49. Ermakov, I. V., Lebedeva, T. L., Litmanovich, A. D. and Platé, N. A., *Polymer Sci.*, 1992, **34**, 513–16.
50. Winnik, M. A., in *Polymer Surfaces and Interfaces*, Eds. W. J. Feast and H. S. Munro, Wiley, New York, 1987.
51. Eisenberg, A., Yokoyama, T. and Sambalido, E., *J. Polymer Sci. A-1*, 1969, **7**, 1717–28.
52. Binder, K. and Sillescu, H., in *Encyclopedia of Polymer Science and Engineering*, 2nd edn, Wiley, New York, 1989, Suppl. Vol., pp. 297–315.
53. Klein, J., *Science*, 1990, **250**, 640–6.
54. Crank, J., *The Mathematics of Diffusion*, Clarendon Press, Oxford, 1975.
55. Kramer, E. J., Green, P. F. and Palmstrom, C., *Polymer*, 1984, **25**, 473–80.
56. Binder, K., *J. Chem. Phys.*, 1983, **79**, 6387–409.
57. Brochard, F., Jouffroy, F. and Levinson, P., *Macromolecules*, 1983, **16**, 1638–41.
58. Jilge, W., Carmesin, I., Kremer, M. and Binder, K., *Macromolecules*, 1990, **23**, 5001–13.

59. Hess, W., Nagele, G. and Akcasu, A. Z., *J. Polymer Sci. B Polymer Phys.*, 1990, **28**, 2233–45.
60. Murshall, U., Fisher, E. W., Herkt-Maetsky, C. and Fytas, G., *J. Polymer Sci. Polymer Lett. Ed.*, 1986, **24**, 191–7.
61. Fytas, G., *Macromolecules*, 1987, **20**, 1430–1.
62. Jones, R. A. L. and Kline, J., *Nature*, 1986, **321**(6065), 16.
63. Jordan, A. E., Ball, R. C., Donald, A. M., Feetters, L. J., Jones, R. A. L. and Kline, J., *Macromolecules*, 1988, **21**, 235–9.
64. Yashin, V. V., Ermakov, I. V., Litmanovich, A. D. and Platé, N. A., *Vysokomol. Soed. A*, 1994, **36**, 955–63.
65. Glansdorff, P. and Prigogine, I., *Thermodynamic Theory of Structure, Stability and Fluctuations*, Wiley–Interscience, New York, 1971.
66. Bazarov, I. P., *Termodinamika (Thermodynamics)* 4th edn, Vysshaya Shkola, Moscow, 1991.
67. Brochard-Wyart, F., in *Studies in Polymer Science*, Elsevier, Amsterdam, 1988, Vol. 2, pp. 249–56.
68. Korshak, V. V., Berestneva, G. L., Bragina, I. P. and Zhuravleva, I. V., *Dokl. Akad. Nauk. SSSR*, 1973, **211**, 598–601.
69. Korshak, V. V., Berestneva, G. L., Bragina, I. P. and Astafiev, S. A., *J. Polymer Sci. Part C*, 1974, **47**, 25–34.
70. Lius, L. A., Bessonov, M. I. and Florinsky, F. S., *Vysokomol. Soed A*, 1971, **13**, 2027–34.
71. Lavrov, S. V., Talankina, O. B., Vorobyov, V. D., Izyumnikov, A. L., Kardash, I. E. and Pravednikov, A. N., *Vysokomol Soed A*, 1980, **22**, 1886–90.
72. Tsapovetsky, M. I., Lius, L. A., Bessonov, M. I. and Kotton, M. M., *Dokl. Akad. Nauk. SSSR*, 1981, **256**, 912–15.
73. Kamogava, H., Kato, M. and Sugiyama, H., *J. Polymer Sci. A-1*, 1968, **6**, 2967–91.
74. Paik, C. S. and Morawetz, H., *Macromolecules*, 1972, **5**, 171–7.
75. Eisenbach, C. D., *Ber. Bunsenges Phys. Chem.*, 1980, **84**, 680–90.
76. Monahan, A. R., *J. Polymer Sci. A-1*, 1966, **4**, 2381–90.
77. Monahan, A. R., *J. Polymer Sci. A-1*, 1967, **5**, 2333–41.
78. Otsu, T., Yasuhara, Y., Shiraishi, K. and Mori, S., *Polymer Bull.*, 1984, **12**, 449–56.
79. Otsu, T. and Shiraishi, K., *Macromolecules*, 1985, **18**, 1795–6.
80. Ermakov, I. V., Yakubovich, O. V., Salamatina, O. B., Fateev, O. V. and Litmanovich, A. D., *Vysokomol Soed A*, 1989, **31**, 793–8.
81. Ermakov, I. V., Thesis, Moscow, 1990.
82. Ermakov, I. V., Yashin, V. V., Litmanovich, A. D. and Platé, N. A., *J. Polymer Sci. Part A Polymer Chem.*, 1993, **31**, 395–401.
83. Antipov, E. M., Ermakov, I. V., Kuptsov, S. A., Litmanovich, A. D. and Platé, N. A., *Dokl. Akad. Nauk. SSSR*, 1990, **311**, 382–4.
84. Delmon, B., *Introduction a la Cinetique Heterogene*, Editions Techniq, Paris, 1969, Chap. 10.
85. Nagasawa, M., Koizuka, A., Matsuura, K. and Horita, M., *Macromolecules*, 1990, **23**, 5079–82.
86. Friedman, B. and O'Shaughnessy, B., *Macromolecules*, 1993, **26**, 5726–39.
87. Painter, P. C., Park, Y. and Coleman, M. M., *Macromolecules*, 1988, **21**, 66–72.

CHAPTER 7

Intermacromolecular Reactions

7.1 INTRODUCTION

Reactions between macromolecules are widely used to synthesize various polymer materials. The processes of polycondensation at high conversions, hardening of resins and varnishes, etc., are based on reactions between end groups of oligomer and polymer chains as well as between the end groups and active groups included in structural units of the chains. Depending on their chemical structure, macromolecules may participate in various reactions producing block, graft, branched and crosslinked polymers. To immobilize physiologically active high-molecular species, e.g. proteins, on macromolecules, intermacromolecular reactions are used as well.

All the reactions mentioned are usual chemical reactions. The reactivity of chemical groups belonging to structural units of macromolecules is more or less affected by chain structure of the macromolecules, resulting in quantitative changes in thermodynamical and kinetical characteristics of the reactions caused by the effects of surrounding, neighboring groups, conformation, etc., regarded in Chapter 2.

Another situation arises in the case of intermacromolecular reactions proceeding with formation of double–stranded structures stabilized by sequences of relatively weak noncovalent bonds between monomer units of interacting chains (Scheme 7.1), where A and B represent monomer units of corresponding polymer chains containing atoms, or groups of atoms, having an affinity with one another:

Scheme 7.1

Macromolecules (—A—) and (—B—) are usually named 'complementary' and the product of their interaction-interpolymer complex' (IPC, IP complex), or 'polycomplex'. As an example, fragments of IP complexes stabilized by bonds of a different nature are shown in Scheme 7.2. IP complex formed by poly(ethylene oxide) (PEO) and poly(methacrylic acid) (PMAA) is stabilized by H bonds; that formed by poly(4-vinylpyridinium) salts (PVPy) and poly(acrylic acid) (PAA), by electrostatic interactions:

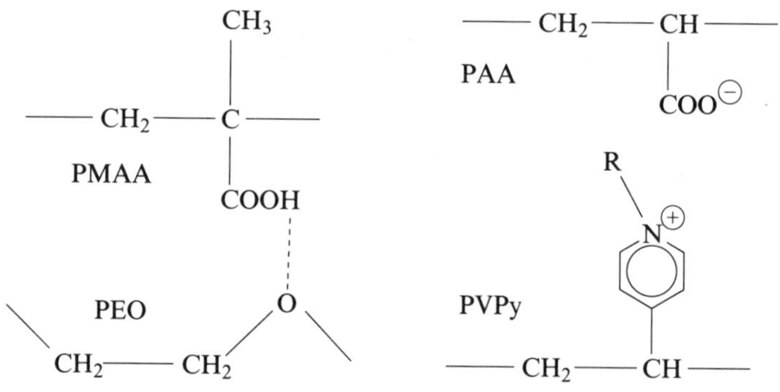

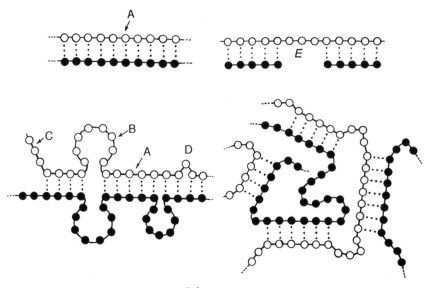

Scheme 7.2

Depending on the structure of polymer components and conditions of interpolymer interaction, the IP complex may represent ideal or quasi-ideal linear double-stranded structures, quasi-linear structures comprising relatively long sequences of intermolecular bonds (A) and imperfections such as loops (B), free tails (C), and single nonbound monomer units (D), block-copolymer-like structures consisting of blocks of double-stranded structures and free segments of

Scheme 7.3

long-chain component (E) (so-called 'nonstoichiometric complexes') or networks comprising double-stranded fragments (see Scheme 7.3).

Two main characteristic features of double-stranded structures are: (a) the cooperativity of the systems of interchain bonds in uninterrupted sequences shown in Scheme 7.1, providing high stability of the systems regardless of low energy of each single bond, and (b) the possibility of structural rearrangements in IP complexes due to realization of equilibria on the level of segments of macromolecules (see Scheme 7.4), thus providing formation of thermodynamically optimal structures and substitution of one macromolecule in the IP complex by another.

Scheme 7.4

The duality is peculiar for the behavior of interacting complementary macromolecules: on the one hand, their behavior is such as to show that a macromolecule participates in the interaction as a whole; on the other hand, the reversible association–dissociation process can proceed inside the IP complex on the level of segments or even monomer units of macromolecules.

These features result in a qualitative change in the thermodynamic behavior of intermacromolecular interactions in comparison with interactions in the systems in which one of the macromolecular reagents (or both) is replaced by its low-molecular analog. New specific thermodynamic properties of the interactions are: strong dependence of the stability of the products on the lengths of interacting macromolecules, their very high stability regardless of low energy related to a single interchain bond and molecular recognition manifesting in extremely high selectivity of intermacromolecular interactions with regard to the length and the structure of interacting polymer chains.

Based on the features of the interactions, very effective macromolecular reagents may be created for extensive practical applications such as fractionation of nonuniform polymers and separation of composite mixtures of macromolecular species of both synthetic and biological origin [1, 2], precise addressing of physiologically active macromolecules to certain molecular structures in living organisms [3, 4], neutralization of the heparine in blood [5], etc.

It is noteworthy that Nature chose intermacromolecular interactions to make the basis for creation of unique composite molecular structures and processes playing a decisive role in living organisms, e.g. providing the storage and reproduction of genetic information. The selectivity of interactions between biological macromolecules (e.g. nucleic acids), i.e. their ability to recognize a certain molecular partner, is known to be brought to perfection. Knowledge of the principles and mechanisms of molecular recognition on the level of simple polymer systems seems to be of key significance for understanding the causes that have determined the speed of pre-biological evolution of macromolecules [6].

Returning to usual synthetic macromolecules, it should be noted that IP complexes represent a special case of individual polymer compounds, usually characterized by definite composition. They may be regarded as ladder-like polymers and blends in the same time due to lability of the systems of interchain bonds providing their rearrangements; moreover, these products, being very stable in some definite conditions, may be separated on to free chain components under special conditions. A variety of IP complexes may be created as precursors in which weak interchain bonds may be replaced via chemical transformation by stable covalent bonds, thus producing stable ladder structures.

In general, usual chemical reactions taking place in IP complexes have their own special properties caused by different reactivities of structural units of one and the same macromolecule situated in double-stranded regions and various imperfections of the structure. This provides spatial separation of different reactions in which the units may participate. Intramolecular stresses arising or decreasing in the course of these reactions may affect their relative rates as well.

In this chapter, the main attention will be paid to thermodynamic aspects of intermacromolecular reactions, molecular recognition and mechanisms of the reactions. Taking into account the specific properties of the reactions proceeding in IP complexes, their features will be regarded as well in a special section of this chapter.

7.2 STATISTICAL THERMODYNAMICS OF INTERPOLYMER INTERACTIONS

One of the general thermodynamical features of interpolymer interactions is practically a complete shift of the equilibria toward the formation of products, i.e. high stability of the products—IP complexes—even in dilute solutions, whereas in similar interactions between low-molecular analogs (or if only one of the reactants is low-molecular) stable products do not form. The reason for the stability is the cooperativity of the system of bonds in the structures shown in Scheme 7.1. In such structures, the rupture of any bond $A \cdots B$ is not counterbalanced by an increase in the entropy of the system, as it is in the case when one of the components, or both, are low-molecular. Appreciable conformational mobility of a chain segment (and, then, some increase in the entropy) appears only at simultaneous breaking of sufficiently long sequence of $A \cdots B$ bonds, but configurational entropy of the system arises at total dissociation of the IP complex to free macromolecular components.

The dependence of the equilibrium constant on chain lengths of interacting macromolecules P_1 and P_2 in the reaction

$$P_1 + P_2 \rightleftharpoons IPC(P_1 P_2)$$

Scheme 7.5

is of key significance in understanding the reasons for all unusual thermodynamical properties of interpolymer interaction, and the high stability of the products as well. For this purpose, the most productive is investigation of the equilibrium

$$\text{oligomer} + \text{polymer} \rightleftharpoons IP \text{ complex}$$

Scheme 7.6

depending on the degree of polymerization of relatively short-chain polymers (conditionally called 'oligomers'). Scheme 7.6 may be regarded as an adsorption of the oligomer chains on a long polymer chain considered as a one-dimensional lattice.

The reversible adsorption of an oligomer (with the d.p. equal to n) on a one-dimensional lattice was considered [6, 7] on the basis of the theory of complex equilibria using the model of adsorption:

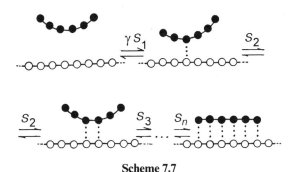

Scheme 7.7

where γS_1 is the probability of the first bond formation between the oligomer and the polymer, γ being dependent on concentration and S_1 on the bond energy; S_i is the probability of the ith bond formation following $i-1$ bonds. The value of S_i depends on the energy Δg_i of the ith bond: $S_i = \exp(-\Delta g_i/kT)$.

If the polymer and oligomer are homopolymers, then it may be assumed that

$$\Delta g_1 = \Delta g_2 = \cdots = \Delta g_n = \Delta g$$

and

$$S_1 = S_2 = \cdots = S_n = S$$

In dilute solutions $\gamma S \ll S$.

According to the theoretical analysis of the model [6, 7], even with the single interchain bond energy only slightly exeeding kT (i.e. $S > 1$) an oligomer chain exists at equilibrium practically in two states, namely free or completely bound, because the statistical weights of any intermediate state is negligibly small. Then the interaction Scheme 7.8 may be represented as:

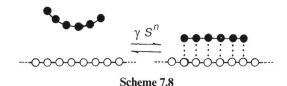

Scheme 7.8

and all the intermediate states of Scheme 7.7 may be neglected.

To derive the adsorption equation for the oligomer–polymer interaction, some assumptions concerning the adsorption model should be adopted:

(a) The IP complex is double-stranded (no difficulties arise, for example, for triple-stranded IP complexes).

(b) Oligomer chain with d.p. n is much shorter than that of the polymer; all the oligomer chains are of the same length.

(c) All the polymer chains present in solution are to be replaced by a unique long chain (i.e. by an uninterrupted one-dimensional adsorptional lattice) to neglect the end effects.

(d) An oligomer chain with the d.p. equal to n occupies a polymer chain segment of equal d.p., i.e. the stoichiometry of the IP complex at mutual saturation of the reactants is $1:1$ with respect to monomer units; two or more oligomer chains can not occupy one and the same monomer unit of the polymer chain. (Deviation from $1:1$ stoichiometry may be easily taken into account in the adsorption equation to be derived.)

(e) The energy of an interpolymer bond does not depend either on the length of adsorbed oligomer chain or on the number of oligomer chains associated with the polymer. Note that there are many reasons why one of the conditions or both are not fulfilled in real systems. The reasons will be discussed later as well as resulting effects in behavior of an oligomer–polymer system.

(f) Under interaction, an oligomer can occupy any free place on the polymer chain, if this place is not shorter than the chain of the oligomer (i.e. the adsorption does not take place on a free segment situated between two occupied polymer segments, if the d.p. of the free segment is less than n). Note that this condition seems to be 'working' but not at high degrees of filling of one-dimensional adsorbent (polymer) by oligomer chains.

Now let us consider a solution containing N_s molecules of solvent, N_{ol}^o chains of oligomer with d.p. n and one polymer chain with d.p. N_p^o which represents all polymer chains joined into one to neglect end effects. If ΔG is the change in free energy of the system after adsorption of $N_{ol(a)}$ oligomers on the polymer chain, the equilibrium condition can be written as

$$\frac{d\Delta G}{dN_{ol(a)}} = 0 \tag{7.1}$$

The total change in free energy can be represented as

$$\Delta G = \Delta H - T(\Delta S_1 + \Delta S_2) = \Delta G_1 - T\Delta S_2 \tag{7.2}$$

where $\Delta G_1 = \Delta H - T\Delta S_1$, ΔH is the change in enthalpy of the system, ΔS_2 and ΔS_1 are the changes in entropy caused respectively by a change in the number of configurational (translational) and all other (conformational, etc.) degrees of freedom under formation of the IP complex.

Combining Eqs. (7.1) and (7.2),

$$\frac{d\Delta G_1}{dN_{ol(a)}} = T\left(\frac{d\Delta S_2}{dN_{ol(a)}}\right) \tag{7.3}$$

Assume ΔH and ΔS_1 to be proportional to the total quantity of IP complex formed, i.e. to the number of unit moles of the oligomer $q_{ol(a)}$ incorporated in the IP complex:

$$\Delta G_1 = q_{ol(a)}\Delta G_1^o \tag{7.4}$$

where ΔG_1^o is the change in free energy of the system, when one unit mole of the IP complex is formed. By definition,

$$\Delta S_2 = k \ln\left(\frac{W_{ol(s)}'' W_{ol(a)}}{W_{ol(s)}'}\right) \tag{7.5}$$

where $W_{ol(s)}'$ and $W_{ol(s)}''$ represent the numbers of arrangements of free oligomers in the solution (before and after part of the oligomer chains has been bound in the IP complex) and $W_{ol(a)}$ is that of bound ('adsorbed') oligomer chains.

All changes in conformational degrees of freedom are taken into account in ΔS_1, so, for N_{ol} oligomer chains in a dilute solution containing N_s molecules of solvent,

$$W_{ol(s)} = \frac{xN_{ol}(N_s/x + N_{ol})!}{(N_s/x)!N_{ol}!} \tag{7.6}$$

where x is the number of elements in the oligomer, each of which occupies a volume equal to the volume of a solvent molecule.

As assumed, an oligomer chain occupies on a one-dimensional adsorbent (the polymer) any sequence of free n monomer units. The sequence (segment of the polymer) occupied by one oligomer chain may be regarded as a new single structural unit, comprising n monomer units:

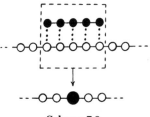

Scheme 7.9

Then after adsorption of $N_{ol(a)}$ chains, the polymer chain may be represented as a chain consisting of $N_p^o - nN_{ol(a)}$ free monomer units and $N_{ol(a)}$ occupied ones.

Taking into account the fact that one and the same monomer unit of the polymer cannot be occupied by two or more oligomers,

$$W_{ol(a)} = \frac{[N_p^o - (n-1)N_{ol(a)}]!}{(N_p^o - nN_{ol(a)})!N_{ol(a)}!} \tag{7.7}$$

This way of calculating $W_{ol(a)}$ was proposed in reference [8]. Substituting in Eq. (7.5) the appropriate thermodynamic probabilities for the expressions (7.6) and (7.7) and then combining Eqs. (7.3), (7.4) and (7.5) one obtains, after suitable transformations and differentiations,

$$\frac{\beta[1 - (1 - 1/n)\beta]^{n-1}}{m_{ol}(1 - \beta)^n} = \exp\left(\frac{-n\Delta G_1^o}{RT}\right) \tag{7.8}$$

where m_{ol} is the unit mole fraction of free oligomer in solution in the equilibrium state and β is the degree of saturation of the polymer (adsorbent) by the oligomer. With the assumed 1:1 stoichiometry of the IP complex,

$$\beta = \frac{nN_{ol(a)}}{N_p^o} = \frac{q_{p(a)}}{q_p^o} = \frac{q_{ol(a)}}{q_p^o} \tag{7.9}$$

where $q_{p(a)}$ and q_p^o are the numbers of unit moles of the polymer incorporated in the IP complex and total respectively.

When $n \gg 1$ and $n(1 - \beta)/\beta \gg 1$, i.e. if the oligomer chains are sufficiently long (remember that the term 'oligomer' is conditional and indicates only that n is much less than the d.p. of the polymer) and β is not too high, Eq. (7.8) can be transformed [9] to

$$\frac{\beta}{m_{ol}(1 - \beta)}\exp\left(\frac{\beta}{1 - \beta}\right) = \exp\left(\frac{-n\Delta G_1^o}{RT}\right) \tag{7.10}$$

The right-hand side of Eq. (7.10) represents the equilibrium constant for the reaction (7.6) with the d.p. of the oligomer equal to n:

$$K_n = \exp\left(\frac{-n\Delta G_1^o}{RT}\right) \tag{7.11}$$

or

$$K_n = K_1^n \tag{7.12}$$

where $K_1 = \exp(-\Delta G_1^o/RT)$ is the equilibrium constant in a formation of each next ith bond after $(i-1)$ already formed bonds between monomer units of a polymer and an oligomer chain under complexation. (In general K_1 is the effective value, see below.)

In comparison with the usual equation for the adsorption of low-molecular species, Eq. (7.10) includes the supplimentary term $\exp[\beta/(1-\beta)]$. This term seems to demonstrate 'sterical hindrance', i.e. a progressive decrease in the ratio of free polymer segments which are long enough to bind next oligomer chains at sufficiently high β.

A theoretical dependence of the equilibrium constant on the length of a macromolecule, analogs to Eq. (7.12), was obtained for fixed β (e.g. for $\beta = 0.5$) by various methods in the 1960s on deriving the equations to estimate the role of stacking interactions between end nucleotide groups in complexation of complementary oligo- and polynucleotides [6, 7, 10, 11].

The consequence of Eqs. (7.10) to (7.12) is a number of general thermodynamical properties of interpolymer interactions, such as:

Consequence 1. The equilibrium in reaction (7.6) must be strongly dependent on the d.p. of the oligomer and shift rapidly to the IP complex formation when a certain interval of the d.p. is reached.

Consequence 2. Very stable to dissociation IP complexes may be obtained even in dilute solutions at very low energy of interaction between monomer units of complementary macromolecules.

Consequence 3. Apart from direct A⋯B, low-energy interactions, others such as volume or hydrophobic interactions may contribute substantially to overall energy of IP complex stabilization, being commensurable with the energy of the interaction between the monomer units.

Consequence 4. Molecular recognition, i.e. very high selectivity of interpolymer interactions with respect to the length and the structure of a macromolecular partner, must be the general property of the interactions, even on the level of the simplest synthetic macromolecules. This is the direct consequence of the strong dependence of K_n both on K_1 (depending on the chain structure) and on n (the chain length).

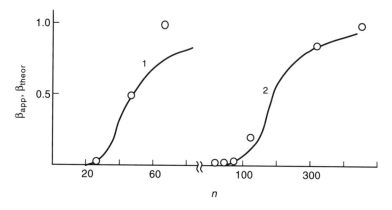

Fig. 7.1   Dependence of the IP complex stability on the oligomer chain length. Points are experimental values of β_{app} using data [1, 12] for the system PMAA–PEO (1) and PAA–PEO (2); solid lines are the theoretical dependences plotted using Eq. (7.10) when $\Delta G_1^\circ = -230$ (1) and -45 (2) J/unit-mol and the other conditions corresponding to the experimental ones. (Reprinted from *European Polymer J.*, **17**, 974, 1981, with kind permission from Elsevier Science Ltd, The Boulevard, Langford Lane, Kidlington OX5 1GB, UK)

The theory and experimental data concerning molecular recognition will be discussed separately in the next section of this chapter.

All of the published experimental data concerning the stability of IP complexes approve the correctness of consequence 1. Much data has been obtained in the 1960s dealing with complementary oligonucleotide–polynucleotide interactions as well as the interactions of common synthetic polymers [1, 2]. As an example, in Fig. 7.1 the dependences of β_{app} (apparent value of β) on the d.p. of PEO (the oligomer) for its interaction with PMAA or PAA (the polymers) are shown in comparison with correspondent theoretical dependences obtained with the use of Eq. (7.10). The values of β_{app} were calculated from the plots of η_{sp} versus n [1, 12] and the values of β_{theor}, using the value of K_1 calculated for $\beta_{app} = 0.5$ using Eq. (7.10). Good agreement between the theory and experiment is seen in Fig. 7.1. The dependence of the equilibrium constant on n described by Eq. (7.12) was confirmed experimentally both for interacting uncharged macromolecules and for polyelectrolytes [13].

Equation (7.12) predicts very sharp dependences of the stability of IP complexes on temperature, concentration and the nature of the solvent (including pH, ionic strength, etc.). It is the term 'melting temperature' that is often used to describe temperature dependence of the stability of IP complexes of complementary poly- and oligonucleotides (e.g. poly-U + oligo-A) (see Fig. 7.2) [14].

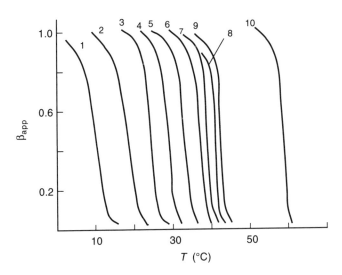

Fig. 7.2 Temperature stability of IP complexes of poly-U and oligo-A in water at d.p. of oligo-A of 4 (1), 5 (2), 6 (3), 7 (4), 8 (5), 10 (6), 11 (7), 12 (8), 13 (9), 300 (10). (Reproduced by permission of John Wiley & Sons Ltd from *Biopolymers*, 1973, **12**, 776)

Some experimental data concerning the reversible formation–dissociation of IP complexes PEO (oligomer)–PMAA (polymer) under variations of concentration, temperature and the nature of the solvent are shown in Fig. 7.3. The viscosity of the solution of noninteracting long-chain PMAA and relatively short-chain PEO is determined by high-molecular PMAA. Formation of the IP complex is accompanied with a decrease in the viscosity; when the interaction becomes complete, the viscosity becomes low and practically independent of molecular weights of the components due to the very compact structure of the IP complex particles [12, 15]. It can easily be seen from Fig. 7.3 that the intervals of IP complex formation–dissociation are relatively narrow regardless of very low absolute values of ΔG_1° (see Table 7.1).

The 'melting' of IP complexes in water with a decrease in temperature indicates the positive sign of the enthalpy of IP complex formation due to the decisive role of hydrophobic interactions in the complex stabilization [15] (see Consequence 3). In water–alcohol solutions, where the role of hydrophobic interactions becomes negligible, the 'melting' proceeds normally with an increase in temperature (compare Fig. 7.3a and c).

The interactions between charged macromolecules with formation of interpolyelectrolyte complexes were shown to be very sensitive to pH, ionic strength and even the nature of small ions [16, 17]. Thus, experimental dependences of IP

Table 7.1. Thermodynamic parameters for the reaction $P_1 + P_2 \rightleftharpoons IPC(P_1P_2)$

P_1	P_2	Solvent	T_{lim}(K)	ΔH_1^o (J/unit-mole)	ΔG_1^o (J/unit-mole)	$\ln K_1^*$
PMAA	PEO	H_2O	260 (lower limit)	$+700 \pm 90^a$ $+1160^b$ $+770^c$	$-140^{b,c}$	0.1^b
		$H_2O:CH_3OH$ 7:3	460 (upper limit)	-400 ± 90^a	-230^d	0.16^d
		$H_2O:CH_3OH$ 1:1	326 (upper limit	-1370^b	-115^d	0.08^d
PAA	PEO	H_2O	309 (upper limit)	$+300 \pm 90^a$	-45^d	0.03^d
		$H_2O:CH_3OH$ 7:3		-420 ± 90^a		
PMAA	PVPd	H_2O		$+230 \pm 90^a$	-700 ± 140^c	

^{a}According to calorimetric data [15].
^{b}According to the data on thermal stability of the IP complex.
^{c}According to the data on potentiometry [15].
^{d}Calculated from Eqs. (7.10) and (7.19).

complex stability on the oligomer chain length, temperature and concentration are, in general, satisfactorily described by Eqs. (7.10) to (7.12). However, there are the reasons why these equations may be used for quantitative treatment of experimental data and calculation of the thermodynamic parameters only if certain conditions are fulfilled; thermodynamic parameters, thus obtained, represent some effective values.

Some of the reasons are as follows:

(a) Defects in the structure of an IP complex, such as different-type loops formed by the unbound-in-a-complex parts of interacting macromolecules (see Scheme 7.3). Such defects may be present in the particles of an IP complex either due to equilibrium in the particle (see Scheme 7.4), which is a kind of microreactor [18, 19], or due to incomplete reaction in the particles. The latter may be caused by a decrease in intramolecular mobility of macromolecules in the process of development of continuous intermolecular bond sequences after formation of primary intermolecular bonds, which occasionally appear when the coils of complementary macromolecules come into contact with each other in solution.

The presence of imperfections such as loops, etc., does not exert an influence on type (7.11) or (7.12) equations. An IP complex composed of long macromolecules can be represented as a chain comprising Z segments sufficiently long to consider the average total energy ΔG_s of intermolecular bonds to be similar for all the

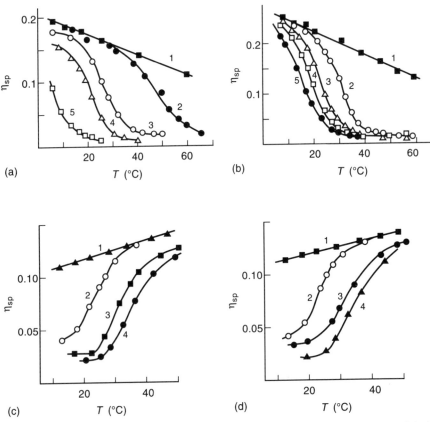

Fig. 7.3 Temperature stability of IP complexes PMAA–PEO in (a, b) water and (c, d) water–methanol 1:1 mixture at (a, c) different PEO chain lengths and (b, d) different PEO concentrations. PMAA concentrations are (a) 0.05 and (b–d) 0.1 g/dl. PEO concentrations are equimolar to PMAA in (a) and (c); in (b, d), 0.05 (2), 0.1 (3), 0.15 (4) 0.2(5) g/dl. PEO d.p. in (b. d) is 45; in (a, c), 45 (2), 70 (3), 90 (4), 140 (5). Curves 1 present solutions of PMAA without PEO. (a and b Reprinted from *European Polymer J.*, **17**, 974 (1981, with kind permission from Elsevier Science Ltd, The Boulevard, Langford Lane, Kidlington OX5 1GB, UK)

segments. Then

$$K_n = \exp\left(\frac{-Z\Delta G_s}{RT}\right)$$

and since Z is proportional to n, i.e. $Z\Delta G_s$ may be replaced by $n\Delta G^*$, then

$$K_n = \exp\left(\frac{-n\Delta G^*}{RT}\right) = K_1^n \qquad (7.13)$$

where $K_1 = \exp(-\Delta G^*/RT)$, ΔG^* and K_1 are corresponding effective values connected with the degree of nonperfection of the IP complex structure which depends not only on the nature of interacting chains but on the conditions of the IP complex formation and even on the length of interacting chains as well.

(b) If at least one of the participants of interpolymer interaction is a weak polyelectrolyte charged in the course of the interaction, $-\Delta g_i$ (and then S_i, see Scheme 7.7) in the general case should be a diminishing function of i due to contribution of the electrostatic energy to the value of Δg_i. The contribution must depend on the number of bonds i, i.e. on the number of charges appearing in the complexing segment of a weak polyelectrolyte chain. Many reactions of this kind have been investigated to date [20, 21]. Moreover, because of the decrease in $-\Delta g_i$ (or S_i) with the increase in i, after reaching some definite i the value of S_i may become equal to or even less than unity. This means that under proper conditions and with a certain length of oligomer chains, a partially bound state may predominate in the reaction products; in addition, the further development of uninterrupted sequences of interpolymer bonds may become thermodynamically unfavorable. For this reason, in the case of interaction between relatively long-chain oppositely charged polyelectrolytes, one of which (or both) being weak and charging during IP complex formation, the degree of conversion Θ under certain conditions is much less than unity and the equilibrium exists between bound and nonbound segments of the macromolecules [18, 19]. In the case when the length of the 'oligomer' becomes relatively high and Θ does not depend on the chain lengths of interacting participants, an IP complex can be represented as a chain consisting of Z segments with all consequences regarded above.

This means that, in reactions with participation of weak polyelectrolytes that one charged in the course of the reactions, K_1 must be a diminishing function of oligomer chain length for relatively short chains. However, after the length becomes sufficiently high, K_1 reaches some definite limit value and does not change practically with a further increase in d.p. of the 'oligomer' (of course, this term is arbitrary).

(c) Changing flexibility of macromolecules under complex formation and changing energy of the interaction of their units with the solvent. For example, if the interaction proceeds in water, the solubility of IP complexes is in general much lower than that of free macromolecules owing to screening of lyophilic groups in an IP complex. As follows from Eqs. (7.10) to (7.12), quite small energies of interaction A $\cdots$ B are sufficient to stabilize IP complexes. Thus, apart from the bond energy $\Delta \varepsilon_1$ (per unit mole), a considerable contribution to ΔG_1^o (which is an effective value) can be made, for example, by the energy of long-range interaction $\Delta \varepsilon_2$ between units or segments of double-stranded chains of an IP complex in a poor solvent, by the energy of hydrophobic interactions $\Delta \varepsilon_3$, if the solvent is water, and so on [1], i.e.

$$\Delta G_1^o = \sum_i \Delta \varepsilon_i \qquad (7.14)$$

For example, in PEO or PVPd (oligomers)–PMAA (polymer) systems the main contribution to stabilization of corresponding IP complexes in water is made by hydrophobic interactions (the formation reaction of these IP complexes in water is endothermic) [15]. The relative contribution of each of $\Delta\varepsilon_i$ to ΔG_1° varies with changing both the reaction medium and the degree of 'filling' of polymer with oligomer, i.e.

$$\Delta G_1^\circ = f(\beta) \qquad (7.15)$$

In some cases the value of ΔG_1° increases with increasing β so rapidly that, with the average $\beta < 1$, oligomer chains turn out to be distributed between polymer chains in accordance with the principle 'all or nothing', i.e. for one part of the polymer chains $\beta \approx 1$ and for the other one $\beta \approx 0$ [22, 23]. This 'self-organization' is rather typical of interpolymer interactions with participation of short and long chains [24, 25].

(d) Stacking interaction between contacting purine and pyrimidine end groups of the oligonucleotides adsorbed on complementary polynucleotide chain. In such a case the adsorption equation can be analytically derived only for certain β, e.g. for $\beta = 0.5$ [10, 11].

Therefore, taking into account the fact that ΔG_1° in general is a function of β, Eq. (7.10) has to be transformed as

$$\frac{\beta}{m_{ol}(1 - \beta)} = f(\beta)\exp\left(\frac{-n\Delta G_1^\circ(\beta)}{RT}\right) \qquad (7.16)$$

where $f(\beta)$ and $\Delta G_1^\circ(\beta)$ are functions of β [1] and their expressions are unknown. Hence, the experimental dependences of β on concentration, chain length of oligomer and temperature may be employed to find thermodynamic parameters only for a fixed value of β. Practically, it is more convenient to use particular equations for $\beta = 0.5$ as below:

(i) at constant temperature T_a:

$$\ln c_{ol}^* = A_1 n + B_1 \qquad (7.17)$$

where $A_1 = \Delta G_1^*/(RT_a)$, $B_1 = -\ln Z^*$; Z^* depends on the value of the function $f(\beta)$ when $\beta = 0.5$ and on how the concentration c_{ol}^* is expressed;

(ii) at constant oligomer length n_a:

$$\frac{1}{T^*} = A_2 \ln c_{ol} + B_2 \qquad (7.18)$$

where $A_2 = R/(n_a \Delta H_1^*)$, $B_2 = 1/T_{lim} + A_2 \ln Z^*$, $T_{lim} = \Delta H_1^*/\Delta S_1^*$;

(iii) at constant oligomer concentration $c_{ol(a)}$:

$$\frac{1}{T^*} = \frac{A_3}{n} + B_3 \qquad (7.19)$$

where $A_3 = R\ln(c_{ol(a)}Z^*)/\Delta H_1^*$, $B_3 = 1/T_{lim}$.

The asterisks in Eqs. (7.17) to (7.19) imply that the value of a given parameter is related to the condition $\beta = 0.5$. ΔH_1^* and ΔS_1^* are the effective values of enthalpy and entropy of IP complex formation, respectively, per unit mole of the oligomer or the polymer. T_{lim} is the temperature of IP complex semi-decomposition at infinite lengths of the participants; it may be the upper or the lower limit temperature of the IP complex stability depending on the sign of ΔH_1^* [1].

Plotting the experimental dependences in the coordinates of Eqs. (7.17) to (7.19), it is possible to determine ΔG_1^*, ΔH_1^*, ΔS_1^* and T_{lim} and to calculate the temperatures of semi-decomposition T^* for any n at given concentrations of the

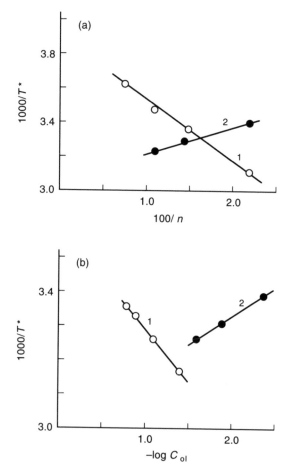

Fig. 7.4 The data given in Figs. 7.3(a) (a, 1), 7.3(b) (a, 2), 7.3(c) (b, 1) and 7.3(d) (b, 2) in the coordinates of (a) Eqs. (7.19) and (b) (7.18). (Reprinted from *European Polymer J.*, **17**, 974 (1981, with kind permission from Elsevier Science Ltd, The Boulevard, Langford Lane, Kidlington OX5 1GB, UK)

oligomer in a given solvent. Analogs to Eqs. (7.17) to (7.19) were derived first for a particular model of interaction between complementary oligo- and polynucleotides. The only difference is that in this model the value of parameter Z^* is dependent only on the stacking interaction energy, but as seen from the general consideration, unambiguous interpretation of the parameter may be invalid due to the possible role of all other factors (e.g. long-range or hydrophobic interactions) influencing the value of $\Delta G_1^0(\beta)$ at $\beta = 0.5$. As an example, the corresponding experimental dependences in the coordinates of Eqs. (7.17) to (7.19) are represented in Fig. 7.4. The experimental data in general satisfy the equations that make it possible to calculate thermodynamic parameters of interpolymer interactions for oligomer–polymer systems.

It is to be remembered, however, that the value of the equilibrium constant K_1 as well as the Gibbs energy, the enthalpy and the entropy of the interaction in general depends on β. This means that the values of corresponding parameters obtained by different methods and related to the formation of one unit mole of IP complex from $\beta = 0$ to $\beta = 1$ (e.g. calorimetry, potentiometric titration, etc.) or at $\beta = 0.5$ (e.g. by using Eqs. (7.17) to (7.19)) may not necessarily coincide, although, as seen in Table 7.1, in some cases their correspondence is satisfactory.

7.3 MOLECULAR RECOGNITION AS A SPECIFIC FEATURE OF INTERPOLYMER INTERACTIONS

7.3.1 GENERAL REMARKS

As was mentioned above, 'molecular recognition', i.e. high selectivity of interpolymer interactions, should be the consequence of exponential dependence of the stability of an IP complex on the length of macromolecules. In a limited case, the recognition may be schematically represented as

$$P + \sum_{i=1}^{n} P_i \rightarrow IPC(PP_k) + \sum_{i=1}^{k-1} P_i + \sum_{i=k+1}^{n} P_i$$

Scheme 7.10

where P is a macromolecule complementary to each of the macromolecules P_1, $P_2, \ldots, P_n$ present in the system, which differ from each other in chemical structure, degree of polymerization, or both.

In the case of nonuniform polymer P, recognition may be represented as

$$P + \tilde{P} \rightarrow IPC(P\tilde{P}_a) + \tilde{P}_b$$

Scheme 7.11

where $\tilde{P}_a$ and $\tilde{P}_b$ are fractions of $\tilde{P}$, bound into IP complex and free respectively. The recognition (selection) of some definite macromolecule P_k (Scheme 7.10) or fraction $\tilde{P}_a$ (Scheme 7.11) is possible if three conditions are fulfilled. The first is that the recognizing macromolecules P should be taken in deficiency, i.e. they must be saturated with a fraction of complementary macromolecules present in the reactive system. The second condition is that the IPC(PP$_k$) must be the most stable. In other words, the equilibrium

$$P + P_i \rightarrow IPC(PP_i)$$

Scheme 7.12

for $i = k$ is shifted to the right much more than for any other i. This is the thermodynamic condition. The third is the kinetic condition: the rates of all the substitution reactions

$$P_k + IPC(PP_j) \rightarrow IPC(PP_k) + P_j$$

Scheme 7.13

(P_j is any of P_i but not P_k) must be sufficiently high. This condition is fulfilled even at high stabilities of participating in the reaction IP complexes due to the existence of the special mechanism (see Section 7.4).

It is obvious that equations describing the equilibrium in the system consisting of recognizing polymer P and the mixture of polymers complementary to P (P_1, $P_2, \ldots$, etc.), are independent of the initial state of the system and should be equally valid for recognition and substitution reactions.

Theoretical analysis of the equilibria is more convenient if systems 'recognizing oligomer–mixture of polymers' and 'recognizing polymer–mixture of oligomers' are regarded separately. The reason is that when obtaining 'working' theoretical equations, different limit conditions have to be assumed for these systems concerning the degree of filling of one-dimensional adsorbent in the equilibrium state.

7.3.2 RECOGNITION WITH REGARD TO STRUCTURE IN THE SYSTEMS 'OLIGOMER–TWO OR MORE POLYMERS'

Consider the interaction of uniform oligomer P (d.p. $= n$) with a mixture of polymers $P_1, P_2, \ldots, P_n$ each of which is complementary to P. Any IP complex (PP$_i$) formed is in equilibrium with free oligomer. Then for any pair P–P$_i$ Eq. (7.10) may be used, i.e.

$$\frac{\beta_i}{m_{ol}(1 - \beta_i)} \exp\left(\frac{\beta_i}{1 - \beta_i}\right) = K_{1(i)}^n \tag{7.20}$$

where m_{ol} is the concentration of free oligomer in solution, β_i is the degree of filling of polymer P_i with the oligomer and $K_{1(i)}$ is the corresponding effective constant for given pair $P - P_i$. Since m_{ol} in equilibrium is the same for any P_i, then it is possible to obtain the value characterizing the relative selectivity of interpolymer interaction with P for any pair P_k and P_j:

$$\frac{B_k}{B_j} = \left(\frac{K_{1(k)}}{K_{1(j)}}\right)^n \tag{7.21}$$

where

$$B_i = \frac{\beta_i}{(1-\beta_i)} \exp\left(\frac{\beta_i}{1-\beta_i}\right)$$

In general, $K_{1(i)}$ is a function of β (see above) and Eq. (7.21) can not be used for quantitative calculations. However, Eq. (7.21) may be used in a limited case $(\sum_i \beta_i) \to 0$ (or, what is the same, each $\beta_i \to 0$) [26]. It corresponds to the addition of a very small portion of the oligomer to a large excess of polymers P_i. Then Eq. (7.21) may be simplified to

$$\lim_{(\Sigma_i \beta_i) \to 0} \left(\frac{\beta_k}{\beta_j}\right) = \left(\frac{K_{1(k)}}{K_{1(j)}}\right)^n \tag{7.22}$$

Obviously, in this equation each value of $K_{1(i)}$ corresponds to the condition where each $\beta_i \to 0$.

If only two polymers, P_1 and P_2, participate in the competition to be recognized by the oligomer P, then

$$\lim_{(\beta_1+\beta_2) \to 0} \left(\frac{\beta_1}{\beta_2}\right) = \left(\frac{K_{1(1)}}{K_{1(2)}}\right)^n \tag{7.23}$$

The right-hand side of the equation was named 'the factor of recognition' (or selection) Ψ [26]:

$$\Psi = \left(\frac{K_{1(1)}}{K_{1(2)}}\right)^n \tag{7.24}$$

Ψ characterizes quantitatively the relative preference of one of the polymers, P_1 or P_2, by the oligomer P under the interaction. Evidently,

$$\frac{m_{ol(1)}}{m_{ol(2)}} = \left(\frac{m_1}{m_2}\right)\Psi \tag{7.23a}$$

where m_1 and m_2 are unit-mole concentrations of the polymers and $m_{ol(1)}$ and $m_{ol(2)}$ are unit-mole concentrations of the oligomer bound with corresponding polymer under the condition $\beta_1 + \beta_2 \to 0$.

Taking into account Eq. (7.11),

$$\ln \Psi = \frac{-n\Delta\Delta G_1^{\circ}}{RT} \tag{7.25}$$

where $\Delta\Delta G_1^{\circ} = \Delta G_{1(1)}^{\circ} - \Delta G_{1(2)}^{\circ}$; $\Delta G_{1(1)}^{\circ}$ and $\Delta G_{1(2)}^{\circ}$ represent the effective free energies of interpolymer interaction calculated per unit mole of the corresponding IP complexes under the condition $\beta_i \to 0$. It follows from Eqs. (7.23) to (7.25)

Table 7.2. Investigated equilibria: IPC(PP$_1$) + P$_2$ ⇌ IPC(PP$_2$) + P$_1$

Number	P^a	P$_1$ ^{a}	P$_2$ ^{a}	Type of system^b	Comments^c	Reference
1	PMAA	PVPd, PEO, PVA		A	Row of reactivity PVPd > PEO > PVA	[27]
2	PMAA	PEO, PAAm, PVPd, PVA		A, C	Row of reactivity PEO (MM 1.4 × 10^6) > PAAm > PVPd > PEO (MM 3 × 10^3) > PVA	[28]
3	PMAA	PVPd, PEO, PAAm			Row of reactivity PVPd > PEO > PAAm	[29]^d
4	PAA	PVPd	PEO	A	↓	[30]
5	PMAA	P4VPy	P2VPy	A	↓	[28]
6	PEO	PMAA	PGA	B	↓	[28]
7	BPFR	PVPd	PEO	B	↓	[31]
8	PVPy	PVSA	PMAA	C	↓	[32]
9	PMAA	PVPd	PEO	A	← Under MM PEO 15 000, MM PVPd 40 000 → Under MM PEO 100 000, MM PVPd 5000	[33]
10	PVPd	PMAA	copolymers AA–MAA	B	← Under growth of AA content in copolymers ← Under growth of MM of PVPd	[26, 34]

11	PVPd	PMAA 6:41:53	PMAA 14:32:54	B	↓	[26,35]
12	PEO	PMAA 14:32:54	PMAA 6:41:53	B	←Under growth of MM PEO	[26,35]
13	PEAD	PVSA	PGA	A	→Under growth of pH	[36]
14	PMAA	PVPd	P2VPy	A	←Under pH = 3 →Under pH = 6	[28]
15	PMAA	PVPd	PEI	A	←Under pH = 2 →Under pH = 5	[28]
16	PVPy	PPh	PMAA	C	←Under growth of MM of PPh ←Under growth of [NaCl], [LiCl] →Under growth of [KCl]	[17]

[a] PVA, Poly(vinyl alcohol); PAAm, poly(acryl amide); PGA, polyglutamic acid); BPFR, *p*-bromphenolformaldehyde resin; PVSA, poly(vinyl sulphuric acid); PEAD, poly(2-diethylamino dextran), PPh, sodium polyphosphate. The ratio of iso-, hetero- and syndiotactic triads is given for PMAA samples in rows 11 and 12.

[b] Types of the systems: A, $n > n_1$, $n > n_2$; B, $n < n_1$, $n < n_2$; C, $n_1 > n > n_2$.

[c] In the row of reactivity, every other polymer is substituted with the preceding one in the IP complex; → and ← indicate partial or full shift of the equilibrium in the corresponding direction.

[d] Disagreement with the preceding rows are probably explained by the d.p. of polymers, which are not given in this work.

that the precision in selection of an optimal macromolecular partner by the oligomer increases very rapidly with an increase in d.p. of the oligomer if $K_{1(1)}/K_{1(2)} \neq 1$. Even at insignificant absolute values of $\Delta\Delta G_1^0$, such as tens of joules per unit mole, practically error-free recognition should occur at sufficiently high d.p. of the oligomer (remember that the term 'oligomer' is conditional; no objections exist to use Eqs. (7.22) to (7.25) even if the lengths of recognizing polymer P and competing for interaction polymers P_1, P_2,... are commensurable).

Some experimental data concerning recognition as well as substitution reactions of the type 'oligomer–two polymers' are represented in Table 7.2. Generally, if different homopolymers P_1 and P_2 are used and d.p. of the oligomer P is high enough, a complete recognition of one of the homopolymers is observed (see Table 7.2).

The recognition manifests itself in matrix polymerization as well. For instance, if macromolecules of two different matrices P_1 and P_2 are present in the polymerization medium, the growing chain P, being complementary to both P_1 and P_2, recognizes only the 'strongest' matrix (the IP complex with which is more stable). After the recognition step, the 'daughter' chain grows under the control of this 'strongest' matrix. For instance, growing chains of PMAA recognize macromolecules of PVPd under free-radical polymerization of MAA in the presence of both PVPd and PEO (or both PVPd and PVA) as matrices, although each of the matrices, taken separately, is capable of being recognized by a growing chain of PMAA [37]. It is obvious that the choice of preliminary prepared macromolecules by PMAA in analogous systems is the same (see Table 7.2).

In general, the molecular recognition in interpolymer interactions is so perfect that only negligible differences in structures of p_1 and p_2 provide a means to observe experimentally real competition between them for P at reasonable d.p. of the latter. Thus, Figs. 7.5 and 7.6 demonstrate the dependence of perfectness of molecular recognition (expressed by either β_1/β_2 at given concentrations of all participants or Ψ at $\beta_1 + \beta_2 \rightarrow 0$) on the chain length of oligomer-recognizer P and the difference in structures of competing polymers respectively. To obtain β_1/β_2 values, the method of polarized luminescence may be used [26, 34, 35]. Both structures labeled with luminescent markers p_1^* and p_2^* were used to study the reactions

$$IPC(Pp_1^*) + p_2 \rightarrow IPC(Pp_2) + p_1^* \qquad \text{Scheme 7.14a}$$
$$IPC(Pp_1) + p_2^* \rightarrow IPC(Pp_2^*) + p_1 \qquad \text{Scheme 7.14b}$$

to be sure that the equilibrium state was indeed reached and that the equilibrium state was not affected by a small concentration of markers covalently bound to macromolecules.

The data concerning the competition between two polymers–stereoisomers of PMAA for 'oligomer'–PEO depending on its d.p. are shown in Fig. 7.5. The microtacticity of the samples (iso : syndio : hetero-triads, %) corresponds to

6:53:41 (P_1) and 14:54:32 (P_2), i.e. the competing polymers are very similar in structure.

It is seen that the perfectness of recognition depends on the d.p. of recognizing oligomer, increasing with an increase in the length of its chains. The linear dependence of $\ln \Psi_0$ on n (the values of $\Psi_0 = \lim_{(\beta_1 + \beta_2) \to 0} (\beta_1/\beta_2)$ are found by corresponding extrapolation of the experimental curves) is in agreement with the

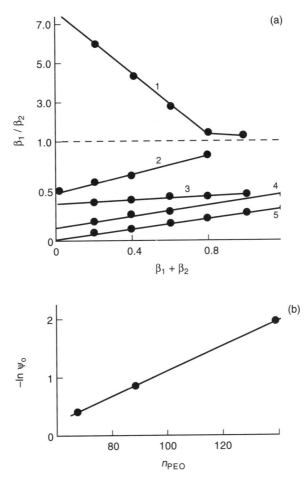

Fig. 7.5 (a) Dependence of degrees of filling macromolecules of PMAA samples of different microtacticity with oligomers PVPd (1) and PEO (2–5) on the total degree of filling of the polymers. β_2, the degree of filling PMAA with I/S/H = 14:54:32; β_1, the same for a sample with I/S/H = 6:53:41. The oligomer d.p.: 80 (1), 70 (2), 90 (3), 140 (4), 450 (5) b) Dependence of the factor of recognition Ψ on the d.p. for selective interactions of PEO with PMAA stereoisomers in the coordinates of Eq. (7.25). (Reproduced by permission of Nauka, Moscow, Russia, from *Dokl,. Akad. Nauk. SSSR*, 1979, **246**, 925)

theory. Note that the value of Ψ_0 for d.p. $= 450$, obtained by extrapolation of the linear dependence shown in Fig. 7.5(b), is of the order of $\exp(-10)$, which is practically zero; the same gives extrapolation of curve 5 in Fig. 7.5(a).

Calculated from Fig. 7.5(b)' the difference in effective free energies, $\Delta\Delta G_1^{\circ}$, of IP complex formation between PEO and the given stereoisomers of PMAA at $\beta_1 + \beta_2 \to 0$ is as small as ~ 40 J/unit-mole. Values of such order, being the reason

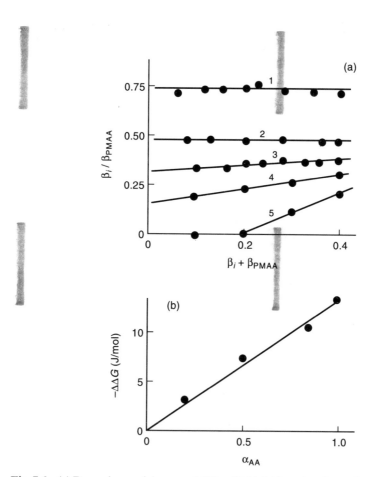

Fig. 7.6 (a) Dependence of degrees of filling PMAA (β_{PMAA}) and copolymers AA–MAA (β_i) in their mixtures with PVPd on the total degree of filling of the polymers. Current of units AA in copolymers: 20 (1), 50 (2), 85 (3, 5), 100 (4), mol %; d.p. of PVPd: 80 (1–4), 360 (5) b) Dependence of the difference between average IP-complex formation free energies per unit-mole, calculated from Eq. (7.25), on the composition of copolymer $\alpha_{AA} = [AA]/([AA] + [MAA])$. (Reproduced by permission of Nauka, Moscow, Russia, from *Vysokomol. Soed. A*, 1980, **22**, 1182)

for practically error-free recognition in interpolymer interactions with participation of sufficiently long macromolecules (up to 100% at $\beta_1 + \beta_2 \to 0$ for d.p. = 450; see Fig. 7.5b), never manifest themselves in reactions between low-molecular compounds and can even hardly be measured by usual methods.

The preference given to one of the stereoisomers (containing less iso-triads) by PEO seems to be caused by structural correspondence of interacting macromolecules since another oligomer, PVPd, under the same conditions preferably binds macromolecules containing more isotactic triads [26, 35].

It is seen from Fig. 7.6 that the perfectness of selection depends not only on the d.p. of the oligomer but also on the difference in structure of competing polymers. The data deal with competition between PMAA and methacrylic– acrylic acid (MA–AA) copolymers of different composition in order to be recognized by the 'oligomer'—PVPd; it is seen that the higher the content of monomer units of AA in the copolymer, the higher is the recognition of PMAA. The plot in Fig. 7.6(b) shows the linear experimental dependence of $\Delta\Delta G_1^0$ on the content of AA units in the copolymer. This means that for this binary copolymer the effective free energy of IP complex formation, $\Delta G_{1(1,2)}$, is the additive function of the copolymer composition, i.e.

$$\Delta G_{1(1,2)} = a\Delta G_{1(1)} + b\Delta G_{1(2)} \qquad (7.26)$$

where a and b are unit-mole fractions of the first and the second monomer units in the copolymer and $\Delta G_{1(1)}$ and $\Delta G_{1(2)}$ are the corresponding increments of free energy.

In principle, the reaction medium may affect the value $K_{1(1)}/K_{1(2)}$, thus influencing the perfectness of recognition and, moreover, changing the selectivity of the reaction with respect to the competitors, P_1 and P_2. In interactions with participance of polyelectrolytes, a change in the nature of mineral salts present in the reaction solution, in some cases changes the preference of one of the competitors (say, P_1) to another (say, P_2) in their recognition by P (see below).

7.3.3 RECOGNITION IN THE SYSTEMS 'POLYMER–TWO OR MORE OLIGOMERS'

In general, the problem of recognition in the systems under consideration is reduced to finding the relative probability of binding macromolecule P (here, 'polymer') with macromolecules ('oligomers') $P_1, P_2, \ldots, P_n$ present in the mixture if each of them, taken separately, is capable of interacting with P. If all macromolecules P_i are brought together under the term 'nonuniform oligomer $\tilde{P}$' and a continuous or discrete distribution function is introduced by nonuniformity parameters $w(\{x_i\})$ then the task is formulated in a different way: how will the distribution function $w(\{x_i\})$ be changed after binding part $\tilde{P}$ in an IP complex with P.

Consider solution $\tilde{P}$ and distribution function $w_0(\{x_i\})$ normalized to the total unit-mole concentration m° of oligomer:

$$\int_{\{x_i\}} w_0(\{x_i\})\,d\{x_i\} = m^\circ \tag{7.27}$$

Add to this solution complementary polymer P which is capable of forming the equimolar IP complex with $\tilde{P}$ (stoichiometry is not important since it must be only constant), its concentration being $m < m^\circ$. Then part of $\tilde{P}$ will be bound in an IP complex and the distribution function $w_s(\{x_i\})$ of $\tilde{P}$ remaining in the solution, normalized to $m - m^\circ$, should not be identical to the initial one, i.e.

$$\frac{1}{m^\circ - m} w_s(\{x_i\}) \neq \frac{1}{m^\circ} w_0(\{x_i\}) \tag{7.28}$$

Assuming that each 'oligomer' among $\tilde{P}$ is either fully bound in an IP complex or is fully in a nonbound state, the following correlation is valid for each fraction of oligomer $\tilde{P}$:

$$\frac{w_c(\{x_i\})}{w_s(\{x_i\})} = A \exp\left[\frac{-\Delta G(\{x_i\})}{RT}\right] \tag{7.29}$$

where $w_c(\{x_i\})$ is the distribution function of $\tilde{P}$ bound in the IP complex, normalized to m, $\Delta G(\{x_i\})$ is the free-complex-forming energy of fraction $\{x_i\}$ of oligomer $\tilde{P}$ with P. Hence, since

$$w_c(\{x_i\}) + w_s(\{x_i\}) = w_0(\{x_i\})$$

then

$$w_s(\{x_i\}) = \frac{w_0(\{x_i\})}{1 + A \exp[-\Delta G(\{x_i\})/RT]} \tag{7.30}$$

Parameter A denotes the normalizing multiplier, reflecting the fact that polymer P is in deficiency, i.e. that the IP complex phase is of limited 'holding capacity' with respect to 'oligomer'. If this capacity is unlimited then, in accordance with Boltzmann's statistics, $A = 1$; therefore $A < 1$ in the given case.

Equation (7.30) shows that if P is in deficiency then the IP complex-formation process is accompanied with factionation of $\tilde{P}$ by nonuniformity parameters. The efficiency of such fractionation or, in other words, the degree of recognition by polymer P of the optimum partners among $\tilde{P}$ will be determined by the type of function $\Delta G(\{x_i\})$.

In reality, the interest is given to resolve particular problems such as fractionation by molecular weight or by composition.

Recognition with Regard to Chain Length

Consider the interaction of polymer P with polydisperse oligomer $\tilde{P}$ under the following conditions:

(a) The d.p. of P is greater than the d.p. of the highest molecular weight fraction of $\tilde{P}$.

(b) $\tilde{P}$ is in excess with regard to P.

(c) Under conditions of an experiment, the degree of filling P with oligomers is close to 1, i.e.

$$(\sum_i \beta_i) = 1 \tag{7.31}$$

In this case, in Eq. (7.30), $\{x_i\} \equiv n$ and

$$\Delta G(n) = -nRT \ln K_1 \tag{7.32}$$

Then the MWD of the nonbound (in an IP complex) oligomer boils down to the follwing equation:

$$w_s(n) = \frac{w_o(n)}{1 + AK_1^n} \tag{7.33}$$

The normalization parameter A is determined by the correlation of initial concentrations of P and $\tilde{P}$, i.e. m and m^o respectively.

An analytical expression may be obtained to describe recognition in the mixture of polymer P and two monodisperse oligomers P_1 and P_2 with similar structures but different lengths n_1 and n_2, their initial concentrations being m_1^0 and m_2^0, respectively [38]. Providing the conditions (a) to (c) are fulfilled, the ratio of degrees of filling chains P with oligomers P_1 and P_2 or, what is the same, the ratio of quantities of these oligomers bound to polymer P, is expressed as

$$\frac{\beta_1}{\beta_2} = \frac{m_1}{m_2} K_1^{\Delta n} = \frac{m_1^o - \beta_1 m_p}{m_2^o - \beta_2 m_p} K_1^{\Delta n} \tag{7.34}$$

where m_p is the initial concentration of the polymer, m_1 and m_2 are equilibrium concentrations of respective oligomers, all concentrations being expressed in unit-mole fractions, and $\Delta n = n_1 - n_2$ [38]. In the particular case, if initial concentrations of all participants are the same (i.e. $m_1^o = m_2^o = m_p$), then only half of the total amount of oligomers can be bound in an IP complex and

$$\frac{\beta_1}{\beta_2} = K_1^{\Delta n/2} \tag{7.35}$$

Equations (7.34) and (7.35) show that the selectivity of interpolymer interactions with regard to chain length depends only on the difference in the d.p. but not on their absolute values. Although comprehensible, this result seems to be nevertheless surprising.

If macromolecules P_1 and P_2, competing for binding with P, are such that the chain lengths correlate as $n_1 > n_p > n_2$, where n_p is the d.p. of polymer P (the term 'oligomer' naturally cannot be applied to P_1), this represents the case when the stability of IPC(PP$_2$) is determined by the length of chain P$_2$ but the stability of IPC(PP$_1$) is determined by the length of chain P (because the stability is determined by the shortest chain component). The 'factor of recognition' may be estimated in this case as

$$\Psi = \frac{K_1^{n_P}(0)}{K_1^{n_2}(1)} \tag{7.36}$$

Note that constants K_1 here should be different because they are attributed to different filling degrees, i.e. $\beta = 0$ and $\beta = 1$ respectively. Therefore the efficiency of recognition depends on the difference between the lengths of chains P$_2$ and P but does not depend on the d.p. of P$_1$. Indeed, only high-molecular weight PEO binds in an IP complex after addition of PMAA with molecular weight of 6×10^5 to dilute solution of a mixture of the PEO samples with molecular weights 1.4×10^6 and 3×10^3 [28].

Calculation of the P-bound fraction in the general case, i.e. in the case of polydisperse oligomer, is impossible analytically but may be carried out on a computer if $w_0(n)$, K_1 and initial concentrations of the participants are given.

As an example, the result of such a calculation for the case of the most probable distribution of the oligomer $(\bar{M}_w/\bar{M}_n = 2)$ and for the value of K_1 which is close to K_1 obtained experimentally for interactions of PAA or PMAA with PEO or

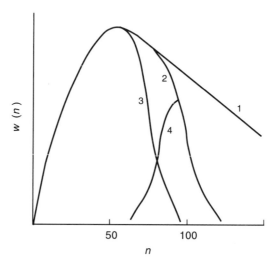

Fig. 7.7 Changing MWD of 'oligomer' under IP-complex formation, calculated from Eq. (7.33) for the initial $w_0(n) = 4 \times 10^{-4} \exp(-0.02n)$: 1, with $\ln K_1 = 0.22$; 2 and 3, MWD of the 'oligomer' not bound in an IP complex after addition of the first $(m = 0.385m^o)$ and the second $(m = 0.115m^o)$ portions of the polymer; 4, MWD of the 'oligomer' bound in an IP complex with the second portion of the polymer. (Reproduced by permission of Nauka, Moscow, Russia, from *Vysokomol. Soed. A*, 1977, **19**, 719)

PVPd in water solutions [9] is shown in Fig. 7.7. The distribution functions in Fig. 7.7 are normalized to the quantity of the oligomer; with such a normalization, the choice of the Y-axis scale is of no significance since it depends only on the total quantity of the oligomer. In this case the changes in the MWD of the oligomer when its fraction is bound in an IP complex are more illustrative. It is clear from Fig. 7.7 that consecutive addition of the polymer to the oligomer solution results in consecutive narrowing of the MWD of the free oligomer due to binding of high-molecular weight fractions in the IP complex. The greater the value of K_1 and the smaller the portion of the polymer, the more efficient is the fractionation [38].

Figure 7.8 shows changes in the MWD of PVPd (oligomer) after binding part of it in an IP complex with PAA (polymer) [39]. It is clear that the experimental pattern (Fig. 7.8) corresponds to the theoretical one (Fig. 7.7) i.e. the high-molecular weight fraction binds in an IP complex while the low-molecular one remains in the solution.

The efficiency of fractionation, being dependent on the value of K_1, must be sensitive to the experimental conditions. In fractionation of PEO with initial $\bar{M}_w/\bar{M}_n = 20.2$ by PAA in water solution, a significant narrowing of the MWD of PEO remaining in the solution was observed with an increase in temperature: from $\bar{M}_w/\bar{M}_n = 3.43$ at 20 °C to 2.46 at 50 °C [40]. This is in good agreement with

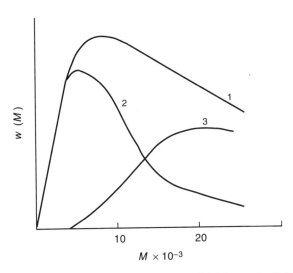

Fig. 7.8 Changing MWD of PVPd: 1, MWD of the initial sample; 2, MWD of PVPd remaining in the solution after binding in an IP complex with PAA and removal of the IP complex; 3, MWD of PVPd bound in an IP complex with PAA. (Reproduced by permission of Nauka, Moscow, Russia, from *Vysokomol. Soed. B*, 1978, **20**, 83)

the theory since the value of K_1 for these two interacting polymers increases with an increase in temperature due to positive enthalpy of the interaction in water.

Recognition with Regard to Chemical Structure

Two particular cases will be considered here. The first is recognition in the case when macromolecules of two or more 'oligomers', each being uniform both in the d.p. and structure, participate in competition for interaction with complementary polymer; the second is fractionation of d.p.-uniform binary statistical copolymer ('oligomer') via binding its fractions in an IP complex with consequently adding portions of complementary polymer. The second case is of practical interest as it represents the most frequently desired operation.

Recognition of an oligomer by a polymer in a mixture of two or more oligomers
The theoretical equation describing the recognition by polymer P of an optimum partner in a mixture of two different in structure oligomers P_1 and P_2 with d.p. n_1 and n_2 respectively, may be derived analytically using a similar procedure to that described in Section 7.2. Some additional conditions should be accepted to avoid difficulties dealing with deriving the equation as well as its application:

(i) oligomers P_1 and P_2 are both in excess with respect to polymer P;
(ii) in the equilibrium state, polymer chains are saturated by P_1 and P_2, i.e. $\beta = \beta_1 + \beta_2 \to 1$;
(iii) the free energy of IP complex formation ΔG is an additive function with regard to composition, i.e.

$$\Delta G = q_p^o(\beta_1 \Delta G_{1(1)} + \beta_2 \Delta G_{1(2)}) \qquad (7.37)$$

where q_p^o is total number of unit-moles of P (remember that, due to $(\beta_1 + \beta_2) \to 1$, all polymer chains are incorporated in the IP complex) and $\Delta G_{1(1)}$ and $\Delta G_{1(2)}$ are respective increments of free energy, i.e. effective free energies referred to the formation of one unit-mole of respective IP complex at $\beta_i \to 1$ and independent of composition of the resulting IP complex.

Condition (ii) arises due to dependence of the effective energy of polymer–oligomer interaction on β in the general case. As was stated in Section 7.2, quantitative calculations are possible at $\beta = $ constant. From a practical point of view, the condition $(\beta_1 + \beta_2) \to 1$ is most convenient because it may easily be reached under condition (i). Condition (iii) is not self-evident, because in real systems the increments may depend on β_1/β_2.

Taking into account conditions (i) to (iii) and using calculations similar to those described in Section 7.2, the equation of recognition may be derived [2] as

$$\lim_{(\beta_1 + \beta_2) \to 1} \left(\frac{\beta_1}{\beta_2}\right) = \frac{m_{1(s)}}{m_{2(s)}} \Psi \qquad (7.38)$$

where $m_{1(s)}$ and $m_{2(s)}$ are equilibrium unit-mole concentrations of free P_1 and P_2, and Ψ is the factor of recognition

$$\Psi = \frac{K_{1(1)}^{n_1}}{K_{1(2)}^{n_2}} \qquad (7.39)$$

Obviously, here the values of $K_{1(i)} = \exp(-\Delta G_{1(i)}/RT)$ correspond to the condition $\beta = \beta_1 + \beta_2 \to 1$.

At the large excess of both P_1 and P_2, the decrease in their concentrations in the free state after complexation may be neglected and

$$\lim_{(\beta_1 + \beta_2) \to 1} \left(\frac{\beta_1}{\beta_2}\right) = \frac{m_1^o}{m_2^o} \Psi \qquad (7.38a)$$

If $n_1 = n_2 = n$, i.e. oligomers P_1 and P_2 are of the same d.p., then

$$\Psi = \left(\frac{K_{1(1)}}{K_{1(2)}}\right)^n \qquad (7.39a)$$

This value represents quantitatively the relative stability of IP complexes (PP_1 and PP_2).

Experimental data concerning the recognition (or substitution, which is a particular case of recognition) in the systems of type under consideration are given in Table 7.2.

It follows from Eq. (7.39) that the value of the factor of recognition depends on the d.p. of the oligomers; an example of model calculation is shown in Fig. 7.9. It is easily seen that the polymer recognizes oligomer P_1 ($\Psi > 1$) or oligomer P_2 ($\Psi < 1$) depending on their lengths or does not give any preference ($\Psi = 1$) at definite values of n_1 and n_2. Thus, the 'recognizability' of an oligomer increases with its d.p. at a constant d.p. of another oligomer.

Inversion of recognition in water solution comprising PVPd(P_1), PEO(P_2) and PMAA(P) was observed experimentally [28, 33]. At a fixed d.p. of PVPd, macromolecules of PMAA did not get sight of PEO if the chains of this 'oligomer' were commensurable in length with PVPd or were shorter, but recognized long-chain PEO. The same situation was observed for the system comprising PVPd(P_1), PAAM(P_2) and PMAA(P) [28].

Inversion of recognition (selection) also takes place under changing external conditions. $K_{1(1)}$ and $K_{1(2)}$ may depend on temperature, pH, ionic strength, etc., in a different way. Therefore, changes in these parameters bring about considerable changes in value of the factor of recognition. The most trivial one is the case when changes in external conditions result in the transition from the range where IPC(PP_1) is stable and IPC(PP_2) is not, to the range where IPC(PP_2) is stable but IPC(PP_1) is not.

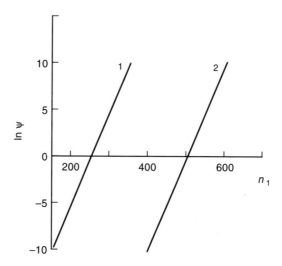

Fig. 7.9 Dependence of the factor of recognition of n_1 calculated from Eq. (7.39) at $\Delta G^\circ_{1(1)} = -140\,\text{J/unit-mole}$, $\Delta G^\circ_{1(2)} = -700\,\text{J/unit-mole}$, $n_2 = 50$ (1) and 100 (2)

Thus, IP complexes stabilized by H bonds are stable in an acid medium and decompose in a neutral one, and IP complexes of weak polyacids (such as PAA or PMAA) with polybases are, on the contrary, stable in a neutral medium but unstable in an acid one. Therefore, along with pH change, an inversion of selection takes place (see examples 14 and 15 in Table 7.2). This effect may be used for preparative separation of one of the macromolecular components from the IP complex [41] whereas other methods of separation, e.g. selective precipitation from a solution of dissociated IP complex, do not give satisfactory results. A similar phenomenon was observed for the system 'polybase + weak polyacid + strong polyacid' (example 13 in Table 7.2).

Even replacement of one salt on to another in common solution of two polyelectrolytes competing for interaction with oppositely charged polyelectrolyte may result in the inversion of recognition. The equilibrium

$$P + P_i \rightleftharpoons IPC(PP_i)$$

where P_i and P are oppositely charged (or charging in the course of the reaction) macromolecules, is sensitive to ionic strength of the reaction medium and shifts to the left with an increase in concentration of low-molecular salt. Due to different affinities of polyanions incorporated in an IP complex for different counter-ions the equilibrium is also sensitive to the nature of cations. Since the recognition in reaction with participation of two different polyanions (P_1 and P_2) and one polycation P is connected directly with relative stabilities of IP complexes (PP_1) and (PP_2), the stabilities being dependent on the nature of low-molecular cation, then the equilibrium

$$IPC(PP_1) + P_2 \rightleftharpoons IPC(PP_2) + P_1$$

may be shifted to the left or to the right depending on the nature of salt present in the system. Such inversion of recognition by means of replacement of KCl on LiCl (or NaCl) was observed in a mixture of PVPy (P), PMAA (P_1) and sodium polyphosphate (P_2): P_1 was recognized by P in the presence of KCl, but P_2 in the presence of LiCl or NaCl [17].

A quantitative description of the system in question within a nonmodified model of unidimensional adsorption is impossible because, first, the forming IP complexes are nonstoichiometric and, second, the polycation is an oligomer in relation to one polyanion (P_1) but a polymer to another one (p_2). It may be assumed, however, that the type of equation for the factor of recognition in this system should not greatly differ from an equation of type (7.36) or (7.39); in any case, it is to contain the factor $K_{1(2)}^{n_2}$ (n_2 is the d.p. of polyphosphate—an oligomer being in excess with regard to complementary polymer PVPy), since the inversion of recognition (selection) has also been observed in work [17] on changing the d.p. of polyphosphate.

Fractionation of binary statistical copolymer ('oligomer') by composition
If, besides nonuniformity in chain length, oligomer $\tilde{P}$ is also characterized by nonuniformity in composition, then to solve the molecular recognition task it is necessary to know the dependence of ΔG on the structure of the macromolecule. To a first approximation, it is possible to accept for the binary statistical copolymer (oligomer) with units A and B,

$$\Delta G(a, b) = a\Delta G_{1a} + b\Delta G_{1b} \tag{7.40}$$

where a and b are the numbers of units A and B in a macromolecule of the copolymer ($a + b = $ d.p.) and ΔG_{1a} and ΔG_{1b} are the respective increments of the free energy. This additive approximation represents a natural generalization of Eq. (7.32), but it does not take into consideration the possible dependence of free energy of the IP complex formation on the fashion of the distribution of monomer units in the polymer chain. Since no experimental data on such dependence are available at present, the analysis is confined here to the additive approximation which is valid in some systems. (For instance, contributions of units of acrylic and methacrylic acids to the free IP complex-formation energy of their copolymers with PVPd are additive; see Section 7.3.2.) The equation describing the change in the function of the compositional and molecular weight nonuniformity of copolymers $\tilde{P}$ upon binding one of its part in an IP complex with P reads in the form of the additive approximation [26] as

$$w_s(a, b) = \frac{w_o(a, b)}{1 + A K_{1a}^a K_{1b}^b} \tag{7.41}$$

This equation is analogous to one describing fractionation of copolymers by fractional precipitation [42]. To obtain uniform fractions, it is necessary to use cross-fractionation [42, 43] in both cases since under a standard scheme of fractionation the macromolecules with different a and b values but with a similar denominator on the right-hand side of Eq. (7.41) will equally distribute between the solution and the IP complex (between the diluted and the concentrated phases respectively). In fact, cross-fractionation is perhaps the only possible direct method of experimental determination of the compositional and molecular weight nonuniformity of the copolymers; the algorithm for calculation of the nonuniformity by cross-fractionation was suggested in reference [44].

It is evident that Eqs. (7.38) and (7.39) describing recognition by polymer P of the optimum partner in the mixture of two oligomers are easily obtained after elementary transformations of Eq. (7.41) taking into account the function of compositional and molecular weight nonuniformity:

$$w(a, b) = \delta_{a,m}\delta_{b,0}m_1^o + \delta_{b,n_2}\delta_{a,0}m_2^o$$

where $\delta_{x,y}$ is Kronecker's symbol

$$\delta_{x,y} = \begin{cases} 1, & \text{if } x = y \\ 0, & \text{if } x \neq y \end{cases}$$

In a particular case of d.p.-uniform copolymer, i.e.

$$n = a + b = \text{constant} \tag{7.42}$$

fractionation takes place only by composition and Eq. (7.41) becomes uniparametrical, analogous to Eq. (7.33):

$$w_s(\alpha) = \frac{w_0(\alpha)}{1 + A'\kappa^\alpha} \tag{7.43}$$

where $\alpha = a/(a + b)$ is the fraction of units A in the copolymer and $\kappa = (K_{1a}/K_1 b)^n$. In this case, the dependence of the average composition of the separated fraction of copolymer (i.e. bound in the IP complex) on unit-mole correlation $\mu = [P]/[\tilde{P}]$ is expressed [26] by

$$\bar{\alpha}_{IPC} = \int_0^1 \frac{\alpha w_0(\alpha)}{\mu + A''\kappa^\alpha}\, d\alpha \tag{7.44}$$

This particular case has been investigated experimentally [26, 34]. It has been shown that under PAA (P) IP complex formation with an excess of partly quaternized PVPy ($\tilde{P}$), the latter is quite efficiently fractionated by composition, macromolecules of the copolymer enriched with nonalkylated units being selected in the IP complex. It is of importance that the preliminary molecular weight fractioned sample of PVPy is quaternized with ethyl bromide; i.e. under the conditions of an experiment Eq. (7.42) is approximately fulfilled. It is clear from Fig. 7.10(a) that Eq. (7.44) leads to a satisfactory description of the

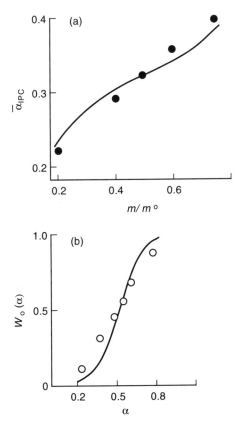

Fig. 7.10 (a) Fractionation of PVPy, alkylated with ethyl bromide to 49 mol%, under IP-complex formation with PAA. Points, experimental data on the average composition of fractions of the copolymer bound in an IP complex with different quantities of PAA; curve, calculation from Eq. (7.44) with $\kappa = 1.65$. (Reprinted from *European Polymer J.*, **17**, 987, 1981, with kind permission from Elsevier Science Ltd, The Boulevard, Langford Lane, Kidlington OX5 1GB, UK). (b) Integral distributions by composition of the same PVPy sample, obtained from the data [34] on alkylation kinetics (curve) and from fractionation data [34] by cumulative routine (dots)

experimental data using the distribution function by composition of the initial copolymer $w_0(\alpha)$ obtained from the data on the kinetics of quaternization [34].

Compare, on the other hand, integral distribution functions by composition obtained from the kinetic data (Gaussian approximation) and from the fractionation data by means of the cumulative routine in Fig. 7.10(b). Dots are close to the curve but indicate that distribution is broader than when it follows from the kinetic data. This result demonstrates once again exceptional selectivity of interpolymer interactions.

7.4   ON KINETICS AND MECHANISM OF INTERMACROMOLECULAR INTERACTIONS

Intermacromolecular interactions in dilute solutions are complex reactions including bimolecular initiation (i.e. formation of the first interchain bond) and repeating monomolecular propagation reactions resulting in formation of an uninterrupted sequence of the bonds (see Scheme 7.3) in the microreactor formed by two interacting chains. If one of the interacting chains, or both, is not too long, 'termination' in the propagation should take place at the end of the interacting chain. However, if the interacting chains are both relatively long, simultaneous or consecutive initiation, i.e. formation of single bonds between structural units belonging to different parts of two encountering macromolecular coils, may take place. This means that the microreactor formed by two interacting macromolecular coils at early stage of the interaction should represent a microgel comprising the segments of occasional lengths situated between interchain bonds. Then two segments, belonging to one and other chains and situated between interchain bonds formed occasionally, will in general be nonequal in length (see Scheme 7.15). The 'destination' of the microreactor should depend on whether the propagation reactions are reversible or not.

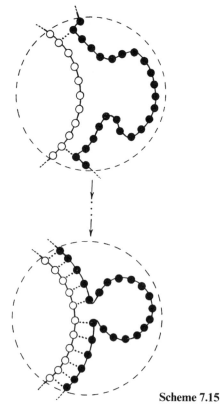

Scheme 7.15

If the reactions are reversible and the rate of structural rearrangement inside the microreactor is sufficiently high to anneal occasionally formed imperfections then the product with an equilibrium structure is formed, the defects in such an equilibrium structure being caused by equilibrium inside the IP complex (see Scheme 7.4). However, if the reactions are irreversible, simultaneous growth of the sequences of interchain bonds should be accompanied, on meeting one another, by appearance of loops containing nonbound-in-the-complex monomer units of one of the chains, thus resulting in formation of a much more imperfect structure than the equilibrium one. Just such a situation takes place in the case of high energy of interchain bonds, namely in the case of direct formation of interchain covalent bonds (see Section 7.5).

It may not be easy to imagine that the propagation reaction is irreversible at low energies of each interchain bond. However, if it is remembered that, in general, the quality of solvent decreases (up to phase separation) as a portion of interchain bonds in the IP complex increases, then the progressive compactization of the IP complex particle (microgel) in the course of propagation reactions seems to be able to lay obstacles to the reversibility of the reactions, especially at high conversions, by 'freezing' imperfections in compact particles due to restrictions in intramolecular motion.

The most favourable conditions for obtaining the most perfect structure of an IP complex are realized in matrix processes of polymerization and polycondensation [1, 2]. For example, in matrix polymerization after mutual recognition of the matrix and growing chain, the elementary stages of chain growth and propagation of the interchain bond sequence proceed simultaneously (Scheme 7.16). Corresponding experimental data concerning perfectness of structure of IP complexes obtained in different ways will be discussed in Section 7.5.

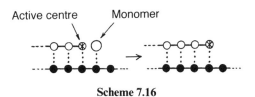

Scheme 7.16

Unfortunately, a kinetic study of intermacromolecular interactions is very difficult since, in general, even in dilute solutions, the reactions are very fast and complete in time of mixing the reagents [1, 2]; a recent estimation of characteristic time of the united coil formation by means of the stopped flow method gives the value as less than 5 ms (the 'dead' time of the equipment) [45]. Investigations in the field of matrix polymerization showed that mutual recognition of matrix macromolecule and growing chain resulting in IP complex formation took place with a period less than the lifetime of a free radical in radical polymerization (~ 1 s) [27, 37].

With reference to recognition and substitution reactions, owing to random encounters, macromolecule P binds occasionally with any macromolecule P_i in the mixture. The choice of the optimal partner P_k (see Scheme 7.10) is possible only due to multiple 'trials and errors' within a reasonable period of time. The possibility of such substitution reactions is not self-evident because in the case of relatively long macromolecules P_i, competing for complexation with P, each $IPC(PP_i)$, or some of them, can be so stable that the equilibria (Scheme 7.12) may be shifted to the right practically completely. Nevertheless, the substitution may take place even in this case due to the existence of two different mechanisms [27] called 'dissociative' and 'contact'.

The dissociative mechanism may be represented as

$$IPC(PP_j) + P_k \rightleftharpoons P + P_j + P_k \rightleftharpoons IPC(PP_k) + P_j$$

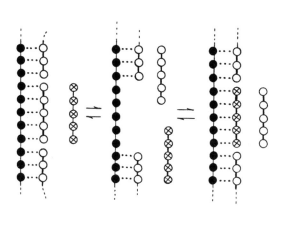

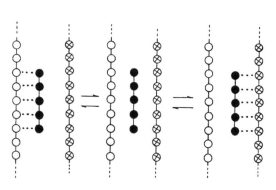

Scheme 7.17

Not too high a stability of IPC(PP$_j$) is the necessary condition to provide dissociation of the complex with liberation of free chains. This mechanism should be peculiar for systems comprising relatively low-molecular P or P$_j$. An increase in stability of IPC(PP$_j$) caused by an increase in the d.p. of P or P$_j$ must be accompanied by a decrease in the rate of substitution by the mechanism under consideration till a full stop of the substitution.

The contact mechanism assumes formation of the intermediate product IPC(PP$_j$P$_k$):

$$IPC(PP_j) + P_k \rightleftharpoons IPC(PP_jP_k) \rightleftharpoons IPC(PP_k) + P_j$$

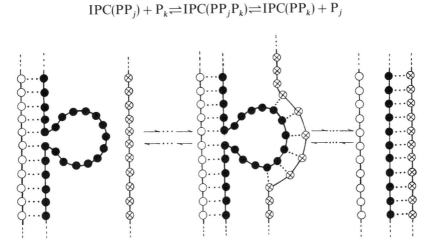

Scheme 7.18

This mechanism seems to be the only one possible when all participating chains are sufficiently long and no dissociation of IPC(PP$_j$) onto free chains occurs, but the presence of imperfections (loops, free tails) is the necessary condition to provide bimolecular initiation with consecutive propagation. Theoretically, the initiation may become the limiting step of the substitution, if the particles of IPC(PP$_j$) are very compact; it may happen when the solution of the IP complex is close to phase separation. For example, the substitution reaction with participation of interpolyelectrolyte complexes practically does not occur in salt-free solutions but proceeds at a high rate after addition of low-molecular salt [45]; it is well known that the stability of interpolyelectrolyte complexes decreases with increasing ionic strength (till dissociation onto free macromolecules) being accompanied by progressive loosening of the complex's particles [20, 21].

Unlike the dissociative mechanism, where the propagation step is similar to that of the usual interchain interaction (IP complex formation), the propagation reaction in the contact mechanism is a complex reaction including one more step, namely the dissociation of the interchain bond belonging to the former IP complex (the reverse reaction with respect to propagation):

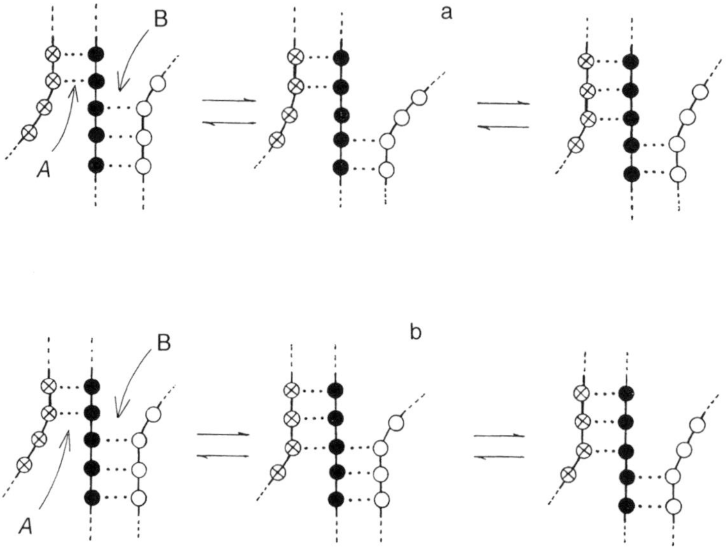

Scheme 7.19

The rate of propagation should depend on the difference in energies of border bonds A and B, since the relative time that a border pair of structural units (belonging to the sequence of interchain bonds) exists in bound and nonbound states is directly connected to the respective equilibrium constant. In the absence of other factors affecting the rate of propagation, the higher the difference in the energies (or, what is the same, in stabilities of corresponding IP complexes), the higher should be the rate of propagation; the lowest rates should be observed in reversible exchange reactions, i.e. in the case when P_k and P_j are identical in structure.

Only assuming the existence of exchange reactions, it is possible to explain reversible processes of structural transformations of IP complexes from microgel particles in dilute solutions to networks in semi-dilute and concentrated solutions and vice versa [46].

The rates of exchange reactions are indeed not too high and may easily be measured, in contrast to substitution reactions which usually proceed with the rate of mixing of the components [1, 2]. The time required for the substitution reaction can be as short as the lifetime of a free radical in radical polymerization; for example, in radical polymerization of methacrylic acid, growing chains of PMAA have time to recognize the presence in dilute solution of matrix chains of PEO in IPC(PEO–PAA), i.e. to initiate substitution by the mechanism represented in Schemes 7.18 and 7.19 and substitute the chains of PAA in the IP complex in the process of growing [27, 37]. In this particular case, propagation steps of polymerization and of substitution coincide.

So far as the structure of chains P_j and P_k in exchange reactions is identical, the kinetics of the exchange can be measured only if labeled macromolecules are used:

$$IPC(PP_1) + P_1^* \rightarrow IPC(PP_1^*) + P_1$$

Scheme 7.20

where P_1 and P_1^* are macromolecules of the same structure but P_1^* is labeled with luminescent markers. As an example, P_1^* and P_1 may be PAA or PMAA with and without luminescent anthryl-containing markers respectively, and P may be PEO [27].

The kinetics of exchange reactions is usually investigated by means of polarized luminescence or fluorescence quenching methods. The idea of the methods is based on the dependence of spectroscopic parameters of a marker, introduced in small concentration into the chains of macromolecules P_1, on whether this macromolecule is in a free state or bound in the IP complex with P. For instance, the value of luminescence polarization of the anthryl marker depends on intramolecular mobility of the chain segment to which it is attached, the mobility being highly dependent on whether P_1^* is free or complexed [27]. If P is the fluorescence quencher, then complex formation–dissociation process is accompanied by fluorescence quenching–inflammation [45]. Both methods give quantitative information concerning the rate of exchange reactions.

Experimental data obtained in investigations of exchange reactions may be summarized as follows:

(a) The rate of exchange in the case of long-chain P_1^* and P_1 ('polymers') and relatively short-chain P ('oligomer'), being high for low-molecular P, decreases rapidly with an increase in the oligomer chain length becoming practically independent of this parameter at high lengths of the chains. This is just an illustration of the transition from the dissociative mechanism to the contact one with increasing chain length [47].

(b) Compactization of IP complex particles in solution caused for any reason results in a decrease in the exchange rate. For example, the compactization may be caused by an increase or decrease in ionic strength in the cases of IP complexes stabilized by H bonds (e.g. between polycarboxylic acids and PEO or PVPd) or by electrostatic interactions (e.g. in the case of interpolyelectrolyte complexes). Correspondingly, for an increase in ionic strength, the rate of exchange in the former systems rapidly decreases [48]; in the latter systems, it increases [45].

(c) In some cases, the kinetic stop of the exchange as well as the substitution is observed in the course of propagation, i.e. a formation of relatively long-life triple complexes (see Scheme 7.18) is fixed, thus approving the contact mechanism [49].

(d) As a rule, the dependence of the exchange rate on time (or degree of conversion) can not be described in the frame of appropriate simple kinetical scheme.

The only known exception is the so-called 'reaction of single-chain transfer' in nonstoichiometric interpolyelectrolyte complexes of polymer–oligomer type [45]. The 'exchange' reaction (Scheme 7.20), where P is oligomer PVPy and P_1 is polymer PMAA, turns to follow the kinetic equation of second-order irreversible reaction, if the initial IP-complex composition corresponds to one chain of PVPy per one chain of PMAA (irreversibility is due to specific interaction between pyridinium cycles and pyrene fluorescent markers in PMAA, so the reaction should be regarded as substitution).

If the initial IP complex composition is more then one chain of PVPy per one chain of PMAA, then the kinetics becomes polychromatic (more complicated) [45]. There should evidently exist a set of rate constants k_{ij} corresponding to transfers of one PVPy chain from a particle of IPC(PMAA–PVPy) containing i chains of PVPy to a particle of IPC(PMAA*–PVPy) containing j chains of PVPy.

One more reason for the complicated character of the exchange reaction kinetics is nonuniformity of samples of exchanging macromolecules. Thus, if P is uniform with respect to composition (or microtacticity) but P_1 (and P_1^*) are not, the fractions of IP complex particles containing different in composition (or microtacticity) macromolecules P_1 are different in their stability and then contain different overall portions of defects. Thus, the rates of initiation both in dissociative and contact mechanisms may be different for each fraction. The rate of propagation, particularly in the contact mechanism, must depend on the composition (or microtacticity) of both chains P_1 and P_1^*. So far as the initiation is occasional, more or less different macromolecules P_1 and P_1^* may meet one another in each given particle of IPC($PP_1P_1^*$). In each IPC($PP_1P_1^*$), the propagation reaction should proceed not only with a different mean rate but the direction may be different as well. Indeed, only in the case of identical P_1 and P_1^* are the right-hand and left-hand states in Scheme 7.20 equiprobable (this reaction represents real exchange); in all other cases, the equilibrium will be more or less displaced to the left or to the right depending on the difference in structure of the competitors P_1^* and P_1 (these reactions therefore actually represent substitution).

It is clear that non-uniformity in the structure of macromolecules P results in nonequal stability of fractions of IPC(PP_1^*) as well as IPC(PP_1) and, then, in nonequal mean rates of propagation in different IPC($PP_1^*p_1$) particles.

SUMMARY   Theoretical conclusions concerning the influence of different factors on recognition in systems 'oligomer–two or more polymers' seem to be very important, mainly in the creation of biologically active polymers comprising macromolecular carriers of active species [50] as well as in the separation of polymer mixtures [2].

In the creation of biologically active polymers intended for recognizing some definite structures in living organisms, attention should be paid to the possible mechanisms of the recognition. The recognizing macromolecule, being able to bind occasionally to any complementary (but not desirable) structure, has to select the target by means of multiple 'trial and error', i.e. participating in multiple substitution reactions. It is evident that if the complementary structures comprising the target are spatially separated, i.e. do not contact with one another, the contact mechanism is prohibited. Then only the dissociative mechanism may be available, providing relatively short-chain recognizing macromolecules for use or, if for some reason recognizing chains are long enough, they must be very weak complexing agents.

By contrast, if the recognizing macromolecule is able to contact with all other complementary structures comprising the target (e.g. all are moving) so that the contact mechanism may be realized, the precision of recognition increases with an increase in the d.p. of recognizing chains. In connection with the separation of polymer mixtures, the quality of an oligomer as recognizer also increases with an increase in its d.p. The best results should be obtained when the chains of recognizer become equal in length or longer than the chains to be recognized.

7.5 REACTIVITY OF FUNCTIONAL GROUPS IN INTERPOLYMER COMPLEXES

7.5.1 GENERAL REMARKS

Identical structural elements of one and the same chain in the IP complex are in quite different states, being included in double-stranded regions (A), loops (B), free tails (C), single unbound monomer units (D) or linear free segments (E) situated between the ends of two chains of another component (Scheme 7.3), as well as in the microsurrounding of the elements; their intramolecular (conformational) mobilities are not identical in the regions mentioned. The structural inhomogeneity should result in a distinctive difference in reactivity of functional groups belonging to the regions mentioned up to spatial separation of inter- and intrachain reactions, each of them proceeding in its own region. Thus, ladder-like structure of IP complexes may be converted into true ladder structures if noncovalent interchain bonds are transformed by appropriate reaction into the covalent ones; such IP complexes may be regarded as precursors of ladder polymers.

On the other hand, the reactions peculiar to macromolecular components of an IP complex may be essentially accelerated or decelerated in different regions of the complex under the influence of intramolecular stresses, i.e. steric factors caused by lowered (in imperfections) or practically frozen (in double-stranded regions) intramolecular mobility, as well as due to microsurrounding effects.

The stresses may increase or decrease in the course of different reactions proceeding in an IP complex, thus providing the decisive role of stress relaxation processes in the reaction kinetics as well as structural conjugation of spatially separated reactions due to their different influence on rigidity of IP complex structure elements.

In this section, we shall consider the features of intra- and interchain reactions in IP complexes proceeding at heating and not requiring the participance of any other additional substances.

7.5.2 INTERCHAIN REACTIONS

So far as spatially separated segments of two macromolecular components of an IP complex (precursor) situated in imperfections of double-stranded structure are not able to participate in the formation of sequences of interchain covalent bonds, the perfectness of the ladder structure to be obtained by chemical transformation of interchain noncovalent bonds into covalent ones depends on the perfectness of the precursor's structure, the rate of rearrangements in the structure, steric and other objections taking place in the initial structure or arising in the course of the transformation process, and kinetic characteristics of the process.

Supposing that no steric hindrances affect the course of the reaction, the perfect ladder structure may theoretically be obtained in two cases: when the ladder-like structure of the precursor is perfect or, if it is not, when the ratio $k_1/k_2 \rightarrow 0$ (k_1 and k_2 are rate constants of the initiation and propagation, see below) and the rate of structural rearrangements in the precursor is high enough. If the structural rearrangements are frozen in the IP complex or their rate is much lower than the rate of the transformation reaction, then the imperfect structure of the ladder-like precursor will be copied in the structure of the resulting ladder polymer.

The reaction of transformation may be qualitatively regarded in the frame of the theory of the neighboring groups effect assuming that the process is determined by three kinetic constants k_1, k_2 and k_3 (see Scheme 7.21); to simplify the consideration $k_2 = k_3$ may be assumed. The corresponding kinetic analysis to calculate the dependence of conversion on time, the dependence on conversion of distribution by length of the sequences of bonds formed, and so on, may be applied only in the cases of ideal double-stranded structure or imperfect structure with known distribution by length of double-stranded regions at the rate of rearrangements equal to zero.

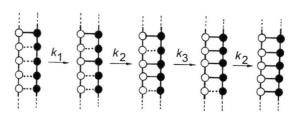

Scheme 7.21

Thus, at $k_1 \ll k_2$, the sequences of covalent bonds (i.e. the elements of ladder structure) will appear at low conversions. If the rate of the rearrangements in the precursor is as high as the rate of propagation, the sequences formed will be commensurate in length with the chains composing the IP complex. At very low rates of the rearrangements, the lengths of the sequences will correspond to the lengths of the noncovalent bond sequences existing in the precursor (note that the probability of the initiation is proportional to the length of a sequence). In the case of IP complexes of 'polymer–oligomer' type, even at relatively high conversion a fraction of oligomer chains will remain in the nonbound (by the covalent bonds) state and may then be separated from the polymer under appropriate conditions; the product should represent block copolymer comprising initial polymer segments and segments of ladder structure (Scheme 7.22a).

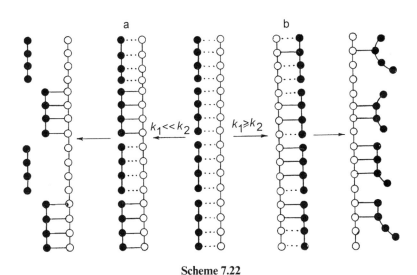

Scheme 7.22

At $k_1 = k_2$, occasional formation of covalent bonds in double-stranded regions of the IP complex should 'freeze' the imperfections caused by occasional and then nonequal lengths of segments of two chains situated in the nearest double-stranded regions separated by imperfection (Scheme 7.15a). Relatively long sequences of interchain covalent bonds can appear in such conditions only at high conversions; even at 100% conversion of noncovalent bonds into covalent ones, the resulting ladder structure will be defective. In the complexes of 'polymer–oligomer' type, at relatively low conversion, a large fraction of oligomer chains or almost all of them will be bound to the polymer by one or a few covalent bonds, and will then become inseparable from the polymer chain (Scheme 7.22b). A similar situation should be peculiar for $k_1 \geqslant k_2$, accompanied by essential deceleration of the reaction at high conversions.

It should be mentioned that the formation of covalent interchain bonds is not necessarily connected with preliminary complexation of the reacting macromolecules. The direct interchain reaction between appropriate monomer units is possible as well, but such reactions have their own features. First of all, the rate of such a reaction and the limit conversion depend on mutual compatibility of reacting macromolecular coils, so far as the reaction may occur only in the zone of their interpenetration. Computer simulations show that the limit conversion reaches no more than 50% in the case of ideal interpenetration [51, 52].

Two effects are evidently capable of changing the above value of the limit conversion, namely incompatibility of reacting polymers which should decrease the value and, contrariwise, preliminar IP complex formation which may increase it. As a very good illustration of the effects, consider interpolymer reactions in solutions between poly-trischlorobutadiene and polystyrene [51, 53, 54] or PEI [52, 55]:

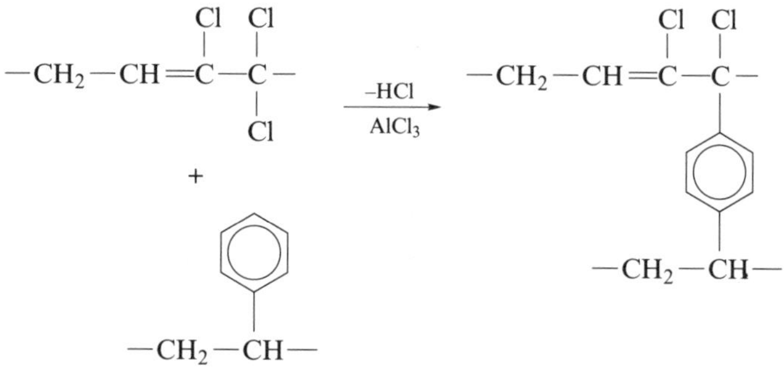

Scheme 7.23a

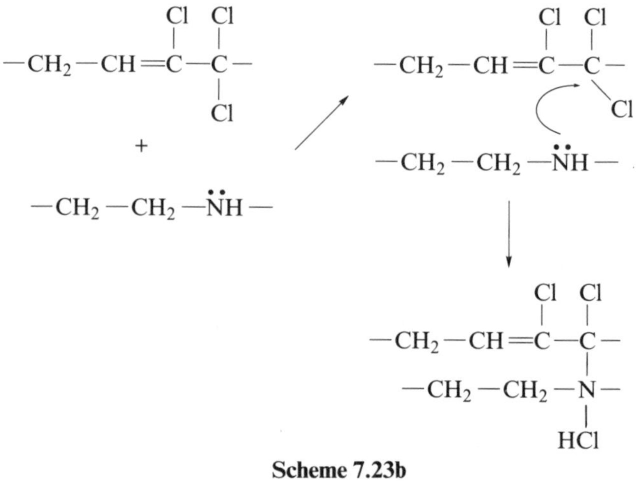

Scheme 7.23b

Participants of the former reaction (Scheme 7.23a) are incompatible, so the limit conversion achieved is *ca.* 10%, though almost all coils are already paired at conversion of *ca.* 1% [53, 54]. Thus, the resulting interpolymer looks like two glued spheres in this case. The reaction of covalent bonding in Scheme 7.23b is preceded by formation of IP complex precursor, so the limit conversion exceeds 80% [55], thus demonstrating the advantages of IP complexes in the formation of more perfect ladder structures.

Among the reactions of interchain covalent bonds formation studied, the interchain amidation reaction in the IP complex formed by PAA and PEI:

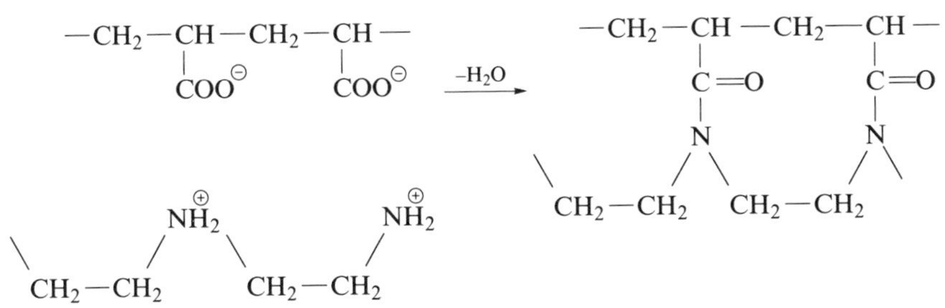

$$-CH_2-CH-CH_2-CH- \qquad -CH_2-CH-CH_2-CH-$$

Scheme 7.24

is not complicated by any accompanying intrachain reaction. It was shown by IR spectroscopy [56, 57] that only those functional groups that had already formed electrostatic bonds in the initial IP complex (precursor) participated in the amidation reaction. All the bonds may be almost entirely converted to amide ones, while nonionized carboxylic groups in defect regions of the IP complex stay unchanged in the course of the reaction at given conditions (heating for 40 min at 150–250 °C). In this system, therefore, the interchain reaction proceeds only in double-stranded regions and the decrease in concentration of electrostatic bonds does not provocate the unbound functional groups to form new ones (though bound and unbound groups are in equilibrium in the foregoing water solution, their balance being unchanged on water evaporation during preparation of the sample).

Both the intrachain dehydration, and the intrachain decarboxylation, being characteristic for pure PAA, do not occur on heating of IP complexes PAA–PEI, which is proved by means of IR spectroscopy (absorbance bands of anhydride groups are absent in the IR spectra of heated samples) and gas chromatography (no CO_2 elimination is determined) [56, 57].

The amidation kinetics was found to follow a first-order equation up to some definite conversion degree and then followed the Kohlrausch equation [57]. This means that the reaction rate is defined at the first stage by concentration of the electrostatic bonds, but then accumulating stresses convert the reaction in the relaxation regime (see Fig. 7.11a).

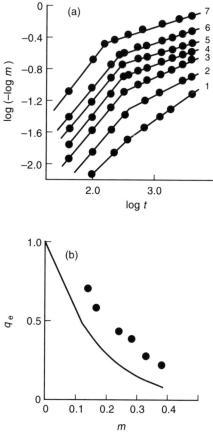

Fig. 7.11 (a) Kinetics of amidation reaction in IPC(PAA–PEI) in double-logarithmic coordinates. m is the degree of conversion of electrostatic bonds to the covalent ones. (Reproduced by permission of Nauka, Moscow, Russia, from *Vysokomol. Soed. B*, 1976, **18**, 785) (b) Dependence of the ratio q_e of tetraethylenepentamine chains extracted from its IPC with PAA after heating on m. Curve, calculated by the Poisson equation; dots, experimental data

The distribution of intermolecular covalent bonds between oligomer chains of PEI is found to be different from the equiprobable one [56]. It was shown (see Fig. 7.11b) that the amount of oligomers extracted from the sample after heating is appreciably greater than that calculated by the Poisson equation at the same conversion; in other words, the situation is just like in Scheme 7.22a. This means that the formation of sequences of interchain amide bonds is advantageous, which is similar to the accelerating effect of neighboring units, i.e. $k_1 > k_2$ (see Scheme 7.21). It is therefore concluded that the first stage is described formally by the first-order equation due to intercompensation of the accelerating effect and deceleration induced by the arising stresses.

The deceleration is strong enough to provide the 'kinetic stop' of the amidation reaction in IPC(PAA–PEI), the limit conversion being dependent on temperature. It is noteworthy that 'kinetic stop' does not occur in the thermolysis of IPC(PAA–PEI) which forms on the destruction of poly-*tert*-butylacrylate in composites with PEI at 180 °C [58]. The destruction is accompanied by the formation of amide bonds between amine groups of PEI and carboxylic groups step by step, formed from the ether groups of poly-*tert*-butylacrylate, and no absorbance bands of either COOH or COO$^-$ groups are observed in IR spectra of heated composites. In this case, either amide bonds are formed from undissociated carboxylic groups or electrostatic bonds are quickly and quantitatively converted into amide ones.

Interchain amidation occurs in water solution of IPC(PAA–PEI) as well [59]. In this case, significant degrees of conversion may be also achieved: up to 50% of electrostatic bonds convert into amide ones at 235 °C (the reaction was carried out in an autoclave). In the whole range of degrees of conversion investigated, the products of amidation are soluble in water and water–salt media in the same pH region, as is initial IPC(PAA–PEI). This fact proves that the amidation reaction proceeds only in double-stranded regions; the unbound groups of the polyelectrolytes in the defect regions, which define solubility of the IP complex in water, do not participate in the reaction.

Formation of covalent bonds in double-stranded structures of IP complexes is not necessarily accompanied by deceleration of the reaction up to its kinetic stop if no considerable stresses arise in the course of development of continuous sequences of the covalent bonds, as it is in the case of IP complexes of PAA with urea–formaldehyde polymer (PFU) [60, 61]. These IP complexes, as well as composites consisting of an IP complex and an excess of one of its macromolecular components, are obtained by matrix polycondensation of urea and formaldehyde on PAA in water solutions [62]. The structure of PFU thus formed depends on reaction conditions being controlled (at pH < 3.7) or not (at pH > 3.7) by the matrix macromolecules, though in both cases (up to pH ≅ 5) at an appropriate ratio of the matrix and comonomers, the product represents an IP complex formed by the matrix and grown chains of PFU. The structure of monomer units shown in Scheme 7.25a predominate at pH < 3.7, being very unusual for polymers obtained from urea and formaldehyde, while at pH > 3.7 the polymer consisting of 'usual' monomer units is formed (Scheme 7.25b). The products of matrix polycondensation in water represent hydrogels due to the specific mechanism of microseparation of a new phase of the IP complex as well as networking as usually happens in the formation of PFU [62]. It was shown by using Stewart models [63] that PFU-I provides formation of regular continuous double-stranded structures in IPC(PAA–PFU) in which each pair of monomer units may form two H bonds (Scheme 7.25). In the case of PFU-II the structure is more crumbly, and formation of two H bonds per one pair of monomer units cannot take place, being only 1.5 maximum (on average). Transformation of both structures into ladder ones by formation of continuous sequences of interchain imide bridge bonds (Scheme 7.25) is not accompanied by rising significant intrachain stresses [63]. As a consequence, both IP complexes are completely converted into structures shown in Scheme 7.25 at a fixed temperature in the range 180–250 °C; this is shown by the IR spectroscopy method [64]. The rate of the reaction is higher in the case of the more perfect IP complex formed by PFU-I (Fig. 7.12). The kinetic curves in this figure correspond to elimination of water only in the imidization reaction, since another possible reaction proceeding with water elimination (viz. formation of cyclic anhydride groups in PAA chains) in the IP complexes does not occur at all (see the next section).

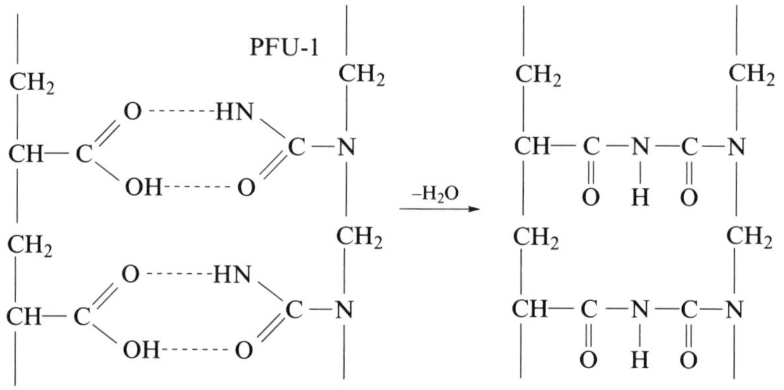

Scheme 7.25a

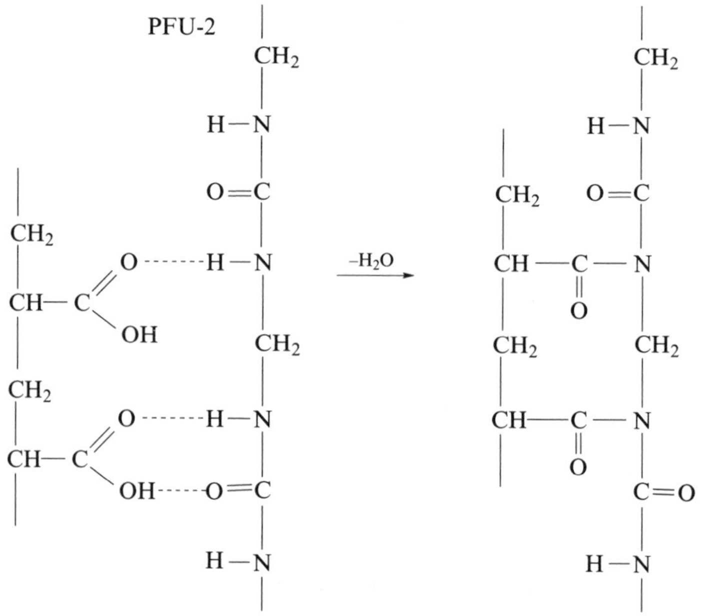

Scheme 7.25b

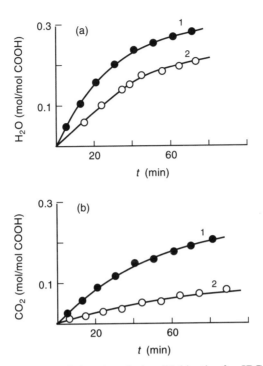

Fig. 7.12 Dehydration (a) and decarboxylation (b) kinetics for IPC(PAA–PFU), obtained at pH = 2.4 (1) and 4.1 (2) [65]

The products of thermal reactions in the IP complexes only slightly swell in water due to the presence of free segments of PAA in initial IP complexes, while the degree of swelling of the complexes is as high as 1000–10 000% [66]. As could be expected, the ladder structures obtained by intermolecular amidation in IPC(PAA–PEI) and imidation in IPC(PAA–PFU) are much more stable at elevated temperatures (up to 300–350 °C) than macromolecular components of these IP complexes.

7.5.3 INTRACHAIN REACTIONS

The reactions requiring conformational (intramolecular) mobility of chain segments should be the most sensitive to the state of segments in different structural regions of an IP complex. Thermal dehydration with formation of cyclic anhydride groups in polycarboxylic acids, being very characteristic of PAA and PMAA, may be considered as an example:

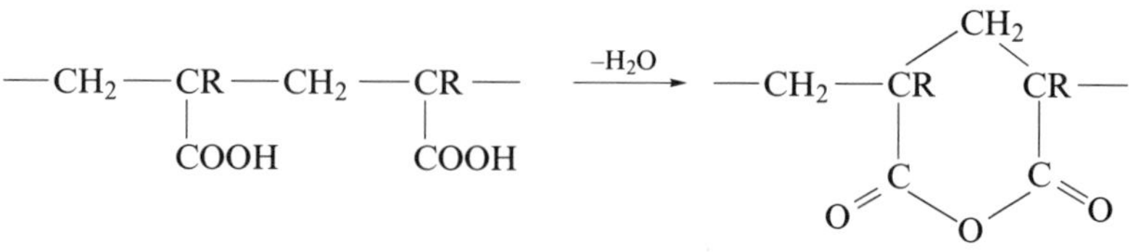

$$-CH_2-\underset{\underset{COOH}{|}}{CR}-CH_2-\underset{\underset{COOH}{|}}{CR}- \quad \xrightarrow{-H_2O} \quad -CH_2-CR\overset{CH_2}{\diagup}\underset{\underset{O}{\overset{C}{\diagdown}}\diagdown_O}{\diagdown}CR-$$

Scheme 7.26

In all IP complexes studied, the reaction under consideration was shown to proceed on heating only in imperfections of double-stranded structure [67, 68]. As an example, Fig. 7.13 shows the changes in optical density of the bonds in the IR spectra of IPC(PMAA–PEO) corresponding to valent oscillations C=O of carboxylic groups bound (1730 cm^{-1}) and unbound (1710 cm^{-1}) by the H bond with PEO (i.e. situated in double-stranded regions and the imperfections respectively), as well as that of cyclic anhydride group (1030 and 1800 cm^{-1}). It is clearly shown that with an increase in conversion in the reaction of Scheme 7.26, the

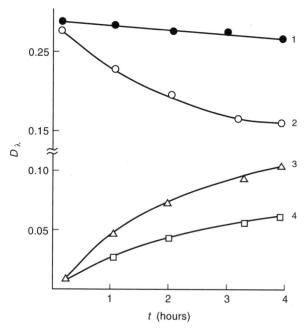

Fig. 7.13 Variation of optical density D with time at wave numbers 1730 (1), 1710 (2), 1030 (3) and 1800 (4) cm^{-1} in the IR spectrum of IPC(PMAA–PEO) with PEO d.p. = 45 at 150 °C. (Reprinted from *European Polymer J.*, **20**, 193, 1984, with kind permission from Elsevier Science Ltd, The Boulevard, Langford Lane, Kidlington OX5 1GB, UK)

concentration of carboxylic groups in the imperfections decreases, while that in double-stranded regions does not.

As a rule, the optical density of bands corresponding to groups participating in complexation, i.e. included in double-stranded regions, does not change in the course of the reaction manifesting an inessential influence of changes in the structure of chain segments belonging to imperfections on the stability of the

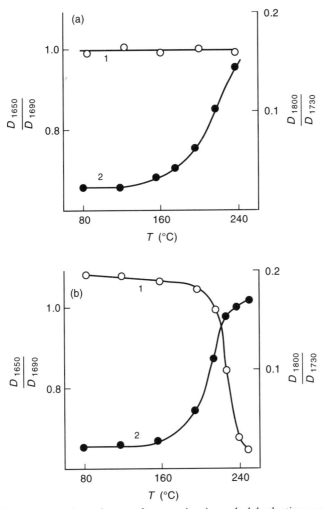

Fig. 7.14 Temperature dependences of conversion in cyclodehydration reaction, characterized by D_{1800}/D_{1730} (1), and degree of binding of PVPd units by H bonds with polycarboxylic acids, characterized by D_{1650}/D_{1690} (2), in (a) IPC(PMAA–PVPd) and (b) IPC(PAA–PVPd). (Reproduced by permission of Nauka, Moscow, Russia, from *Vysokomol. Soed. B*, 1983, **25**, 213)

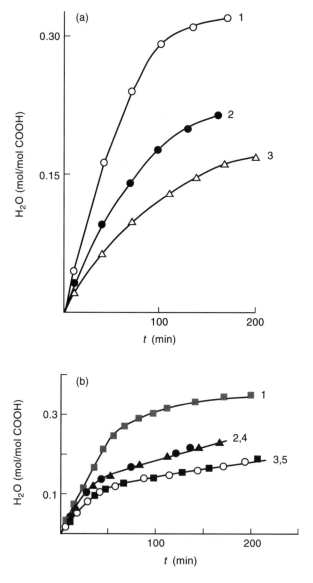

Fig. 7.15 Dehydration (a, b) and decarboxylation (c) kinetics at 200 °C of IP complexes of PVPd with (a) PMAA and (b, c) PAA, obtained in different ways. 1, pure polycarboxylic acids; 2, IP complexes, obtained by mixing the components in common solution; 3, IP complexes, obtained by matrix polymerization; 4, fully destroyed and 'rebuilt anew' matrix IP complexes; 5, partially destroyed and 'rebuilt anew' matrix IP complexes. (Reproduced by permission of Nauka, Moscow, Russia, from *Vysokomol. Soed B*, 1985, **27**, 495)

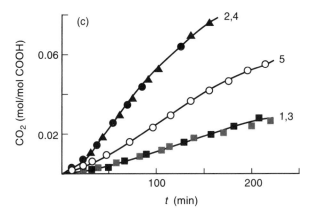

Fig. 7.15 (*Continued*)

double-stranded structure. However, at least in one case, namely at heating of IPC(PAA–PVPd) (see Fig. 7.14), the concentration of interchain H bonds decreases with an increase in concentration of forming anhydride groups, thus indicating consecutive dissociation of double-stranded regions in the course of the reaction [68]. This phenomenon was explained by the destructive influence of increasing stresses in chain segments undergoing the cyclization reaction, taking into account the fact that the stability of respective IP complexes formed by PAA is lower than that of PMAA (see Table 7.2).

The higher the perfectness of the double-stranded structure of an IP complex, the lower is the rate of the reaction. Figure 7.15(a) and (b) demonstrates the influence of the method of preparation of IP complexes (PMAA–PVPd) and (PÅA–PVPd) on the kinetics of dehydration reaction in the complexes. It is clearly shown that the reaction proceeds in free polycarboxylic acids more rapidly than in IP complexes. The limit conversions in free polyacids are higher as well, indicating that only a fraction of a polyacid participates in the reaction; this fraction represents the segments included in imperfections.

Further, the rates of the reactions in the IP complexes obtained by mixing macromolecular components in common solvent are higher than those in the complexes obtained in the course of matrix polymerization of methacrylic or acrylic acids on macromolecules of PVPd. If the differences in the reaction rates are caused by more perfect structure of the 'matrix' IP complexes in comparison with that of 'usual' complexes obtained by mixing, then after complete dissociation of a 'matrix' IP complex onto free macromolecular components (in water solution at pH > 9) and subsequent formation ('rebuilding anew') of the complex (at pH < 3), the structures of this 'rebuilt anew' IP complex and of the 'usual' one must be identical. Indeed, as seen from Fig. 7.15(b), the dehydration curves for 'usual' and 'rebuilt anew' IP complexes coincide, which proves that the double-stranded structure of an IP complex obtained via matrix polymerization may be more perfect than that obtained by direct interaction of previously prepared macromolecules in common solvent by the reasons regarded above.

It is interesting that after only partial destruction of the 'matrix' IP complexes by swelling in neutral media (complete separation of a polyacid and PVPd chains does not take place under the conditions, but dissociation of a fraction of double-stranded structure inside each IP-complex particle), with subsequent 'rebuilding anew' at pH < 3, this 'rebuilt anew' complex 'keeps in mind' the structure of the original 'matrix' IP complex. This follows from the coincidence of curves 2 and 3 in Fig. 7.15(a).

The passiveness in intramolecular anhydridization of carboxylic groups of polyacids included in double-stranded regions of IP complexes is not surprising. High rigidity of the double-stranded structure should lay obstacles to mutual orientation of neighbor carboxylic groups favorable for the reaction. A certain intramolecular (i.e. conformational) mobility of polyacid segments is obviously the necessary condition to provide this favorable orientation. Hence, the rate of the reaction under consideration should depend not only on the fraction of carboxylic groups in imperfections of the double-stranded structure but also on the type of imperfections and other features of the given IP complex structure. Thus, intramolecular mobility which can affect the rate of intramolecular reactions seems to be dependent on the kind of imperfections in which the groups are situated: in a loop B (the mobility being dependent on its size), in a free tail C or in a linear free segment E (evidently, 'linear' does not mean 'rigid rod') situated between the ends of two chains of a second component of the IP complex (Scheme 7.3). It is evident that the relative fraction of a definite kind of imperfection depends on relative chain lengths of macromolecules forming the IP complex. Thus, loops should predominate in the case when both macromolecules (e.g. PMAA and PEO) are long-chain 'polymers', free tails—in the case when the polyacid is short-chain and the second component is long-chain- and linear free segments, in the case of a long-chain polyacid and a short-chain second component. It is evident that the worst conditions to realize the mobility as well as to relax the stresses arising in the course of formation of rigid cyclic elements in the main chain should be characteristic for loops. This means that the lowest rates of intramolecular dehydration of polycarboxylic acids should be realized in IP complexes formed by long-chain macromolecules. In polymer–oligomer types of IP complexes, where the polyacid is long-chain, the conditions for the reaction should be more auspicious due to a higher degree of freedom of the ends of free polyacid segments than in loops, especially if it is remembered that such complexes are less stable and free 'oligomer' chains, liberated on heating due to dissociation of the IP complex, may, in principle, play the role of plasticizer for free segments of the 'polymer' chains, thus accelerating the reaction.

Such a phenomenon was observed when investigating the reaction under consideration in IP complexes formed by high-molecular PMAA and PEO of different d.p. As a qualitative characteristic of the reaction rate, the temperature T^* was used at which heating of an IP complex resulted in the appearence of cyclic anhydride group bands in the IR spectrum [67, 69]. It was shown that for an IP complex formed by high-molecular PMAA and relatively long-chain PEO ($M = 20\,000$), as well as for free PMAA, $T^* = 160\,°C$, while for an IP complex of the same PMAA and short-chain PEO ($M < 4000$) $T^* = 100\,°C$, i.e. in this IP complex the reaction proceeded even easier than in free PMAA, probably due to the plasticizing effect of free PEO.

Conversely, the intramolecular mobility may be essentially decreased in networked IP complex structures. For example, in IPC(PAA–PFU) and composites consisting of the IP complex and the excess of one or other macromolecular component, the intramolecular reaction in PAA chains with formation of cyclic anhydride groups does not proceed at all, regardless of the presence of free segments of PAA chains providing the swelling of the IP complex in water [60]. It is the three- or fourfold excess of PAA in the composite with respect to the quantity necessary for mutual saturation of PAA and PFU in the IP complex that provides the appearance of cyclic anhydride group bands in the IR spectrum of heated composite. Apparently, the networked structure of the IP complex as well as corresponding composites and internal stresses arising in the process of network formation in the course of matrix polycondensation create very unfavorable conditions for realizing conformational rearrangements in free segments of PAA in the structure.

It is noteworthy that another intramolecular reaction peculiar to PAA, namely decarboxylation:

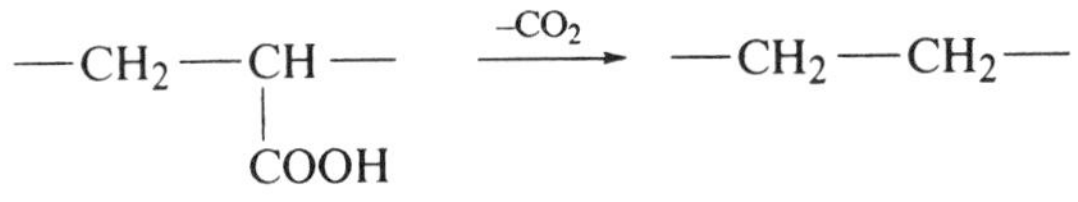

$$-CH_2-\underset{\underset{COOH}{|}}{CH}- \xrightarrow{-CO_2} -CH_2-CH_2-$$

Scheme 7.27

is accelerated in IPC(PAA–PFU) and its composites as well as in other IP complexes formed by PAA. The mechanism of this reaction is not known for certain, but the most probable one consists of decarboxylation of carboxylic groups which do not form either intra- or intermolecular H bonds [70].

In contrast to dehydration with formation of cyclic anhydride groups, this thermal reaction results in an increase in flexibility of PAA chains. Available experimental data allow us to assume that acceleration of this reaction in IP complexes is connected with striving of PAA chains to decrease the stresses arising due to limitations in the conformational mobility of the chains. For example, an increase in the PFU fraction in the composites comprising IPC(PAA–PFU) (PFU is responsible for rigidization of PAA due to both formation of double-stranded structures by intermolecular H-bonding and networking) results in essentially increasing the decarboxylation rate (Fig. 7.16).

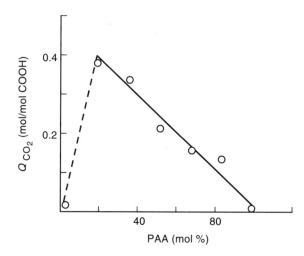

Fig. 7.16 Dependence of limit yield of CO_2 at 170 °C per unit-mole of PAA on unit-mole content of PAA in IP-complex composites with PFU, obtained at pH = 2.4. (Reproduced by permission of Nauka, Moscow, Russia, from *Vysokomol. Soed A*, 1991, **33**, 642)

More crumbly IP complex formed by PAA and PFU comprising a large fraction of PFU-II (see Scheme 7.25) is characterized by a twofold lower rate of decarboxylation than more perfect IP complex of the same composition but formed by PFU-I (Fig. 7.12b). Finally, in IPC(PAA–PVPd), the decarboxylation reaction proceeds not only with a much higher rate than in free PAA but also accelerates with increasing conversion of another intrachain reaction, resulting in the rigidization of free PAA segments, namely dehydration with the formation of cyclic anhydride groups (Fig. 7.15c).

It is interesting that decarboxylation is more sensitive to the structure of the IP complex than dehydration. Thus, the rate of this reaction in IPC(PAA–PVPd) after partial destruction and consequent re-formation of the matrix IP complex is intermediate between the rates in the matrix and 'usual' IP complexes [71].

7.6 CONCLUSIONS

Investigations in the field of intermolecular reactions were started in the early 1930s by Bungenberg de Jong [72], but were extensively developed in many laboratories in the world only in the 1960s due to stimulation by well-known discoveries in genetics and molecular biology. The most valuable contribution to the investigations of cooperative interactions between macromolecules of both synthetic and biological origin was made by scientific school headed by

Kabanov and by Tsuchida. It was not the purpose of this chapter to give exhaustive information concerning intermacromolecular reactions, IP complexes and their applications; for more detailed information see cited reviews and books [1–4, 13, 16, 20, 21, 50 and also 73–80].

It should be mentioned that a number of theoretical conclusions concerning interpolymer interactions may be, at least qualitatively, applied to interactions of macromolecules with the surface of different types of particles. There are definite grounds for believing that under adsorption of macromolecules on the 'infinite' two-dimensional lattice (surface), the dependence of the adsorption constant on the chain length, with low filling degrees of the surface and not too large chain length, will be expressed by an equation close to Eqs. (7.10) to (7.12) with all the following consequences.

The assumptions developed above may also be applied to the interactions of macromolecules with small particles, in particular with globulae and particles of sols, if the length of the macromolecule is sufficiently high to be capable of binding simultaneously with more than one particle. One of the expected effects, for instance, is fractionation by size of very small particle dispersions with the help of macromolecules interacting with them.

Listing these almost random examples, we intend to underline the fundamental character of the features of intermacromolecular reactions as well as almost unlimited possibilities to develop both theoretical and practical investigations in this branch of polymer physicochemistry.

REFERENCES

1. Kabanov, V. A. and Papisov, I. M., *Vysokomol. Soed A*, 1979, **21**, 243–81.
2. Papisov, I. M. and Litmanovich, A. A., *Adv. Polymer Sci.*, 1989, **90**, 139–79.
3. Kabanov, V. A., *Vysokomol. Soed*, 1994, **36**, 183–97.
4. Torchilin, V. P., *Vysokomol. Soed*, 1994, **36**, 279–97.
5. Skorodinskaya, A. M., Kemenova, V. A., Chernova, O. V., *et al.*, *Khim-Farmacevt J.*, 1983, **12**, 1463–7.
6. Gladilin, K. L., Orlovsky, A. F., Kirpotin, D. B. and Oparin, A. I., *Origin of Life*, 1978, 357–62.
7. Applequist, J. and Dumle, V. *J. Am. Chem. Soc.*, 1965, 1450–7.
8. Khodakov, Yu. S., Berlin, A. A., Kalyaev, G. I. and Minachev, Kh. M., *Teor. i Exper. Khimiya*, 1969, **5**, 631–8.
9. Baranovsky, V. Yu., Litmanovich, A. A., Papisov, I. M. and Kabanov, V. A., *Europ. Polymer J.*, 1981, **17**, 969–80.
10. Magee, W., Gibbs, J. and Zimm, B., *Biopolymer*, 1963, **1**, 133–40.
11. Magee, W., Gibbs, J. and Newell, R., *J. Chem. Phys.*, 1965, **43**, 2115–32.
12. Antipina, A. D., Baranovsky, V. Yu., Papisov, I. M. and Kabanov, V. A., *Vysokomol. Soed A*, 1972, **14**, 941–9.
13. Tsuchida, E. and Abe, K., *Adv. Polymer Sci.*, 1982, **45**, 1–119.
14. Blake, R. and Fresco, J., *Biopolymer*, 1973, **12**, 775–9.

15. Papisov, I. M., Baranovsky, V. Yu., Sergieva, E. I., Antipina, A. D. and Kabanov, V. A., *Vysokomol. Soed A*, 1974, **16**, 1133–8.
16. Zezin, A. B. and Rogacheva, V. B., in *Uspehi Khimii i Fiziki Polymerov* (*Advances in Chemistry and Physics of Polymers*), Moscow, Nauka, 1973, pp. 3–30.
17. Izumrudov, V. A., Bronich, T. K., Saburova, O. S., Zezin, A. B. and Kabanov, V. A., *Dokl. Akad. Nauk. SSSR*, 1988, **301**, 634–8.
18. Lutsenko, V. V., Lopatkin, A. A. and Zezin, A. B., *Vysokomol. Soed A*, 1974, **16**, 2429–34.
19. Lutsenko, V. V., Zezin, A. B. and Kalyuzhnaya, R. I., *Vysokomol. Soed A*, 1974, **16**, 2411–17.
20. Kabanov, V. A., *Pure and Appl. Chem. Macromol. Chem.*, 1973, **8**, 121–36.
21. Kabanov, V. A. and Zezin, A. B., *Macromol. Chem. Suppl.*, 1984, **6**, 259–75.
22. Papisov, I. M., Baranovsky, V. Yu., Chernyak, V. Ya., Antipina, A. D. and Kabanov, V. A., *Dokl. Akad. Nauk. SSSR*, 1971, **199**, 1364–6.
23. Baranovsky, V. Yu. and Papisov, I. M., *Dokl. Akad. Nauk. SSSR*, 1974, **217**, 133–7.
24. Birshtein, T. M., El'yashevich, A. M. and Morgenshtern, L. A., *Vysokomol Soed B*, 1972, **14**, 487–8.
25. Birstein, T. M., Eliashevich, A. M. and Morgenstern, L. A., *Biophys. Chem.*, 1974, **1**, 242–5.
26. Litmanovich, A. A., Papisov, I. M. and Kabanov, V. A., *Europ. Polymer J.*, 1981, **17**, 981–8.
27. Papisov, I. M., Nedyalkova, Ts. I., Avramchuk, N. A. and Kabanov, V. A., *Vysokomol. Soed A*, 1973, **15**, 2003–7.
28. Abe, K., Koide, M. and Tsuchida, E., *Macromolecules*, 1977, **10**, 1259–64.
29. Baranovsky, V. Yu., in *III Vsesoyuznaya Konferentsiya* on *Vodorastvorimye Polymery i ih Primenenie* (3rd All-Union Symposium on *Water-Soluble Polymers and Their Applications*), Irkutsk, 1987, p. 126.
30. Korugic-Perkovich, L. and Ferguson, J., *Polymery SFRI*, 1983, **4**, 301–8.
31. Chatterjee, S. K. and Sethi, K. R., *Polymer*, 1984, **25**, 1367–70.
32. Izumrudov, V. A., Bronich, T. K., Zezin, A. B. and Kabanov, V. A., *Dokl. Akad. Nauk. SSSR*, 1984, **278**, 404–8.
33. Litmanovich, A. A., *Vestnik Mosk. Univ., seriya khimiya*, 1978, **19**, 617–19.
34. Litmanovich, A. A., Papisov, I. M. and Kabanov, V. A., *Vysokomol Soed A*, 1980, **22**, 1180–4.
35. Litmanovich, A. A., Anufrieva, E. V., Papisov, I. M. and Kabanov, V. A., *Dokl. Akad. Nauk. SSSR*, 1979, **246**, 923–7.
36. Kikuchi, Y. and Kubota, N., *Makromol. Chem. Rapid Commun.*, 1985, **6**, 387–9.
37. Papisov, I. M., Nekrasova, N. A., Pautov, V. D. and Kabanov, V. A., *Dokl. Akad. Nauk. SSSR*, 1974, **214**, 861–4.
38. Papisov, I. M. and Litmanovich, A. A., *Vysokomol Soed A*, 1977, **19**, 716–22.
39. Litmanovich, A. A., Kirsh, Yu. E. and Papisov, I. M., *Vysokomol Soed B*, 1978, **20**, 83–4.
40. Kokufuta, E., Yokota, A. and Nakamura, I., *Polymer*, 1983, **24**, 1031–4.
41. Aleksina, O. A., Zezin, A. B. and Papisov, I. M., *Vysokomol Soed A*, 1971, **13**, 1199–200.
42. *Fraktsionirovanie Polymerov* (*Polymer Fractionation*), Moscow, Mir, 1971, p. 321.
43. Posental, A. J. and White, B. B., *Industr. Engng Chem.*, 1952, **44**, 2693–8.
44. Litmanovich, A. A., *Vysokomol Soed B*, 1985, **37**, 350–4.
45. Bakeev, K. N., Izumrudov, V. A., Kuchanov, S. I., Zezin, A. B. and Kabanov, V. A., *Macromolecules*, 1992, **25**, 4249–54.

46. Higgs, P. G. and Ball, R. C., *J. Phys. France*, 1989, **50**, 3285–308.
47. Anufrieva, E. V., Pautov, V. D., Papisov, I. M. and Kabanov, V. A., *Dokl. Akad. Nauk. SSSR*, 1977, **232**, 1096–7.
48. Chen, H.-L. and Morawets, H., *Macromolecules*, 1982, **15**, 1445–9.
49. Listova, O. V., Izumrudov, V. A., Bronich, T. K., *et al.*, *Vysokomol. Soed B*, 1986, **28**, 724–5.
50. Kabanov, A. V. and Kabanov, V. A., *Vysokomol. Soed*, 1994, **36**, 198–211.
51. Korshak, V. V., Suprun, A. P., Vointseva, I. I., *et al.*, *Vysokomol. Soed A*, 1984, **26**, 111–18.
52. Korshak, V. V., Suprun, A. P., Slonimsky, G. L., *et al.*, *Makromol. Chem.*, 1986, **187**, 2153–78.
53. Vointseva, I. I., Shashkov, A. S. and Suprun, A. P., *Vysokomol. Soed A*, 1978, **20**, 1640–5.
54. Korshak, V. V., Askadsky, A. A., Vointseva, I. I., Mustafaeva, B. B., Suprun, A. P. and Slonimsky, G. L., *Vysokomol. Soed A*, 1981, **23**, 1002–9.
55. Voinstseva, I. I., Lebedeva, T. L., Evstifeeva, I. I., Askadsky, A. A. and Suprun, A. P., *Vysokomol. Soed A*, 1989, **31**, 416–20.
56. Komarov, V. S., Rogacheva, V. B., Bezzubov, A. A. and Zezin, A. B., *Vysokomol. Soed B*, 1976, **18**, 784–7.
57. Komarov, V. S., Rogacheva, V. B. and Zezin, A. B., *Vysokomol. Soed A*, 1978, **20**, 1629–33.
58. Cherkezyan, V. O., Litmanovich, A. D., Godovsky, Yu. K., Litmanovich, A. A. and Hromova, T. N., *Vysokomol Soed B*, 1984, **26**, 112–15.
59. Kabanov, V. A., Zezin, A. B., Rogacheva, V. B. and Ryzhikov, S. V., *Dokl. Akad. Nauk. SSSR*, 1982, **267**, 862–6.
60. Papisov, I. M., Kuzovleva, O. Ye. and Litmanovich, A. A., *Vysokomol. Soed B*, 1982, **24**, 842–3.
61. Bolyachevskaya, K. I., Litmanovich, A. A., Litmanovich, A. D., Markov, S. V. and Papisov, I. M., *Vysokomol. Soed B*, 1987, **29**, 845.
62. Kuzovleva, O. E., Etlis, V. S., Shomina, F. N., Davidovich, G. N., Papisov, I. M. and Kabanov, V. A., *Vysokomol. Soed A*, 1980, **22**, 2316–21.
63. Litmanovich, A. A., Markov, S. V. and Papisov, I. M., *Dokl. Akad. Nauk. SSSR*, 1984, **278**, 676–9.
64. Papisov, I. M., Kuzovleva, O. E., Markov, S. V. and Litmanovich, A. A., *Europ. Polymer J.*, 1984, **20**, 195–200.
65. Bolyachevskaya, K. I., Intra- and intermolecular reactions in interpolymer complexes of polycarboxylic acids stabilized by hydrogen bonds, PhD Thesis, Moscow, 1992.
66. Litmanovich, A. A., Markov, S. V. and Papisov, I. M., *Vysokomol. Soed A*, 1986, **28**, 1271–8.
67. Litmanovich, A. A., Kazarin, L. A. and Papisov, I. M., *Vysokomol. Soed B*, 1976, **18**, 681–4.
68. Kazarin, L. A., Baranovsky, V. Yu., Litmanovich, A. A. and Papisov, I. M., *Vysokomol. Soed B*, 1983, **25**, 212–4.
69. Baranovsky, V. Yu., Kazarin, L. A., Litmanovich, A. A. and Papisov, I. M., *Europ. Polymer J.*, 1984, **20**, 191–4.
70. Bolyachevskaya, K. I., Litmanovich, A. A., Markov, S. V., Molotkova, N. N., Pshenitsyna, V. P. and Papisov, I. M., *Vysokomol. Soed A*, 1992, **33**, 638–41.
71. Bolyachevskaya, K. I., Litmanovich, A. A., Papisov, I. M., Litmanovich, A. D. and Cherkezyan, V. O., *Vysokomol. Soed B*, 1985, **27**, 494–7.
72. Bungenberg de Jong, H. G., in *Colloid Science*, Ed. H. Kruyt, Chemical Publishing Co., New York, 1949, Vol. II, Chap. 10.

73. Zezin, A. B. and Kabanov, V. A., *Uspekhi Khimii*, 1982, **51**, 1447–83.
74. Borue, V. Yu. and Erukhimovich, I. Ya., *Macromolecules*, 1990, **23**, 3625–32.
75. Haronska, P. and Seidel, Ch., *J. Phys. Sec. I (Fr.)*, 1992, **2**, 1645–55.
76. Zezin, A. B., Izumrudov, V. A. and Kabanov, V. A., *Frontiers Macromolecular Science*, Blackwell Science Publishers 1989, pp. 219–25.
77. Anufrieva, Ye. V. and Pautov, V. D., *Vysokomol. Soed A*, 1992, **34**, 41–7.
78. Ibragimova, Z. H., Ivleva, Ye. M., Pavlova, N. V., *et al.*, *Vysokomol. Soed A*, 1992, **34**, 134–9.
79. Kabanov, V. A., *Macromol. Chem. Macromol. Symp.*, 1986, **1**, 101–24.
80. Bekturov, E. A. and Bimendina, L. A., *Adv. Polymer Sci.*, 1981, **41**, 99–147.

CHAPTER 8

Experimental Characterization of Units Distribution and Compositional Heterogeneity in the Products of Macromolecular Reactions

All experimental approaches to the analysis of the distribution of units are associated with measuring relative contents of chain fragments having a definite length. Today the maximum length of such a fragment rarely exceeds several monomeric units. Therefore experimental data describe the distribution of units only in rather short chain segments. For a complete analysis of the distribution of units along the macromolecular chain, one should assume that the processes controlling the structure of macromolecules are random in character and can be approximated by suitable probability models which depend on a small number of parameters. When describing a probability model by equations interrelating the relative quantities of various experimentally observed chain fragments, a possibility arises of estimating the values of corresponding parameters from the experimental data and of checking the adequacy of the probability model.

The examples discussed in this chapter apply mostly to the products of copolymerization. However, the same approaches may be suitable for the products of macromolecular reactions that in essence are also copolymers.

For the experimental determination of relative contents of various chain fragments the common approach is to measure some property conditions by the presence of units A in a copolymer containing units A and B, which depends on the environment of a given unit (A or B). Such a property can be frequencies and line intensities of IR and UV spectra, NMR line intensities, various thermodynamic characteristics, melting and vitrification points, the presence of characteristic chemical groupings, etc. The most complete quantitative information in such investigations is obtainable through the use of the NMR method and through the use of thermal decomposition and some chemical methods.

We give here a brief introduction into possibilities of a modern NMR technique to investigate the structure of copolymers also including stereo-sequences (remember that polymer microtacticity may affect the kinetics of its chemical transformation). The necessary details may be found in the books and reviews cited below. Gas chromatography and destructive mass spectrometry methods as well as some chemical transformations in the chain will also be discussed.

8.1 THE USE OF NMR SPECTROSCOPY FOR INVESTIGATING THE UNITS SEQUENCE DISTRIBUTION

Nuclear magnetic resonance (NMR) spectroscopy has been firmly established as a primary technique for polymer analysis [1–4]. The use of NMR spectroscopy is based on a strong dependence of resonance frequency on the chemical environment of the nucleus and on the possibility of recording spectra under conditions ensuring proportionality of the signal areas to the number of corresponding nuclei in a specimen.

The merit of polymer NMR spectra is their ability to present directly all collections of structural chain parameters (a particular unit structure and differences in the close and distant environment) through the spectral line location, its intensity and multiplicity. Strong dependence of the NMR frequency on the chemical environment leads to a shift of signals due not only to the alternation of chemically different units but also as a result of the alternation of configurations of asymmetric or pseudo-asymmetric carbon atoms of chemically identical units.

The application of NMR spectroscopy for polymer structure investigation has begun with broad-band NMR and relaxation measurements followed by high-resolution NMR spectroscopy. There are some advantages in employing the high-resolution NMR spectroscopy method, which consist of the ability to elucidate fine structural effects, in particular the quantitative analysis of polymer chain stereoregularity [5].

At present NMR spectroscopy, and in particular high-resolution NMR, is the main structural–analytical method for determination of chemical and stereochemical molecular structure, polymer chain heterogeneity, copolymer structure, molecular dynamic, polymer phase composition, reaction kinetic and mechanism [1–3, 6–8], despite the difficulty of creating a strong consistent theory of NMR spectra in polymers by the existence of numerous variants for spectral parameter calculations [9–12]. The development of such a theory is fraught with a number of problems and must take into account the fact that the observed NMR spectra of polymer substances significantly depend of the movement type in macromolecular chains and spatial structures.

The great success of NMR spectroscopy applications is caused by instrumentation improvement and development of new methods for spectrum registration and processing as well as much empirical data accumulation (correlation of spectral and structural parameters). Among the above-mentioned factors the most important are the working frequency increase and transfer from continuous to pulse spin-system excitation with Fourier transformation of free induction decay (FID). As the result of an increase in the working frequency (up to 750 MHz for proton resonance) we have a higher sensitivity and the possibility of facilitating interpretation of the spectrum. FID registration and aquisition essentially improve the concrete NMR sensitivity and make it possible to apply the resonance of the rather rare nuclei (^{13}C, ^{29}Si, ^{15}N, etc.). NMR spectroscopic data are of great use because they make it possible to set up some correlations between chemical shifts or other spectral parameters and reaction ability (in particular, polymerization rate constants) [13].

In recent years ^{13}C NMR has mostly displaced ^{1}H in studies of polymer microstructure. Because of its larger chemical shift range ^{13}C provides more detailed structural information. However, ^{1}H NMR spectroscopy, being less popular for polymers, has the advantage of higher sensitivity and relative ease in determining quantitative values.

8.1.1 ^{1}H NMR SPECTROSCOPY

Compositional and stereochemical information may be important when the physical or mechanical properties of copolymers are under consideration. Determination of the units distribution is the first spectral problem concerning polymers. The second problem associated with the stereoisomerism of macromolecules is more complicated to solve, since alternation of the chain configurations results in a smaller contribution to the chemical shift than alternation of chemically different units. This leads to a poorer resolution of the spectra. Thus, for example, methylene protons of polymethyl methacrylate (if we consider all possible tetrads: *mmm, mmr, rmr, rrm, mrm, rrr*) gives superposition of eighteen lines [14]. For vinyl polymers with one side substituent (such as polyvinyl chloride) the NMR spectra of the methylene protons reflects the interaction of six nonequivalent nuclei having up to 800 individual components.

From this point of view the best studied systems are polymers and copolymers of methacrylates. As long ago as in the 1960s copolymers of methyl methacrylate with styrene or methacrylic acid were investigated by Harwood and Ito [15, 16]. In these works a Markovian first-order distribution of units was postulated, and Bernoulli's scheme with equal probabilities of isotactic addition of units was used to describe stereoisomerism. The case of nonequal probabilities is considered in references [17] and [18] by Platé, Stroganov and co-workers.

Among numerous works devoted to the NMR investigation of polymetha-crylates [i.e. 19–27] the detailed study [28, 29] by Klesper and co-workers in-cludes a large number of specially synthesized compounds (methyl methacrylate–methacrylic acid copolymers) and has a great value. First of all it extended the list of model systems that enable the stereochemical and compositional effects to be examined. One of the latest is extremely demonstrative, giving ^1H spectral information about block and random copolymers of methyl methacrylate (MMA) and ethyl methacrylate (EMA) [30]. For block copolymers there are two sets of signals indicating the presence of highly isotactic poly-(MMA) and poly-(EMA) blocks (Fig. 8.1b and c): two AB quartets for methylene protons at 1.6–2.5 ppm and two sharp singlets at 1.33 and 1.38 ppm for α-methyl protons (it follows from comparison with the spectra of both isotactic homopolymers).

In the random copolymer spectrum (Fig. 8.1a) one can see a complicated splitting and broadening due to different types of monomer sequences. As

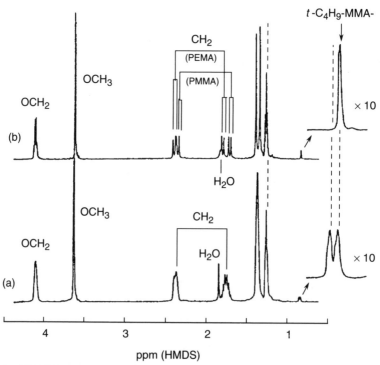

Fig. 8.1 ^1H NMR spectra of highly isotactic copolymers of methyl methacrylate and ethyl methacrylate in nitrobenzene-d_5: (a) random copolymer; (b) block copolymer of poly(MAA) with EMA. (Reproduced by permission of John Wiley & Sons Ltd from *Polymer Bull.*, 1991, **25**, 686)

Table 8.1. Triad concentration in acrylic acid–methyl methacrylate copolymers

[MMA] in feed	Triad	By NMR spectroscopy	From Bernoullian statistics
0.40	121	0.54	0.55
	122	0.40	0.38
	222	0.06	0.07
0.70	121	0.15	0.22
	122	0.47	0.50
	222	0.38	0.28

(Reproduced by permission of the Society of Polymers Science, Japan, from *Polymer J.*, 1989, **21**, 691)

mentioned earlier [31], α-methyl proton signals can give a simple and clear indication of block and statistical MMA copolymers with other substances. Methoxy proton resonance at ~3.6 ppm is less sensitive to sequence distribution as compared to other ones.

In addition to the foregoing, the important thing is the choice of solvent. For example, methylene signals of such types of polymers are insensitive to monomer sequence in $CDCl_3$ [30] or overlap each other in DMSO-d_6 [26]. In the last case the only opportunity to calculate the unit distribution by signal splitting exists in the α-methyl region. Three signals are observed there for acrylic acid–methyl methacrylate copolymers and attributed to 121 (1.05 ppm), 122 (0.933 ppm) and 222 (0.892 ppm) triads, the relative concentrations of which are given in Table 8.1. Indexes 1 and 2 represent acrylic acid unit and methyl methacrylate unit respectively.

The indicated triad concentration calculated from the Bernoullian statistics can be given as:

$$[222] = F_2[P_{22}]^2 \quad [122] = 2F_2[P_{22}(1 - P_{22})] \quad [121] = F_1[(1 - P_{11})(1 - P_{22})]$$

where F is the mole fraction of comonomer in copolymer determined by NMR. A good agreement of first and second results shows a preferably random monomer distribution for such polymers prepared in bulk.

The increased work frequency enables the distribution of units in chlorinated polyolefins in particular to be analyzed. For example, the products of chlorination of isotactic polypropylene, which decreases the crystallinity and affects other thermodynamic properties, must be controlled, and a 220 MHz magnetic field is sufficient to solve this problem [32]. In the proton spectra of chlorinated samples five new peaks appear at 3.4–4.2 ppm; their intensities may be successfully used for quantitative calculations. Assignment of the chemical shifts comply with the spectral parameters of model compounds and correspond to the following:

$$CH(CH_2\,{}^*Cl)\!-\!CH_2\!-\!\qquad -\!CH(CH_2\,{}^*Cl)\!-\!CHCl\!-$$
$$3.42\ \text{ppm}\qquad\qquad\qquad 3.54\ \text{ppm}$$

$$-\!CH(CH_3)\!-\!CH\,{}^*Cl\!-\!\qquad -\!CCl(CH_2\,{}^*Cl)\!-\!CH_2\!-$$
$$3.74\ \text{ppm}\qquad\qquad\qquad 3.74\ \text{ppm}$$

$$-\!CH(CH_2Cl)\!-\!CH\,{}^*Cl\!-\!\qquad -\!CH\,{}^*Cl\!-\!CH(CH_3)\!-\!CH\,{}^*Cl\!-$$
$$3.94\ \text{ppm}\qquad\qquad\qquad 3.94\ \text{ppm}$$

$$-\!CCl\!-\!(CH_3)\!-\!CH\,{}^*Cl\!-$$
$$4.12\ \text{ppm}$$

As a result the degree of chlorination, chlorine distribution and chlorinated structure with unchlorinated triad probability were defined and compared with the theoretical predictions provided by Frensdorff and Ekiner [33]. There Markov chains formalism has been applied to linear nearest-neighbor systems, where only the states $n+1$ and $n-1$ affect the substitution probability of the position n, and transition probabilities were derived in terms of the triad probabilities.

The high-resolution ^{1}H nuclear magnetic resonance applied for determination of copolymer composition and sequence distribution in references [34] and [35] gives rather sufficient information. In the first paper a question deals with the diad peak assignments for sebacinic anhydride (S) and 1,3-bis(p-carboxy-phenoxy)propane (C) copolymers:

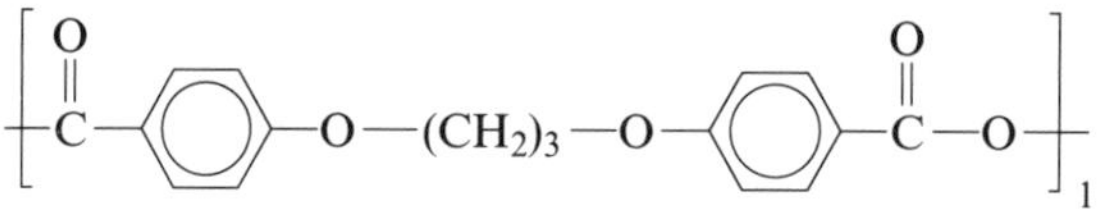

C

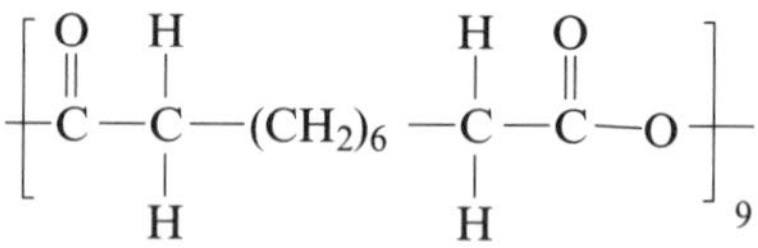

S

Examination of the proton spectra revealed two doublets at 8.1 and 8.0 ppm (CC and CS diads respectively) and two triplets at 2.6 and 2.4 ppm (SC and SS diads respectively) (Fig. 8.2). The assignment was made by comparison with ^{1}H spectra of homopolymers and explained by long-range shielding/deshielding effects. From integral intensities of the above spectra one can calculate the following: probabilities $P(S)$, $P(C)$, $P(SS)$, $P(SC)$, $P(CC)$, degree of randomness $h = P(CS)/P(S)P(C)$ ($0 < h < 2$ from block to completely alternating copolymer) and average chain length with SC sequence $L = 1/P(SC)$.

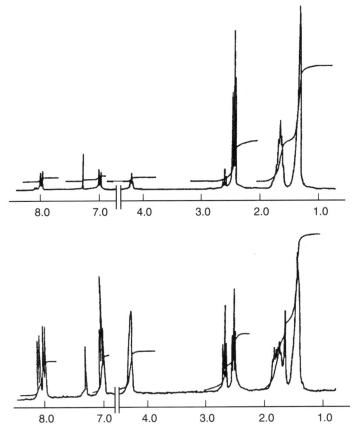

Fig. 8.2 ^{1}H NMR spectra of poly[1,3-bis(p-carboxyphenoxy) propane-sebacic acid]: CPP/SA = 90:10 (upper) and 50:50 (lower). (Reprinted with permission from *Macromolecules*, 1991, **24**, 2280. Copyright 1991 American Chemical Society)

In reference [35] the copolyesters poly(neopenthyl isophthalate-fumarate) are in consideration. The signals of methyl and methylene protons were split into five and six peaks, reflecting the linkages between neopenthyl glycol (N) and two types of acids, i.e. isophthalic acid (I) and fumaric acid (F). All the peaks were assigned and the triad sequence (FNF, FNI, INI, etc.) was clarified. This copolyester was found to be of a random polycondensation type, though there was a large difference between I and F concerning the rate of esterification. The detailed description of mathematical techniques for such calculations is described in reference [36] with numerous examples.

One of the most complicated problems solved for polymer objects using ^{1}H NMR experiments is the identification of amino acid residues in the chain of helical domain, which is the dominant structural motif in proteins [37]. As spectral criteria one considers the nuclear Overhauser enhancement (NOE) effect between neighbor NH protons; the NOE effect between NH and H_α three residues away; three bond coupling constants ^3J; and H_α chemical shift values. Without details we should only mention that the two-dimensional homonuclear Hartmann–Hahn (HOHAHA) spectral experiment was used [38]. The method relies on the principle of homonuclear cross-polarization obtained by switching on a single coherent radiofrequency (RF) field for a mixing time τ_m. For a proton A (offset Δ_A from the carrier frequency) the effective RF field is $v_A = v_A^2 + \Delta_A^2/2v$ by the nominal RF field $v \gg \Delta_A v$. Thus effective magnetization transfer between

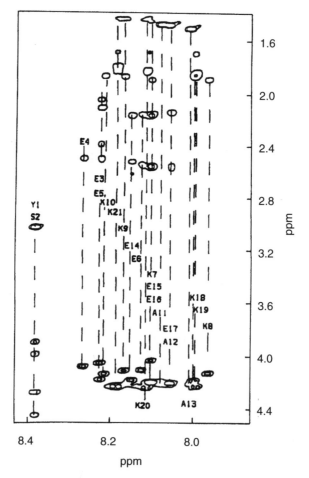

Fig. 8.3 Section of HOHAHA spectrum of EAK peptide, where E-Glu, K-Lys, S-Ser, Y-Tyr, A-Ala in H_2O/D_2O, mixing time 48.8 ms. (Reprinted with permission from *J. Am. Chem. Soc.*, 1991, **113**, 1015. Copyright 1991 American Chemical Society)

protons A and X is possible if $|\Delta_A^2 - \Delta_X^2| = |J_{AX}|$. For example, the magnetization transfer btween NH and H_α protons in the peptides mentioned above can be obtained by positioning the carrier frequency midway between the spectral regions of interest. Figure 8.3 shows one region of such a spectrum of EAK peptide: succinylTyrSerGlu$_4$Lys$_4$Ala$_3$Glu$_4$Lys$_4$NH$_2$ with Ala$_3$ central trimers as an example. Each line corresponds to one residue, and each cross-peak belongs to the same residue, which allows all the protons of a given spin system to be located by verifying a mixing time.

The presence of a large cross-relaxation component in the magnetization relaxation process in addition to the spin-lattice relaxation is a characteristic property of macromolecules. This process is responsible for two-dimensional NOE effects in NOESY experiments [39]. In turn, the $1/r^6$ distance dependence of the intensity of cross-relaxation peaks allows internuclear distances to be determined, which can be combined to produce three-dimensional structures. In most cases in NOESY spectra there are deviations from the intensity ratios predicted on the basis of first-order theory ($1/r^6$); this is the result of (a) the spin

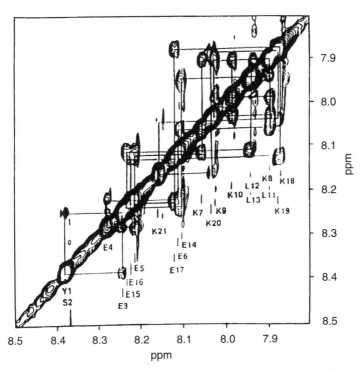

Fig. 8.4 Amide–amide region of NOESY spectrum of ELK peptide, where E-Glu, K-Lys, S-Ser, Y-Tyr, A-Ala in H$_2$O/D$_2$O, mixing time 250 ms. (Reprinted with permission from *J. Am. Chem. Soc.*, 1991, **113**, 1016. Copyright 1991 American Chemical Society)

diffusion process and (b) coherent transfer of the magnetization process. A careful consideration of some aspects of interpretation of cross-peak intensities is presented in references [40] and [41]. An example of the spectral expression of amid–amid connections in the NOESY spectrum of peptide ELK: succinylTyr-SerGlu$_4$Lys$_4$Leu$_3$Glu$_4$Lys$_4$NH$_2$ having trimer Leu$_3$ as the central part (Fig. 8.4).

8.1.2 ^{13}C NMR SPECTROSCOPY

Many problems inherent in proton magnetic resonance do not exist in the carbon NMR method. The natural content of the ^{13}C isotope is 1.1%. Hence, there is a negligibly low probability of a close neighborhood of two ^{13}C nuclei, so the carbon–carbon interaction can be left out of the account. The carbon–proton spin–spin interaction can be suppressed by means of special techniques (spin decoupling). Such spectra have no complicated multiplets and usually consist of individual signals. Furthermore, the range of the carbon chemical shift is larger than the proton one. This fact, combined with the smaller gyromagnetic ratio, γ_C, and the possibility of recording spectra under conditions of complete ^{1}H broadband decoupling makes ^{13}C NMR spectroscopy especially suitable and the most powerful method for investigating polymer and copolymer microstructure, which determines the physical properties of the material. This method continues to

Table 8.2. Identification of ^{13}C NMR signals of the crosslinked polyethylene

$$\overset{\gamma}{-C}-\overset{\beta}{C}-\overset{\alpha}{C}-\overset{}{C}-\overset{\alpha}{C}-\overset{\beta}{C}-\overset{\gamma}{C}-$$

with C_n, C_2, C_1 branch below the central carbon.

ppm	Carbon[a]	ppm	Carbon[a]
11.05	E-1	30.02	Backbone
14.07	A,B,L-1	30.51	A,B,E,L-γ
22.87	A,L-2	32.19	L-3
23.38	B-2	32.73	A-3
26.93	A-4,E-2	34.24	B-4,E-α
27.37	A,B,E,L-β	34.65	A,B,L-α,A-5
29.61	B-3,L-4	38.27	A,B,L-branch

[a]A = amyl, B = butyl, E = ethye, L = long chain.
Reproduced by permssion of Hüthig & Wepf Verlag, Basel, Switzerland from *Die Makromol. Chem.*, 1985, **186**, 133

expand rapidly. Since the early 1960s ^{13}C nuclear magnetic resonance has provided detailed information on the distribution of monomer units in copolymers, the stereochemical configuration and stereoregularity of polymers, the head-to-head addition by polymerization and other matters for a large number of chemical compounds. More recently this technique has also been applied to the investigation of solid polymers and has contributed much to our understanding of polymer morphology [1, 3, 36, 42–47].

As a first example let us consider the distribution of branches and degree of crosslinking in polyethylenes (PE) determined by ^{13}C NMR spectroscopy in reference [48]. A qualitative correlation was observed between bandwidth of the main signal at 30.02 ppm and degree of crosslinking; furthermore, fourteen signals were assigned in spectra according literature data (Table 8.2, Fig. 8.5), which made it possible to calculate the relative number of branches by direct intensity measurements.

This resulted in 22–24% ethyl branches, 31–46% butyl branches, 19–21% amyl branches and 19–26% long-chain branches for different crosslinked polyethylenes prepared by heating at 180 °C with bis(2-phenyl-2-propyl)peroxide. The activation energies for the intramolecular segmental motions were calculated from measurements of spin-lattice relaxation times T_1 using the following formulae:

$$\frac{1}{T_1} = (2\pi h)^2 \gamma_C^2 \gamma_H^2 \tau_R \sum_i r_i^{-6} \qquad \tau_R = \tau_0 \exp\frac{-\Delta E}{RT}$$

where T_1 = spin-lattice relaxation time; h = Planck constant; $r_1 = 1.10\,\text{Å}$ (C—H bond length); γ_C, γ_H = gyromagnetic ratios for C and H; τ_R = correlation time for interaction. It is known that T_1 does not depend on PE molecular weight and crystallinity and characterizes the segmental motions in PE amorphous regions.

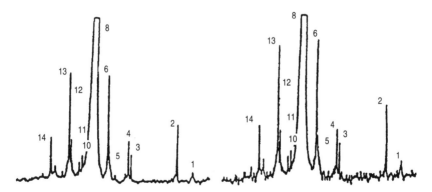

Fig. 8.5 ^{13}C NMR spectra of crosslinked polyethylene in o-dichlorobenzene/benzene-d_6 solution: figures correspond to the index of Table 4.3. (Reproduced by permission of Hüthig & Wepf, Verlag, Basel, Switzerland, from *Die Makromol. Chem.*, 1985, **186**, 133)

The increased value of ΔE (3.55 kcal/mol) for the gel fraction compared with 2.40 kcal/mol is explained by restricted segmental motions due to solvent locking in the gel matrix.

Another type of polymer, namely polydienes, has been studied by many investigators. The set of resonances due to cis-1,4, trans-1,4 and 1,2 units may be assigned to various compositional and conformational features. Let us give an example from reference [49], one of the latest dealing with the NMR study of polybutadienes and using the empirical chemical shift calculations, which gives a comparison of obtained data with those of the model, specially prepared samples and previous data. Figure 8.6 contains characteristic fragments of ^{3}C NMR spectra with attributed resonances. Polybutadiene samples include 50–98% cis-1,4 units.

Figure 8.7 presents a fragment of the synthetic polyisoprene ^{13}C NMR spectrum (methylene part) with peaks assignment made by application of multipulse NMR techniques [50], where c and t designate cis and trans configurations respectively, and the capital letter is related to the nuclei in view.

The reliable division of the closest lines (Table 8.3) and confident spectral fixation of rare structures in solutions made it possible to find an interesting composition peculiarity, the increase of relatively tc-triad number and reduction of inverse adding with molecular weight growth (from about 10 000 to > 50 000) for polymers prepared on Ziegler catalyst in the presence of hydrogen.

Analysis of such polymers via ^{13}C NMR spectra gives a correlation between average numbers of trans-1,4 links and inverse attachments. The dependence of polyisoprene microstructure on its molecular mass confirms the presence of different active centers (on start and during the polymerization) or the sequential regulation of the chain growth mechanism.

The assignment of the ^{13}C NMR peaks to the polymer tacticity is a difficult task. The preparation and separation of isomers, accurate model compounds, polymers of known structure, modified or ^{13}C labeled ones and the analysis of their chemical shifts have led to an almost complete analysis of this problem, in particular when it is concerned with polyolefins. However, it takes much time and is not generally used. For the cases of polypropylene (PP), poly(1-butene) (PB) and series of polyolefins with longer linear side chains the tacticity assignment was made and redetermined [51] by the ^{13}C chemical shift calculation using the γ-effect contributions, which depend on the probability of bond conformations; upfield shift accounts for about -5 ppm for the gauche arrangement compared with the trans arrangement [52, 53]. To solve this problem the authors of reference [51] took into account a downfield shift tendency for C1 and C2 resonances and upfield shift tendency for C3 resonance [54]. However, the tendency of shift is very complex, and it was difficult to attribute the peaks in detail from only chemical shift data.

One can interpret experimental peaks in terms of pentad tacticity of the main chain (Table 8.4) on the basis of such calculation (up to the heptad level) with

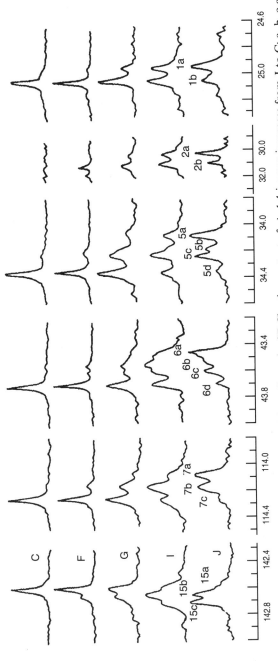

Fig. 8.6 Fragments of ^{13}C NMR spectra of polybutadienes in CDCl$_3$: the content of *cis*-1,4-isomer increases from J to C; a, b, c and d designate *tvt*, *tvc*, *cvt* and *cvc* sequences respectively. (Reprinted with permission from *Macromolecules*, 1993, **26**, 719. Copyright 1993 American Chemical Society)

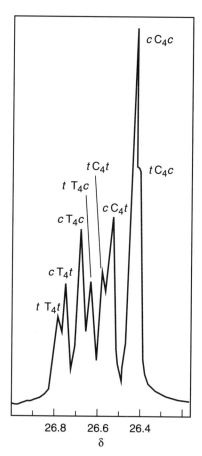

Fig. 8.7 Methylene region of ^{13}C NMR spectrum of polyisoprene: c and t designate *cis* and *trans* configurations; T_4 and C_4 relate to the 4-CH$_2$ group of *cis*-1,4 and *trans*-1,4 units. (Reproduced by permission of Khachaturov, A. S. from *J. Vsesoyuznogo khim. obshchestva im. D. I. Mendeleeva*, 1991, **36**, 231. Copyright J. Vsesoyuznogo khim. obshchestva imeni D. I. Mendeleeva)

spectral simulation using a Lorentzian line form [54]. The ^{13}C NMR assignments for main isotactic peaks are summarized in Table 8.5. In Fig. 8.8 some spectra are presented with resonant peak identification. The relative pentad peak intensities and parameters of the bicatalytic sites model were determined and successfully used to declare a similar polymerization mechanism for these polyolefins.

When copolymers of the olefins mentioned above are considered, a similar approach via γ-effect calculations enable the pentad comonomer sequences to be predicted, which were observed, for example, for 1-butene–propylene

Table 8.3. Identification of ^{13}C NMR aliphatic signals of polyisoprene

Sequence[a]	δ(ppm)	Sequence[a]	δ(ppm)	Sequence[b]	δ(ppm)
tT_5t	16.06	tC_1t	32.01		38.55
tT_5t	16.03	cC_1t	32.03	T_1T_1	38.40
cT_5c	16.00				38.30
tC_5c	23.48	ctC_1c	32.36		
cC_5c	23.44		32.24		
tT_4t	26.77				30.75
cT_4t	26.73	cC_1v	29.85	C_1C_1	30.70
cT_4c	26.67				30.66
tT_4c	26.62	ctT_1c	40.06		
tC_4t	26.57	ctT_1t	39.79		
cC_4t	26.53				28.53
cC_4c	26.43	ctT_1v	40.03	C_4T_4	28.38
tC_4c	26.40	vT_1c			28.18

[a] $c = cis$-1,4 configuration; $t = trans$-1,4 configuration; $v = $ 3,4 configurations.
[b] Inverse attachment.
(Reproduced by permission of Khachaturov, A. S., from *J. Vsesoyuznogo khim. obshchestva im. D. I. Mendeleeva*, 1991, **36**, 231. Copyright J. Vsesoyuznogo khim. obshchestva imeni D. I. Mendeleeva)

Table 8.4. ^{13}C NMR chemical shifts for C3 of polyolefins[a] (relative to *mmmm*)

Pentad	PPE	PHEX	PHEP	PO	PN
mmmm	0.00	0.00	0.00	0.00	0.00
mmmr	-0.18	-0.21	-0.22	-0.24	-0.25
rmmr	-0.18	-0.21	-0.22	-0.24	-0.25
mmrr	-0.18	-0.21	-0.22	-0.24	-0.25
mmrm	-0.35	-0.46	-0.50	-0.54	-0.55
rmrr	-0.35	-0.46	-0.50	-0.54	-0.55
mrmr	-0.59	-0.72	-0.67	-0.69	-0.69
rrrr	-0.59	-0.72	-0.86	-0.89	-0.88
mrrr	-0.8	-0.9	-0.87	-0.89	-0.88
mrrm	-0.9	-1.0	-1.06	-1.06	-1.04

[a] PPE = poly(1-pentene), PHEX = poly(1-hexene), PHEP = poly(1-heptene), PO = poly(1-octene), PN = poly) (1-nonene).
(Reprinted with permission from *Macromolecules*, 1991, **24**, 2334. Copyright 1991 American Chemical Society)

copolymers in reference [55]. Visual spectral information is exhibited in Fig. 8.9, showing the propylene methyl region of the carbon spectrum.

To continue, we present a detailed investigation of the sequence structure of another kind of polymer: poly(propylene oxide)—poly(PO) [56]. The absence of spin multiplets makes the application of ^{13}C NMR spectroscopy extremely effective for stereochemical analysis of homopolymers. Ring-opening polymer-

Table 8.5. ^{13}C NMR chemical shifts of isotactic polyolefins^a (ppm from internal TMS)

Carbon	PP	PB	PPE	PHEX	PHEP	PO	PN
C1	46.37	40.13	1.30	40.90	40.85	40.85	40.85
C2	28.30	34.93	33.33	33.09	33.06	33.04	33.05
C3	21.76	27.69	38.01	34.91	35.06	35.30	35.31
C4		10.79	19.94	28.86	26.22	26.59	26.68
C5			14.69	23.20	32.47	29.91	30.28
C6				13.86	22.58	31.91	29.35
C7					13.81	22.57	31.90
C8						13.79	22.53
C9							13.79

^{a}PP = polypropylene, PB = poly(1-butene), PPE = poly(1-pentene), PHEX = poly(1-hexene), PHEP = poly(1-heptene), PO = poly(1-octene), PN = poly(1-nonene).
(Reprinted with permission from *Macromolecules*, 1991, **24**, 2334. Copyright 1991 American Chemical Society)

ization can result in four head-to-tail (HT) stereochemical triads plus twelve additional regioirregular triads with (TH, TT), (TT, HH) and (HH, TH) addition during cleavage of both C—O bonds in propylene oxide. ^{13}C NMR might be expected to be more successful in such microstructure determination compared to proton resonance methods with selective deuteration and the two- dimensional technique. Stereo sensitive γ-effects were applied to the calculations together with the rotational isomeric state model (RIS model). In HT, poly(PO) methine carbons have two γ-substituents (2 CH), while methylene carbons have three γ-substituents and resonate about 2 ppm upfield from methine ones. In regioirregular poly(PO):

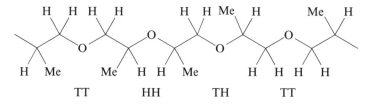

three γ-substituents (2 CH$_2$ + CH$_3$ or CH + CH$_2$ + CH$_3$) correspond to HH methine carbons; HT and TT methylene carbons, like HT methine carbons, have two γ-substituents (2 CH or CH + CH$_2$) with accompanying overlap of resonance signals. In such cases the multipulse techniques (i.e. DEPT or INEPT) help to separate observed NMRs into methine, methylene or methyl.

The 'insensitive nuclei enhanced by polarization transfer' (INEPT) procedure does not result in a full proton-decoupled ^{13}C NMR spectrum, but produces signals with different phases depending of their multiplicity or only signals with a certain multiplicity [57–59]. The distortionless enhancement by polarization

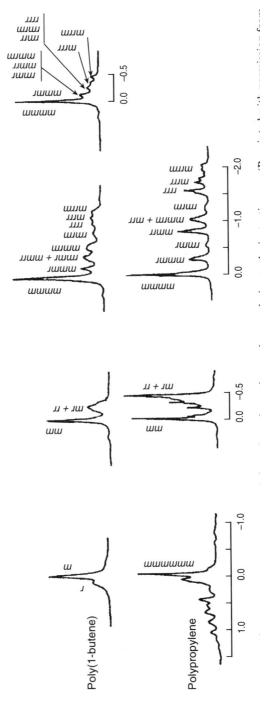

Fig. 8.8 ^{13}C NMR spectra of two polyolefins, where the peaks are shown relative to the isotactic ones. (Reprinted with permission from *Macromolecules*, 1991, **24**, 2334. Copyright 1991 American Chemical Society)

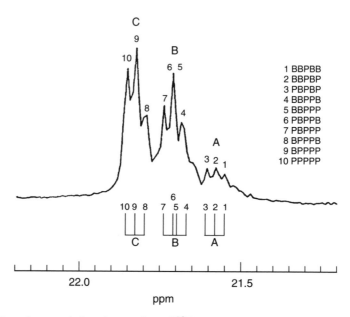

Fig. 8.9 Propylene methyl carbon region of ^{13}C NMR spectra of 1-butene(B)- propylene(P) copolymer: calculated chemical shifts are shown at the bottom. (Reproduced from *Polymer Commun.*, 1990, **31** (April), 131, by permission of the publishers, Butterworth Heinemann Ltd ©)

transfer' (DEPT) method has phase control by the used decoupler pulse angle and the delay time. It yields a series of primary, secondary or tertiary carbon resonances. Compared with the INEPT technique, the DEPT method is less dependent on the coupling constants J(CH) [60].

The most amazing result of the above experiments [56] is the presence of some methine carbon resonances in the upfield region (73.82, 73.30, 72.97, 72.93 and 72.87 ppm), previously thought to contain methylene resonances exclusively. The second step is the identification of end-group signals by variation of molecular weight (all of them occur between 75 and 76.5 ppm), followed by a quantitative analysis of stereo differences and defect additions using the above calculations.

A similar approach applied to copolymers of ethylene oxide (E), propylene oxide (P) and 1,2-butylene oxide (B) can be found in references [61] and [62]. For example, in the latter work, concerning B–E copolymers, one can use the auxiliary spectral information for homopolymers, block and statistical copolymers in $CDCl_3$ and C_6D_6.

The signal at 71.26 (70.33) ppm was assigned to the main-chain E block resonance; four lines at signal splitting in the a-methyl region, additional resonances not present in the block copolymers were assigned as indicated in Fig. 8.10. There is just one more result of this work, the analysis of the substituent

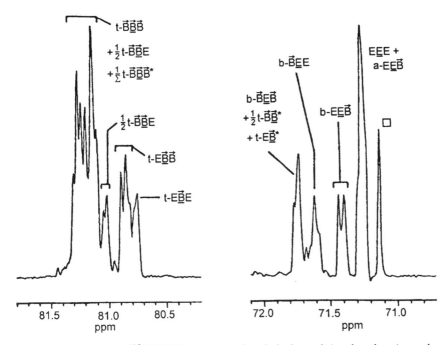

Fig. 8.10 Fragments of ^{13}C NMR spectrum of statistical copoly(oxybutylene/ oxyethylene) in C_6D_6: t designates CH resonances; a and b designate left and right CH_2 resonances. (Reprinted from *European Polymer J.*, **26**, 589, 1990, with kind permission from Elsevier Science Ltd, The Boulevard, Langford Lane, Kidlington OX5 1GB, UK)

additivity scheme, that is important for the following investigations. The complete quantitative determination for studied statistical copolymers takes into account the inverse addition of B units and the contribution of end groups. The triad sequence distribution changes as follows: 0.805 to 0.056 (EEE), 0.009 to 0.086 (BEB), 0.050 to 0.055 (EBE) and 0 to 0.396 (BBB) by variation of E/B comonomer ratio from 210:22 to 30:110.

By attaching ethylene oxide (E) and propylene oxide (P) to long-chain alkanols, important nonionic surfactants including poly(ethylene glycol) and poly(propylene glycol) fragments are obtained in reference [63]. The ^{13}C chemical shift is, in general, influenced by structural differences up to four bonds away from the observed C atom. This affirmation served as the basis for sequence analysis of the polymers in view. First, the measured signals were compared with the theoretical ones and model data; then determination of the individual triads was carried out. The results of the quantitative analysis show that E reactivity is higher (the overall ratio is approximately E/P = 4:1) the greater probabilities are 0.56 for EEE and 0.57 for PPP triads with other ones 0.19 (EPP), 0.10 (EEP), < 0.02 (EPE, PEP).

Apart from everything else, the opportunity to observe such resonances as —CN [43], —SCN [64], $\mathrm{>C=O}$, —COO [65, 66], —CONR$_2$ [66] is one of the benefits of ^{13}C NMR spectroscopy employment. Sometimes such data together with quarternary carbon resonances give exhaustive information, for example stereoregularity of 1-substituted 1,3-poly(bicyclobutanes) in reference [43] or styrene–ethyl acrylate (S–E) copolymers in reference [67]. Figure 8.11 shows the carboxyl (COO)$_E$ carbon spectral region and those of (C1)$_S$ quarternary carbon of phenyl from the latter work.

The signal splitting indicates its sensitivity toward different monomer sequences and tacticity in these carbon resonances. Three sets of signals, the intensities of which are changed with copolymer composition, are assigned to the carbonyl of SES, SEE and EEE triads ($\delta = 174.7$, 175.0 and 175.5 ppm respectively). The attribution of quarternary carbons is shown in the figure. The relaxation times T_1 for atoms of the same type have been found to be of the same order; thus quantitative calculations can be conducted and reactivity ratios can be evaluated:

$$r_{EE} = \frac{2[EEE]}{[EES] + [SEE]} \frac{f_s}{f_E} \qquad r_{SE} = \frac{[SEE] + [EES]}{2[SES]} \frac{f_s}{f_E}$$

as another example, conformational pentads were determined via proton-decoupled ^{13}C NMR spectra of 4-((2-(methacryloyloxy)ethyl)oxy)acetanilide (M)

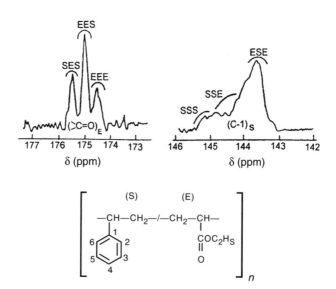

Fig. 8.11   Fragments of ^{13}C NMR spectrum of styrene–ethyl acrylate (S/E) copolymer: spectral regions of carboxyl and quaternary carbons. (Reproduced by permission of Hüthig & Wepf Verlag, Basel, Switzerland, from *Die Makromol. Chem.*, 1993, **194**, 1714)

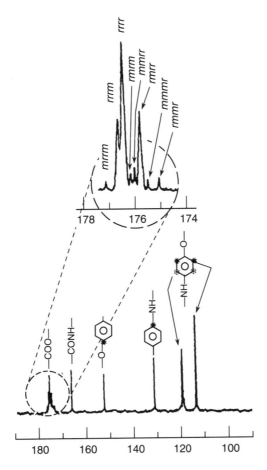

Fig. 8.12 Fragment of ¹³C NMR spectrum of copolymer of 4-((2-(methacryloyloxy)-ethyl)oxy)acetanilide (M) with 2-hydroxyethyl methacrylate (molar ratio 32:68) in DMSO-d_6 with expended carboxyl region. (Reprinted with permission from *Macromolecules*, 1991, **24**, 6086. Copyright 1991 American Chemical Society)

copolymers with 2-hydroxyethyl methacrylate in the carboxyl region (Fig. 8.12) [68]. The assignment of the resonance peaks to the indicated chemical structures was carried out according to resonances of corresponding homopolymers.

In a more complicated case, the esterification reactions of syndiotactic poly(methylallyl alcohol) by N- or/and O-protected amino acids are apparent, for instance, by the appearance of certain $>$C = O resonances in the 150–175 ppm region, though the quantitative characteristics have been obtained from the ¹H NMR spectra [69, 70].

8.1.3   NMR OF RARE NUCLEI

The possibility of using magnetic resonance for many other nuclei is an additional advantage of the method under review. The investigations on such nuclei as [11]B, ^{15}N, ^{19}F, [29]Si and ^{31}P are included in a wide experimental practice thanks to the high sensitivity of modern equipment. On the other hand, deuterium, being an integer spin nucleus ($I = 1$), is used extensively to characterize the dynamics and orientation of polymers and as a mark to control the reaction direction and to elucidate mechanistic and stuctural features, an interaction in the spin system [71–73]. This isotope is especially useful because of its minimal effect on the structure and reactivity of organic molecules. [2]H is easy incorporated into a structure, but has some disadvantages: broad spectral lines and a narrow frequency range. The use of ^{2}H to label the site of interest combined with {[2]H}^{13}C polarization transfer NMR experiments is more informative. Thereby, only carbons directly bounded with ^{2}H are detected, simplifying the spectrum. In reference [73], for example, an additional radiofrequency channel was used to carry out these INEPT experiments for polymer reactivity study.

^{29}Si NMR serves as a source of wide structural information for siloxane polymers [50]. Another interesting example is an investigation of anionic ring opening polymerization of 1-silacyclopent-3-ene by ^{29}Si NMR in reference [74]. Silicon signals observed at 17.60 and -1.92 ppm were assigned to heterocyclic fragments with quaternary and tertiary silicon atoms (low molecular model compounds were previously examined). Such groups are present in the polymer chain structure with numerous branch points, to which [29]Si resonances at -10.66 and -11.87 ppm may correspond:

^{29}Si NMR data also reflect different comonomer addition, as was observed for poly(dimethylsilylene-co-di-n-hexylsilylene) with nonrandom structure, including homosequences >5 di-n-hexyl units (-28.85 ppm from hexamethyldisiloxane in solution, -25 ppm in solid) and mobile, conformationally disordered dimethylsilylene units [75].

Figure 8.13 [76] contains the CP MAS ^{15}N NMR spectra. This is very easy to illustrate, but not to execute and interpret, and indicates the presence of one or more forms of synthetic nylon-12: 84.1 ppm of the α polymorphous component with a broad peak near 87 ppm of the amorphous one and 91 ppm peak of γ crystals. The relaxation parameters confirm these deductions.

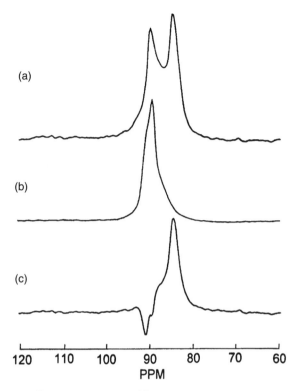

Fig. 8.13 CP MAS ^{15}N NMR spectra of ^{15}N-labeled nylon-12: (a) $\alpha + \gamma$ forms, (b) γ form and (c) subtraction of (b) from (a). (Reprinted with permission from *Macromolecules*, 1991, **24**, 6119. Copyright 1991 American Chemical Society)

As for halogen nuclei resonances, Vogl and co-workers found that purely isotactic polymers with one-handed helicity can be prepared from haloacetaldehydes [77, 78], and the optical purity can be determined by ^{19}F NMR spectroscopy because of chemical shift sensitivity to the diastereostructure. In the above-mentioned case of poly-(fluorochlorobromoacetaldehyde) there are combinations of two resonance signals at 98.1 and 98.2 ppm (C_6F_6 as a standard), assigned to (+) and (−) structures respectively.

The exotic experiments on ^{23}Na, ^{39}K and ^{35}Cl nuclei are described in reference [79]. The synthetic amphoteric metal methacrylate-co-(dimethylamino)ethyl methacrylate hydrogels containing both acidic and basic groups were studied. Line widths and chemical shifts were determined as a function of water content. Alkali metal and halide NMR were found to be sensitive to immediate chemical environment and mobility of ions. For instance, a rapid decrease in the ^{23}Na line width and downfield resonance shift in the range 1–5(6) water molecules per sodium ion suggest the presence of 5(6) water molecules in the first hydration sphere. The upfield shift changes (about 20 ppm) are strong evidence for the formation of contact ion pairs with the polyions.

8.1.4 2D NMR SPECTROSCOPY

Traditional techniques were not always suitable for unequivocal line assignment. As mentioned above, proton NMR spectroscopy is potentially an extremely powerful technique; however, extensive overlap of resonances yields the broad, difficult-to-interpret proton spectra of complex polymer systems. In many cases the power of two-dimensional NMR (2D NMR) spectroscopy has been demonstrated in polymer microstructure characterization [80–85]. 2D NMR spectra are acquired simultaneously along two identical or different frequency axes, and correlated spin states give rise to cross-peaks at the intersections of the respective peak positions. For example, two-dimensional NMR spectroscopy was used in reference [80] for poly-(*tert*-butyl acrylate)(PtBuA) analysis. The combination of three techniques (^{1}H and ^{13}C one-dimensional methods at 499.8 and 125.7 MHz respectively and two-dimensional spectroscopy) considerably simplifies the NMR problems and permits assignments to be made with subsequent structure characterization. In accordance with previous works authors could state that the ^{13}C NMR methine peak of stereoisomeric triads appear in the sequence *rr–rm–mm* up in the field. By direct quantitative analysis of two PtBuA samples prepared under different conditions (PtBuA I should be atactic, while PtBuA II – isotactic) the triad populations shown in Table 8.6 were determined.

In addition to this, the two-dimensional homonuclear COSY spectra show cross-peaks in the methine proton region (2.4–2.6 ppm), which reflect an even pentad structure (Fig. 8.14). The cross-peaks of methylene protons yield a determination of hexad polymer chain sequences.

In heteronuclear correlation (HETCOR) spectra the methine carbons give signals having cross-peaks with their own protons; this enables us to make and revise a corresponding interpretation for overlapped resonances (particularly, *mr*-centered units, resonating between *mm* and *rr* in the ^{13}C spectrum and at a higher field in the ^{1}H one). From the inverse HETCOR experiment (Fig. 8.15)

Table 8.6. Configuration triad and tetrad contents for PtBuA

	rr	*rm*	*mm*	*mmm*	*mrm*	*mmr*	*mrr*	*rrr*	*rmr*
PtBuA I	0.26	0.49	0.25	0.12	0.12	0.24	0.26	0.13	0.13
PtBuA II	0.08	0.42	0.50	0.37	0.15	0.29	0.11	0.02	0.06

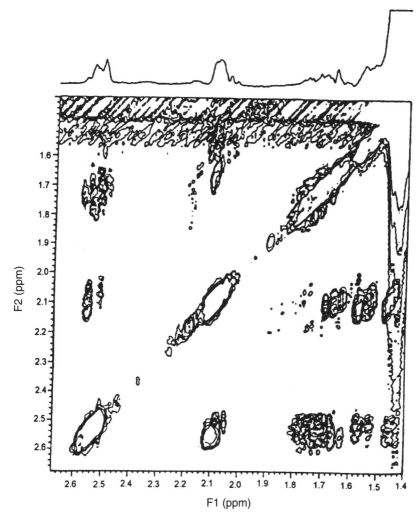

Fig. 8.14 Section of two-dimensional F_1-decoupled COSY spectrum of poly(*tert*-butyl acrylate) in C_6D_6. (Reprinted with permission from *Macromolecules*, 1993, **26**, 104, 105. Copyright 1993 American Chemical Society)

one can see six tetrad sequences and calculate their contents using the estimated probabilities $P_{r/m}$ and $P_{m/r}$. Chain propagation then obeys the Bernoullian statistics (Table 8.6). On the basis of the combined approach mentioned above the complete spectral assignment was made and confirmed.

Another example is the determination of the size of the ring generated during the cyclopolymerization reaction, which yields different ring structures in solution

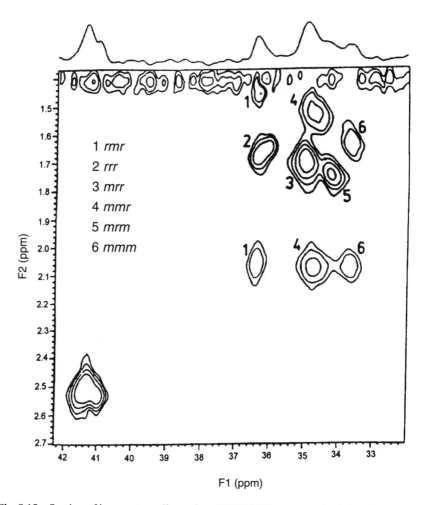

Fig. 8.15 Section of inverse two-dimensional HETCOR spectra of poly(*tert*-butyl acrylate) in C_6D_6. (Reprinted with permission from *Macromolecules*, 1993, **26**, 104, 105. Copyright 1993 American Chemical Society)

and in bulk. This involves the new cyclopolymer poly-[ethyl α-[(allyloxy)-methyl]-acrylate] [86] studied by the two-dimensional INADEQUATE (incredible natura abundance double quantum transfer experiment) NMR method involving a series of FIDs by variable evolution time, during which the double-quantum coherence is developed and observed indirectly via transformation into single-quantum coherence as a usual two-dimensional spectrum (this allows the carbon–carbon connection through the registration of cross-peaks to be defined). This is a significant problem since most systems can cyclopolymerize to give more than one ring structure. In the present case previous examination of two possible structures:

shows that only a five-member ring formation can result in the direct connection of the tertiary and quaternary carbons (d and c), which was really observed by means of two-dimensional spectroscopy. The spectral attribution was made using the combination of ^{13}C spectroscopy with full decoupling and with the decoupler off. There are two peaks for almost each carbon due to the existence of *cis* and *trans* isomers, i.e. for ring tertiary carbon d, 45.7 and 47.1 ppm, ring quaternary carbon c, 55.4 and 56.8 ppm respectively, such that the *trans* isomer is approximately twice that preferred.

The following example presents the application of DEPT, homonuclear 2D *J*-resolved, COSY and ^{13}C–^{1}H correlated spectroscopy to the interpretation of extremely complex spectral data of poly(vinylbutyral)(PVB), which is essentially a copolymer of butyraldehyde rings and unreacted vinyl alcohol [87]. There are two kinds of sequences: stereochemical and compositional. The multiplicity of carbon signals is determined from a series of DEPT spectra with reverse and nullified intensities (Fig. 8.16).

Then ^{13}C resonance assignments are transferred to the proton spectrum via 2D ^{13}C–^{1}H correlated spectroscopy. The proton assignment can be verified via a ^{1}H–^{1}H couplings map thanks to the proton COSY spectrum. After that proton coupling constants can be measured and stereosequences containing *meso* and racemic rings can be observed from the 2D *J*-resolved spectrum (Fig. 8.17). In particular six tetrads (*rrr, mrr, mrm, rmr, rmm, mmm*) were found for the butyral-dehyde part and *mm, mr, rr* triads for the vinyl alcohol portion.

The application of NMR requires only that the different structural and tactic sequences by assigned properly and, preferably, be analyzed properly in the context of propagation statistics. In studies of copolymerization kinetics and polymer microstructure, the use of reaction probability models can provide a convenient framework whereby the experimental data can be organized and interpreted, and can also give insight into reaction mechanisms [88].

Simultaneous investigation of the alternation of chemically and stereo- chemically different units in the chain leads to a drastic increase in the number of unequivalent chain fragments. As a consequence, the number of NMR signals that must be resolved within the same frequency range increases. NMR spectral signals corresponding to different polymer microstructures are often severely

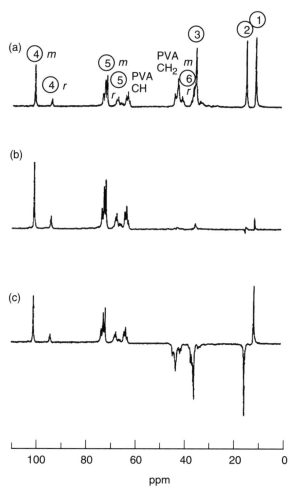

Fig. 8.16 DEPT spectra of poly(vinylbutyral): (a) all protonated carbons have positive signals, but quaternary carbon signals are nulled; (b) only methine signals are present; (c) methine and methyl signals are positive, but methylene signals are negative, quaternary carbons have zero intensity. (Reprinted with permission from *Macromolecules*, 1986, **19**, 1624. Copyright 1986 American Chemical Society)

overlapped. Thus, by the investigation of polymer microstructure one faces the problem of subdividing the overall spectrum into subspectra (each of the latter is a complex spectrum, containing sometimes dozens of components). Only in a few cases can one interpret spectral data unambiguously and carry out sequence determination beyond the triads. There are a few techniques that have been developed to overcome these difficulties, examples being shift reagents, multi-dimensional NMR techniques, deuteration and empirical additive shift rules.

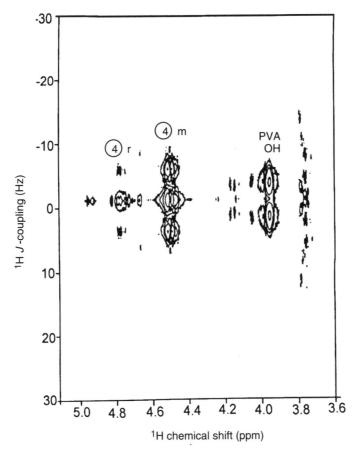

Fig. 8.17 Section of 2D *J*-resolved spectrum of poly(vinylbutyral): expanded CH, OH region. (Reprinted with permission from *Macromolecules*, 1986, **19**, 1625. Copyright 1986 American Chemical Society)

Moreover, some fundamental reasons have resulted in broadening of polymer NMR spectra. The first is macromolecular chain polymorphism, resulting in the partial overlap of multiple picks. The second is unresolved lines splitting. The third is the local mobility of the units or segments at issue. Various experimental procedures have been suggested for the artificial narrowing of spectral lines, among which are reduction of the solution concentration, temperature raising and suppression of the broadening interaction with experimental technique. Thereby much information contained in the spectral line width is lost. One of the possible methods of extracting information from unresolved spectra is computer spectrum simulation.

8.1.5 COMPUTATION ANALYSIS

Computation analysis utilizing all spectral information minimizes experimental errors to provide more precise data and gives quantitative results when polymer microstructures are very complex. One of the possible methods of recovering information from poorly resolved spectra is to simulate the form of the spectral line and to select the parameters (type of spin system, line form for individual components, chemical shifts, constants of spin–spin interactions, contributions of subsystems corresponding to diad, triad or tetrad concentrations) of the best model with minimum differences between experimental and calculated spectra. Such an approach was used for the first time by Bovey and Heatley [5, 89] and is successfully applied at present [4, 50, 54, 90–94]. Although the analytical expression of the individual spectral components may be simple, the reverse task is multiparametric. A sufficiently general procedure for analyzing poorly resolved spectra, which allows a large number of unknown parameters to be determined using a minimum of additional assumptions, was elaborated by Stroganov, Platé and coworkers [95–98]. In the described procedure the studied spectrum $F(x | P, A)$ is represented as the sum of several poorly resolved components $A_k f_k(x | P_k)$ with the Lorentzian or Gaussian form. The positions of the maxima, line half-widths and peak intensities are used as parameters. The sum of square deviations,

$$SSD(P, A) = \sum_i [F^{exp}(x_i) - F^{mod}(x_i | P, A)]^2$$

is a formal criterion of correct simulation. The best coincidence of the model with the experiment corresponds to the minimum SSD. The problem is thus reduced to one of searching the combination of model parameters P and A that minimizes the above sum. By the elaboration of minimizing algorithms there were a number of difficulties associated with the possibility of falling into a local minimum that were successfully overcome. The authors proposed an adaptable system of computation programs for solutions covering a wide area of such questions (including simulation of polymer chain growth, chemical transformation and dynamic of macromolecules) having a great volume of information. To illustrate the operation of the system, let us consider, for example, its application to the analysis of a poorly resolved spectrum of syndiotactic copolymer of methyl methacrylate (M) with methacrylic acid (A). The model of the spectrum is the sum of six Lorentzian components with arbitrary values of the line intensities, positions and widths (Fig. 8.18).

Spectral peaks are attributed to AAA, AMA, MAA, MMA, MAM and MMM triads respectively, from left to right. The obtained model spectrum does not differ practically from the experimental one. The adopted organization of the system makes it possible to use any models of the spectrum components including complicated superpositional ones (for instance, spin multiplets). Figure 8.19

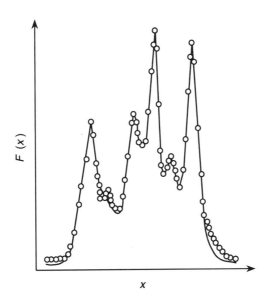

Fig. 8.18 ^{1}H NMR spectrum of methyl groups of cosyndiotactic copolymer of methyl methacrylate and methacrylic acid: experimental (continuous curve) and model (points)

illustrates the simulation of a poorly resolved spectrum of CH_2 protons of polymethyl methacrylate (PMMA) with splitting of the heterosteric tetrad signals. The figures designate spectral components that belong to different tetrad configurations: 1, *mmm*; 2, *rmr*; 3, *mmr*; 4, *mrr*; 5, *mrm*; 6, *rrr*.

On the other hand, this system includes the theory enabling the individual kinetic constants for irreversible polymeranalogous reactions from experimental data to be determined [99, 100]. To estimate the only correct set of kinetic parameters for the multistage process and to propose the most probable reaction mechanism is the essential task of polymerization kinetics. This approach includes the search for the constants set and minimizes the sum of square deviations between the measured and model integral characteristics (the conversion rate or triad concentration determined, for example, by NMR).

The use of this tool has made it possible to obtain a number of important results: to elaborate on a procedure for analyzing the microblock structure of methyl methacrylate–styrene copolymers by a poorly resolved doublet of phenyl protons and to offer the proper polymerization mechanism, to study the specific features of the stereochemical structure of poly-n-alkyl methacrylates, to investigate the distribution of units in syndiotactic methyl methacrylate–methacrylic acid copolymers and to estimate individual constants of hydrolysis and a quantitative relationship between the distribution alteration and neighboring group effect during hydrolysis of syndiotactic PMMA.

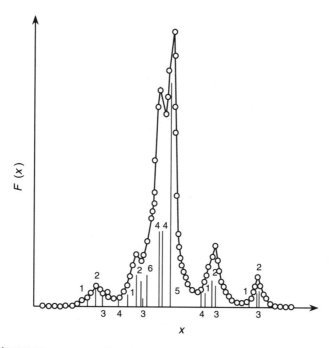

Fig. 8.19 ¹H NMR spectrum of methyl groups of polymethyl methacrylate: experimental (continuous curve) and model (points)

The model spectra can be obtained if the main spectral parameters are known or supposed: the chemical shifts from the experiments with deuteration, some interaction constants from the spectra of model compounds and the relative multiplet contributions from the proposed reaction probability and mechanism. In recent years the major part of the NMR spectral works dealing with polymer systems includes or takes into account the spectrum computation. There are many computer programs for such calculations, which were created on the basis of new and previous algorithms and postulates.

In the works of Cheng and Lee [i.e. 84, 88, 101] it has been noted that most applications of NMR to polymer problems follow this logical scheme: spectrum–interpretation–computational treatment–reaction probabilities ('analytical approach'). Authors use the probability models of polymerization reactions and fit the observed spectral line intensities to the theoretical values in order to test a given polymer for chain-end control or catalytic-site control mechanism. In reference [102] the master table is presented for diad, triad and N-ad calculations from NMR data, using the formalisms of Bovey and Price where the polymerization reactions are approximated by Bernoullian and Markovian first- or

Table 8.7. Calculated reaction probabilities of isospecific polypropylenes

Sample	1	2	3	4
mmmm	90	96.5	93.6	76.5
mmmr	4	1.7	2.7	6.2
rmmr	0	0	0	0
mmrr	4	1.7	2.7	8.7
rrmm + mmrm	0	0	0	2.0
rmrm	0	0	0	0
rrrr	0	0	0	0
rrrm	0	0	0	0
mrrm	2	0	1.0	5.0

second-order statistical models. The particular case, reaction probabilities of polypropylenes made with isospecific catalysts, is shown in Table 8.7.

Furthermore, in these works a general program has been written which simplifies this process and provides the complete ^{1}H and ^{13}C spectral analysis and the optimal reaction probabilities. Every spectral peak intensity is then associated with a theoretical expression involving reaction probability parameters, and the best fit is adjusted between observed and theoretical values by comparison. This permits the polymer system to be described by mathematical expressions.

8.1.6 SOLID STATE NMR SPECTROSCOPY

In recent years the methods of NMR spectroscopy have been used widely to investigate the microscopic and supermolecular organization of polymer chains in the glassy and semicrystalline states and the related dynamic behavior of these polymers, which determines their macroscopic properties including reactivity or modification ability. A classical example of such a relationship is the partial crystallinity of polymers, for instance polyethylene, when the mechanical strength depends on order and mobility of the polymer chains in the crystalline regions and amorphous layers. Among available techniques are the broad-band NMR and different relaxation experiments [93, 103, 104]. Furthermore, the CP MAS (cross-polarization technique with magic-angle spinning, mechanical rotation of the sample in view at 54.7 °C) by high-power coupling and/or the computer difference technique [105] (mutual comparison and corresponding subtraction of the spectra observed for solutions and amorphous, semicrystalline and crystalline samples) usually provides high-resolution spectra of solids.

The macroscopic orientation or preferential packing of polymer molecules is significant for following chemical and mechanical treatment and can be controlled by means of solid state NMR spectroscopy [i.e. 92, 93, 106–108]. Some basic principles, theoretical aspects and general methods available for solid material analysis are described in detail in the works of Spiess, Blumich and co-workers [60, 85, 91, 109–112] with rather full bibliography in references [113] and [114].

As an example of such investigations we can adduce the series of the works of Cholli and co-workers [115–117 and corresponding references], concerned with identification of resonance peaks due to crystalline and amorphous polymers (specifically, high-density polyethylene, poly(ethylene oxide)), with simultaneous detection of cross-linking, redistribution of the cross-links throughout the amorphous phase or unsaturation in gamma-irradiated materials. In Fig. 8.20 spectra are shown using dipolar dephasing when the proton RF field was turned off after the cross-polarization (DD CP MAS).

The main-chain resonance peaks at 34.1 ppm (narrow) and 32.3 ppm (broad) are attributed to the crystalline and amorphous states respectively. The line-broadening factors of amorphous regions reflect a multiplicity of conformational isomers and irregularities in the chain structure and packing. From this spectra the complete isolation of amorphous resonance is seen by the variation of delay time. These characteristics with a conclusion about the preferential place and

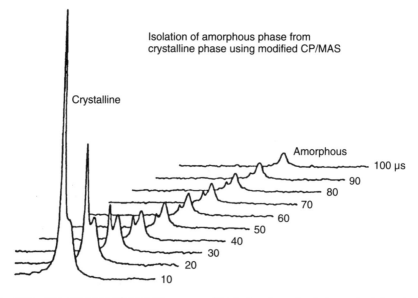

Fig. 8.20   DD CP MAS ^{13}C NMR spectra of high-density polyethylene: the delay times are indicated on the right side for each spectrum. (Reproduced by permission of John Wiley & Sons Ltd from *Spectroscopy Lett.*, 1988, **21**, 524)

mechanism of intramolecular conversion depend on the chemical and morphological changes.

^{13}C CP MAS spectroscopy can also be of use in analyzing the microstructure of insoluble products with determination of the triad sequences. Such an example is reference [118], which deals with insoluble maleic anhydride (M)–styrene (S) copolymers and their derivatives:

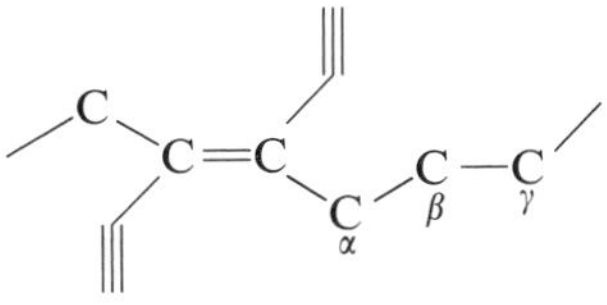

Available model compounds and the resolved spectrum of phenyl quaternary carbon (139, 141, 145 and 147 ppm for MSM, SSM, MSS and SSS triads respectively) favor the successful action. The structural, conformational and dynamic information for insoluble polydiacetylenes, an unusual class of polymers, which can produce macroscopic single crystals, is presented in an extremely interesting work [119]. The urethane groups in the side chains of poly-5,7-dodecadiyne-1,12-diol-bis[((butoxycarbonyl)methyl)urethane] form hydrogen bonds between neighboring chains, resulting in a polymer network. Being heated from 23 to 90 °C, the polymer molecules undergo a transformation, resulting in a large upfield shift of alkyne carbon resonance from 107.6 to 102.9 ppm (DD CP MAS) due to localization of π-electron density (Fig. 8.21); subsequent cooling gives two alkyne peaks at 104.9 and 101.9 ppm corresponding to irreversible transition with partial melting at 90 °C and formation of two (ordered and amorphous) phases. Such heating caused 27.3 and 30.9 ppm chemical shifts for β- and γ-methylene carbons instead of a single initial value of 24.9 ppm, which is believed to accompany the thermochromic transition by the side-chain conformational change (*gtg* to *ttt*). The latter affects the chemical shift via the *γ-gauche* effect. Furthermore, hydrogen bonding of side chains exists in all of the above polymer forms and is characterized with similar urethane carbonyl resonance at about 158.5 ppm.

The two-dimensional solid state NMR spectroscopy is a more informative and conclusive technique for structural investigation [85, 120]. Many simple two-dimensional experiments are used today as routine methods to assign proton and carbon resonances. To solve more complicated problems one can use 2D ^{13}C MAS with rotor synchronization by fixing the rotor position relative to the magnetic field and starting the acquisitions at various angles in two-dimensional exchange spectroscopy; then an exchange of two molecular states during the 'mixing time' (period between two observations) results in the intersectional signal in the two-dimensional spectrum. The technique of combined rotation and multipulse spectroscopy (CRAMPS) being developed for solid samples and

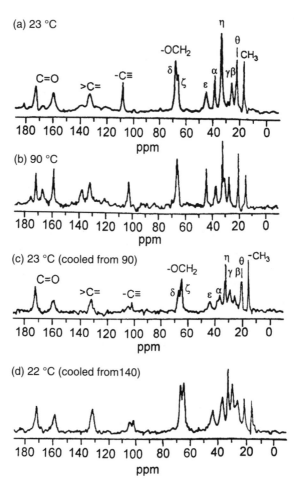

Fig. 8.21 DD CP MAS ^{13}C NMR spectra of poly-5,7-dodecadiyne-1,12-diol bis[((butoxycarbonyl)methyl)urethane]. (Reprinted with permission from *Macromolecules*, 1990, **23**, 3058. Copyright 1990 American Chemical Society)

abundant nuclei, e.g. protons, can sometimes provide structural information not evident from the corresponding ^{13}C spectra [121]. Finally, we would like to mention a three-dimensional NMR experiment for complex solid systems, illustrated by the analysis of isotactic polypropylene [122]. The anisotropy of chemical shifts can be used if it is incorporated into a solid state version of two-dimensional exchange NMR experiment. Thereby one can avoid the loss of information because of averaging anisotropic parts of NMR interactions (chemical shift and quadrupolar coupling). More detailed information is out of place owing to the complexity of the above problems.

8.1.7 SUMMARY

Returning to the above narration let us sum up: the existing level of technique and methodology enables reliable investigations of the units distribution, stereochemical configurations and structural modifications to be carried out for a number of practically important polymers. Depth and authenticity of spectral information depend on peculiarities of the polymer material in question and demand the individual experimental approach. The NMR data of the whole polymer include contributions from all its components, and a proper treatment of these data is not a trivial task. Proton NMR can provide identification of compositional tetrad sequences for different types of copolymers (triad distribution giving considerable insight into the mechanism of chain propagation can practically always be estimated when the model compounds are available); at 400–500 MHz one can extract meaningful information concerning stereochemical tetrad distribution in polymer chains. ^{13}C NMR is more informative from this point of view. Determination of conformational pentads as well as compositional ones is a potentially feasible task for a wide list of macromolecular compounds. In some cases if simulation methods are applied determination of stereo heptads can be achieved. For multicomponent systems the use of reaction probability models and the computerized optimization technique provide a convenient base for analysis. Moreover, it is often possible to detect multiple comonomer additions (up to twenty, as has been found for some biological macromolecules) via resonances of functional group nuclei. Thereat we should have in mind indispensable investigation of auxiliary systems.

Finally, it must be noted that this field of physical chemistry is not limited by represented examples and meets the extremely complex problems with a powerful arsenal of various continuously improved experimental methods and theoretical approaches.

8.2 DETERMINATION OF UNIT SEQUENCE DISTRIBUTION IN POLYMERS BY PYROLYSIS GAS CHROMATOGRAPHY/MASS SPECTROMETRY

As shown above, high-resolution NMR is extensively used in the determination of polymer primary structure. However, methods combining pyrolysis, gas chromatography and mass spectrometry are also very promising. One of these methods, pyrolysis gas chromatography/mass spectrometry (GC/MS), which was developed quite recently, combines the pyrolysis of a polymer, gas chromatographic separation of the products of thermal decomposition of this polymer and mass spectrometric identification of the structure of products in a single-run experiment. Using quantitative gas chromatography and mass spectrometry supported by the extended possibilities offered by automated scientific instrumentation, this method allows rather accurate evaluation of the concentrations of short sequences of monomer units, the block parameters of the copolymers, the degree of disorder of the alternating copolymers, branching of macromolecules and arrangement of monomer units in the chain. Further discussion will concern the possibilities of determining the monomer unit sequence distribution in copolymers by this method [123].

8.2.1 BASIC PRINCIPLES OF PYROLYSIS GC/MS

Three individual methods form the basis of pyrolysis GC/MS. These are pyrolysis, gas–liquid chromatography (GC) and mass spectrometry (MS). The two latter methods are the instrumental techniques most widely used in the analysis of the structure of organic molecules and the qualitative and quantitative analysis of complex mixtures. However, they are applied to the examination of macromolecular compounds, these methods are strongly limited by the nonvolatility of polymers. This problem may be solved by the involvement of thermal decomposition, which has contributed much to the development of modern concepts of polymers structure and composition [124, 125]. Although pyrolysis may be carried out in a separate experiment, its on-line mode is most effective. In this case, the pyrolyzer may be directly connected to the instrument analyzing the composition of the pyrolysis products and their structure. Interfacing a pyrolyzer with a mass spectrometer or a gas chromatograph produced pyrolysis MS and pyrolysis GC, which may be considered to be predecessors of pyrolysis GC/MS. Specific features and potentials of the former two methods were discussed in detail in relatively recent monographs [126, 127].

The type of pyrolyzer may be selected according to the problem to be solved and the method used to identify the products. However, they all should meet the following three requirements [128]:

(a) reproducibility of the decomposition process;
(b) low inertia of the heating and reproducibility and stability of the temperature regime;
(c) rapid removal of the primary products of pyrolysis from the heating zone in order to avoid secondary processes involving the reactions of radicals with the sample or the material of the pyrolysis unit.

Three types of pyrolyzers conform to these requirements: filament pyrolyzers, Curie point pyrolyzers and furnace-type pyrolyzers. The first two are the pulse-mode-heating pyrolyzers, whereas the third is of the continuous-heating type.

In pyrolysis MS instruments, the pyrolyzer (commonly, the furnace-type pyrolyzer) is connected to the ion source of a mass spectrometer. Pyrolysis MS is widely used for the qualitative analysis of polymeric materials, especially commercial polymers, and only rarely to investigate their microstructure.

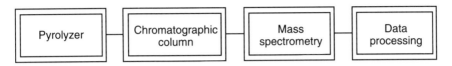

Fig. 8.22 Block diagram of the pyrolysis chromato-mass-spectrometric setup

The unique potential of GC separation and quantitative analysis of various mixtures gave the impetus for combining it with pyrolysis, which is interesting as a method for identification of the microstructure of homo- and copolymers. Pyrolysis GC is a rather 'old' method; however, it is still widely used in polymer analysis [127, 129]. The most serious drawback of pyrolysis GC is that it offers very little information on the structure of the pyrolysis products; it requires laborious syntheses to compare the pyrolysis products with authentic compounds. This is, in general, not a disadvantage of pyrolysis GC/MS because the method involves the mass spectrometer as a detector, which allows efficient identification of pyrolyzate components. Pyrolysis GC/MS may be considered a modification of the reaction GC/MS that was recently developed [130]. Figure 8.22 shows the block diagram of the corresponding setup.

Obviously, pyrolysis GC/MS uses the same types of pyrolyzers as pyrolysis GC, i.e. pulse-mode filament and Curie point pyrolyzers and furnace pyrolyzers with very fast heating. In addition to the three requirements listed above, the pyrolyzers used in pyrolysis GC and pyrolysis GC/MS must conform to two more: (a) they must ensure pulsed injection of the pyrolysis products into the column and (b) the dead volume between the pyrolysis zone and chromatographic column must be as small as possible (this is necessary in order to avoid losses of high-boiling products, which may condense on the cool parts of the instrument).

The first of these conditions is of primary importance because the pulsed injection of pyrolysis products into the column corresponds to the regime of column performance, ensuring good separation of the products. The pulsed injection is provided by the very high heating rate of the analyzed sample to the pyrolysis temperature (for synthetic polymers, this temperature is usually 500–600 °C), which ensures the shortest time of polymer decomposition. In pyrolyzers used in pyrolysis GC and pyrolysis GC/MS, the heating time of the sample to a required temperature is usually several microseconds, but not greater than 5–10 s.

The advantages of pyrolysis GC/MS are high sensitivity, simple and fast sample preparation, the possibility of examining insoluble polymers in small quantities and the vast potential of elucidating the qualitative and quantitative composition of pyrolyzates. Note that, with polymers, the same structural tasks can be solved on a qualitative and quantitative basis using pyrolysis GC. However, the advantages of pyrolysis GC/MS listed above allow a faster, more effective and more sensitive analysis.

8.2.2   QUALITATIVE ASSESSMENT OF POLYMER MICROSTRUCTURE USING PYROLYSIS GC/MS

Identifying the pyrolysis products and examining the pyrograms with pyrolysis GC/MS makes it possible to relate the products to the elements of the polymer microstructure. The principal features of the cleavage of the macromolecular chain are controlled by the chemical structure of the polymer; it is this factor that enables the use of pyrolysis for the elucidation of polymer microstructure. In this connection, it is essential to have a clear understanding of the major trends in the degradation of macromolecules. The mechanisms of degradation of major polymers have been investigated extensively [131, 132] and studies on this topic are ongoing. When conducting pyrolysis at moderate temperatures, the weak bonds are initially split. According to the type of bond scission reaction and the character of the products formed, the processes of thermal degradation of polymers are divided into those occurring via free radical chain mechanisms and nonradical chain mechanisms. The latter involve various rearrangements, Diels–Alder intramolecular cyclization, condensation reactions, concerted cleavages and elimination of small molecules. These reactions are common for polyesters, polyurethanes, polysiloxanes and polymers with polar side chains. Thermal degradation of polyolefins, polydienes, polystyrenes and their derivatives, and of many other similar polymers, proceeds via free radical reactions, which are governed by the well-known principles of free radical chemistry. For example, the formation of tertiary radicals is more favorable than that of secondary and primary radicals because the former are more stable, etc. Degradation of polymers containing aromatic rings favors the formation of benzyl radicals. In macromolecules containing double bonds, it is the α-carbon atom that is preferred as a radical site: $C{=}C{-}C$. The most common reactions of macroradicals occurring in the course of pyrolysis are β-cleavage of the $C{-}C$ bonds of the chain, hydrogen elimination and hydrogen addition. The reactions involved in thermal degradation of macromolecules are controlled by the nature of the decomposed polymer and may be divided into the following groups:

(a) Depolymerization. This route leads to monomers, dimers, etc. This reaction is the predominant route of degradation of polymers containing quaternary and tertiary carbon atoms.

(b) Random chain cleavage followed by chain transfer and stabilization of the fragment formed, for example, by cyclization or β-elimination. This pathway is common for polyolefins.

(c) Elimination of side groups, yielding small molecules, and the subsequent cleavage of the main chain, yielding cyclic products or monomer, dimer and other products. This reaction is typical for vinyl polymers containing polar side groups (e.g. polyvinylcyclohexane, polyacrilonitrile and various polyacrylates).

When investigating polyolefins with pyrolysis GC, the pyrolysis products are often subjected to hydrogenation, which is usually conducted on-line in a micro-reactor placed between the pyrolyzer and chromatographic column. The pyro-gram may thus be reduced to a simpler pattern allowing more reliable product identification. When conducting the investigation with pyrolysis GC/MS, hydro-genation is not always necessary because the method offers extensive opportuni-ties for identifying the pyrolysis product. However, hydrogenation may be very useful when a reliable mass spectrometric identification of pyrolysis products is impossible (this may be the case with unsaturated compounds). The micro-hydrogenator may be placed between the pyrolyzer and the column. However, hydrogenation between the chromatographic column and a mass spectrometer offers greater structural and analytical efficiency [133]. The latter procedure allows separate identification of those components of pyrolyzate that produce identical saturated products when hydrogenated.

Application of pyrolysis GC and pyrolysis GC/MS to the qualitative analysis of the primary structure of copolymers, such as the character of unit alteration, sequence distribution and the extent of block sequences, involves the identifica-tion of true homo- and hetero-dimers, -trimers, etc., as well as of the products resulting from scission of chain C—C bonds at the sites of branching.

Pyrolysis GC/MS appeared to be very informative in distinguishing block and random copolymers. For example, this method was used to identify block, random and random block copolymers of ethylene with vinylcyclohexane (VCH) [134]. It was found that the copolymers synthesized under steady state condi-tions (constant ethylene pressure) did not give VCH dimers during pyrolysis, and the pyrolyzate contained only hydrocarbons related to polyethylene (PE) and alkylcyclohexanes. Hence, these copolymers were assumed to be random. Specially synthesized ethylene–VCH block copolymers, when subjected to pyro-lysis, gave products characteristic of PE and polyvinylcyclohexane [135]. The copolymers prepared under non-steady-state conditions were attributed to random block copolymers because their pyrolysis produced both hydrocarbons characteristic of PE and polyvinylcyclohexane and also alkylcyclohexanes.

8.2.3 QUANTITATIVE DETERMINATION OF SEQUENCE DISTRIBUTION IN COPOLYMERS BY PYROLYSIS GC/MS

When the requirements concerning reproducibility and specificity of the analyti-cal pyrolysis are fulfilled, quantitative treatment of chromatographic data be-comes possible in pyrolysis GC and pyrolysis GC/MS. Consequently, polymer structure may be examined on a quantitative basis. Quantitative analysis by pyrolysis GC/MS involves selecting characteristic compounds in pyrolyzate that reflect the microstructure of the sample, area measurement of the corresponding zones in pyrograms and quantitative processing of this data [136]. By finding a correlation between the yields of corresponding fragments and microstructure of the macromolecule examined, one may identify the arrangement of monomer units, their distribution along the chain, sequence distribution (usually of diads and triads) and other features descriptive of polymer microstructure. For dealing with these aspects, pyrolysis GC/MS may present a strong alternative and addition to high-resolution NMR spectroscopy.

When examining the ethylene–propylene copolymers with pyrolysis GC, Voigt [137] found that a block copolymer and the propylene blend of the same composition gave similar pyrograms. This feature was used to evaluate the content of propylene units in block copolymers, which was based on evaluating the relative areas of 2,4-dimethylhept-1-ene and 1-heptene. To assess the content of propylene units in random copolymers, relative areas of 4-methylhept-1-ene and 1-octene were used. Van Schooten and Evenhuis [138] reported the investigation of copolymers of ethylene with propylene, butene, partially unsaturated ternary copolymers of ethylene with propylene and cyclopentadiene, and others using pyrolysis, hydrogenation and chromatographic separation of the hydrogenated products; the structure of copolymers was established, monomer unit alternation was characterized and the head-to-head and tail-to-tail arrangements were identified. Pyrolysis, hydrogenation of pyrolyzate and vapor-phase chromatography were used to investigate ethylene–propylene copolymers in a broad range of copolymer compositions [139]. The amount of propylene units incorporated in a head-to-head arrangement and distribution of the units along the chain were assessed.

Generally, the pyrolysis of copolymers comprising monomers whose homopolymers are conspicuously different in their thermal properties is controlled by the composition and microstructure of the copolymer and the experimental conditions.

It is a rather old observation that, in the thermal decomposition of copolymers, the yield of monomers is lower than when the blends of the corresponding homopolymers are degraded. This phenomenon is related to the influence that the neighboring monomer unit has on the reactions effecting the decomposition. This influence has generally been called the 'boundary effect'. For vinyl copolymers comprising the monomers $CH_2{=}CHX$ and $CHY{=}CHY$ parameter Q was suggested [124], which characterizes the monomer yield as a function of unit distribution. This parameter is defined as the ratio of the number of scissions of the bonds inherent in initial monomers, which lead to the formation of hybrid monomers $CHX{=}CHY$ and $CH_2{=}CHY$, to the total number of bonds in the chain:

$$Q = 2\frac{\text{internal monomer bonds broken}}{\text{internal monomer bonds in polymer}}$$

When $Q = 0$, only external (i.e. the bonds between the monomer units) bonds break and the initial monomers are recovered. At $Q = 2$, only the internal bonds are ruptured and no initial monomers evolve. In the intermediate case, bonds of both kinds are ruptured at random. Note that only a few copolymers are known that degrade by depolymerization to monomers. Taking these assumptions into account, Zigel *et al.* [140] examined the distribution of units in the copolymers of tetrafluoroethylene with ethylene or hexafluoropropylene.

Shibasaki [141] modified the boundary effect theory by taking into consideration the degradation of the copolymers of the AB type, where A and B denote monomer units. It is well known that, at relatively low temperatures, it is the weak bonds that break first. This process is largely controlled by the stability of the emerging radicals, which are strongly affected by the penultimate unit of the polymer chain. For example, the probabilities for the occurrence of the reactions

$$\cdot AA^{\sim} \rightarrow A + \cdot A^{\sim}$$

$$\cdot AB^{\sim} \rightarrow A + \cdot B^{\sim}$$

are different. Shibasaki introduced parameters β_A and β_B, which are defined as the ratios between the probabilities of the corresponding reactions P_{AB}, P_{AA}, P_{BA} and P_{BB}: $\beta_A = P_{AB}/P_{AA}$, $\beta_B = P_{BA}/P_{BB}$. Having determined the fractional contents of various bonds (A—A, B—B, A—B and B—A) via parameter R, giving the number of homodiads per 100 chain units [142], one may deduce parameters for copolymers of various compositions. For copolymers of styrene with acrylonitrile or methyl methacrylate as examples, Shibasaki [141] demonstrated that parameters were independent of copolymer composition, the pyrolysis temperature and the nature of the monomers. Minin *et al.* [143] applied a similar approach to elucidate the microstructure of commercial formaldehyde–dioxolane copolymers.

Tsuge and Takeuchi [144] suggested that β does not necessarily have to be constant, but rather may vary depending on the monomer unit distribution. A both-side boundary effect theory was developed and successively applied to the analysis of sequence distributions in some vinyl copolymers. The theory is based on the assumption that the dimers evolving in the course of degradation of a copolymer comprising monomers A and B, e.g. AA, BB and AB, may be cut out of different sequences offering different environments of neighboring units. For example, the AA dimer may be cut out of the AAAA, BAAA and BAAB tetrads. It is evident that the probabilities of cutting the AA dimer out of different environments are not equal.

If the yields of the dimers are proportional to the concentrations of the corresponding diads in the copolymers, the following relationships must hold:

$$Y(AA) = K_1 P(AA) \tag{8.1}$$

$$Y(AB) = K_2 [P(AB) + P(BA)] \tag{8.2}$$

$$Y(BB) = K_3 P(BB) \tag{8.3}$$

where $Y(\ldots)$ are the relative yields of dimers measured from the pyrograms (note that $Y(AA) + Y(AB) + Y(BB) = 1$) and $P(\ldots)$ are the concentrations of the corresponding diads in the copolymers, whereas K_1, K_2 and K_3 are the adjusting

Macromolecular Reactions

coefficients, whose distinct values depend on the probabilities of dimer formation. For copolymers whose monomer reactivity ratios are not very different and whose corresponding homopolymers are degraded at an almost equal rate, the values of K_1, K_2 and K_3 should be similar. Such is the situation, for example, with the copolymers of styrene with m-chlorostyrene and styrene with p-chlorostyrene [145]. These monomers are similar in reactivities and their homopolymers depolymerize to a similar extent, forming approximately 12–14% of dimers. On the assumption that K_1, K_2 and K_3 are approximately equal, Okumoto *et al.* [145] deduced the concentrations of corresponding diads solely on the basis of the relative yields of the dimers. A comparison of these data with those from copolymerization theory [146] confirmed the applicability of pyrolysis GC/MS to the elucidation of sequence distribution of these copolymers. A similar approach was used to investigate methyl methacrylate–styrene copolymers obtained at high monomer conversions [147]. The diad and triad distributions were in good agreement with those predicted by the theory.

However, in the case of copolymers, for which the relevant homopolymers are strongly different in thermal properties and reactivity ratios, the probabilities of formation of the products of thermal decomposition are controlled by the units on both sides of the evolving fragment [148]. Because the dimer, e.g. AA, may be cut out of different tetrads, the probabilities of cutting this dimer out of the corresponding tetrads are different and the relative yields of the dimer $Y(\text{AA})$ is a function of the tetrad concentrations $P(...)$:

$$Y(\text{AA}) = k_1 P(\text{AAAA}) + 2k_2 P(\text{BAAA}) + k_3 P(\text{BAAB})$$

where k_1, k_2 and k_3 are the probability constants for cutting the dimer out of corresponding tetrads. Similar probability constants are introduced for the BB and AB dimers. The concentrations of tetrads are related to the concentrations of the corresponding diads in the following way:

$$P(\text{AAAA}) = P_{\text{AA}}^2 P(\text{AA})$$
$$P(\text{BAAA}) = P_{\text{AA}} P_{\text{AB}} P(\text{AA})$$
$$P(\text{BAAB}) = P_{\text{AB}}^2 P(\text{AA})$$

where P_{AA} and P_{AB} are the probabilities of either an A or a B unit following monomer unit A [149]. Copolymerization theory gives the relationships for P_{AA} and P_{AB} as a function of monomer reactivity ratios r_{A} and r_{B} and the ratio between the initial monomer feed concentrations $z = [\text{M}_\text{A}]/[\text{M}_\text{B}]$ [146, 149]. By combining these equations with similar equations for BB and AB dimers the relationships for K_1, K_2 and K_3 as functions of composition and unit distribution are obtained:

$$K_1 = \frac{r_A^2 z^2 k_1 + 2 r_A z k_2 + k_3}{(r_A z + 1)^2} \tag{8.4}$$

$$K_2 = \frac{r_A z^2 k_4 + (1 + r_A r_B) z k_5 + r_B k_7}{\frac{1}{2}(r_A z + 1)(z + r_B)} \tag{8.5}$$

$$K_3 = \frac{z^2 k_s + 2 r_B z k_9 + r_B^2 k_{10}}{(z + r_B)^2} \tag{8.6}$$

where k_1 to k_{10} are the probability constants for cutting the dimers out of corresponding tetrads. For simplicity, the constants k_5 and k_6 of forming the AB dimer in AABA and BABA tetrads are assumed to be equal.

Because each of equations (8.4) to (8.6) involves three probability constants k, each of the k constants may be calculated using three values for adjusting the coefficient K, which are obtained with copolymers of three different compositions. The adjusting coefficients are calculated from Eqs. (8.1) to (8.3) using the dimer yields determined from pyrograms and theoretically calculated concentrations of the corresponding diads. This case may be treated within the framework of copolymerization theory relying on the first-order Markov chain statistics [149]. According to this theory, the concentrations of diads P_t are given by the following relationships:

$$P_t(AA) = \frac{r_A z}{2 + r_A z + r_B/z} \tag{8.7}$$

$$P_t(AB) + P_t(BA) = \frac{2}{2 + r_A z + r_B/z} \tag{8.8}$$

$$P_t(BB) = \frac{r_B/z}{2 + r_A z + r_B/z} \tag{8.9}$$

Three copolymers with the highest diad concentrations are usually used in the initial calculation. The calculated coefficients K are then used to calculate constants k from Eqs. (8.4) to (8.6). These equations are used to calculate new values of adjusting coefficients K_1^{calc}, K_2^{calc} and K_3^{calc}, which are valid, however, for the entire range of compositions. The diad concentrations P_{calc} are then calculated using Eqs. (8.1) to (8.3) and the calculated adjusting coefficients. The fact that the diad concentrations calculated in this way are close to the theoretically calculated values suggests a random monomer unit sequence distribution in the copolymers examined.

Let us exemplify the described treatment by considering the microstructure of vinylcyclohexane (VCH)–styrene copolymers [150, 151]. The pyrolysis of these copolymers produces the same monomeric, dimeric and trimeric products that are typical of individual poly-VCH [135] and polystyrene and also the hybrid

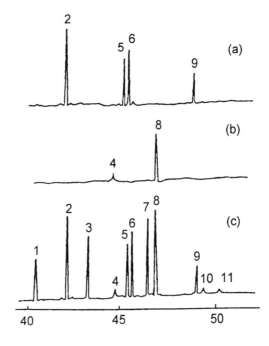

Fig. 8.23   Typical pyrograms in the dimer regions for (a) poly-VCH, (b) polystyrene and (c) VCH–styrene copolymer (40 mol % styrene). (Reproduced by permission of Nauka, Moscow, Russia, from *Vysokomol. Soed. A*, 1993, **35**, 1561)

dimers and trimers that are characteristic of the copolymers. Only the dimers were used to assess sequence distribution because the chromatographic resolution of the trimers was poor. Figure 8.23 shows the dimer regions of the pyrograms of poly-VCH, polystyrene and VCH-styrene copolymer. The structures identified in the mass spectra are presented in Table 8.8.

Relative yields of homodimers $Y(VV)$ and $Y(SS)$ and heterodimers $Y(VS)$ were used in the calculations (V and S denote the monomer units of VCH and styrene respectively). The quantities proportional to these yields are determined from the pyrograms by summing the areas of individual dimers and of the compounds that differ from one another by one carbon atom. For example, $Y(VV)$ is the sum of the areas of zones 2, 5, 6 and 9, whereas $Y(SS)$ is the sum of areas 4 and 8 (Fig. 8.23, Table 8.8). Taking into account the fact that, for compounds which are similar in structure and molecular mass, the sensitivity of a mass spectrometric detector is almost the same, no coefficients adjusting for sensitivity were introduced.

Table 8.9 shows the probability constants of diad formation of VCH–styrene copolymers. One can see that the probabilities of the formation of homodimers are essentially controlled by the nature of neighbor units; they are higher if the

Table 8.8. Structures of dimers[a] (Fig. 8.23)

Zone number	Compound	Zone number	Compound
1	CH_2—CH=CH (phenyl / cyclohexyl)	7	CH_2—CH_2—CH—CH_3 (phenyl / cyclohexyl)
2	CH_2—CH_2—CH_2 (cyclohexyl / cyclohexyl)	8	CH_2—CH_2—C=CH_2 (phenyl / phenyl)
3	CH_2—CH_2—CH_2 (phenyl / cyclohexyl)	9	CH_2—CH_2—CH—CH=CH_2 (cyclohexyl / cyclohexyl)
4	CH_2—CH_2—CH_2 (phenyl / phenyl)	10	CH_2=C—CH_2—C=CH_2 (phenyl / cyclohexyl)
5	CH_2—CH_2—CH—CH_3 (cyclohexyl / cyclohexyl)	11	CH_2=C—CH_2—CH—CH_3 (phenyl / cyclohexyl)
6	CH_2—CH_2—C=CH_2 (cyclohexyl / cyclohexyl)		

[a] Presumed position of the double bonds in shown.
(Reproduced by permission of Nauka, Moscow, Russia, from *Vysokomol. Soed. A*, 1993, **35**, 1561)

dimer neighbors the units of the second comonomer. The probability of formation of hybrid dimers is relatively low and is almost independent of the unit sequence.

Figure 8.24 compares relative dimer yields with theoretical and calculated diad concentrations in VCH–styrene copolymers. As can be seen, the theoretical and calculated diad concentrations are very close within the entire range of compositions, suggesting the applicability of the boundary effect theory to these copolymers and the random distribution of monomer units in the chain.

Okumoto et al. [148] determined diad concentrations in binary copolymers of acrylonitrile with m-chlorostyrene or p-chlorostyrene. The calculated diad concentrations agreed well with those deduced on the basis of copolymerization theory. The boundary effect theory was also applied to determine the triad concentrations in acrylonitrile–styrene copolymers; the calculated diad and triad concentrations showed good agreement with those predicted by the theory [152].

Table 8.9. Dimer probability formation constants for VCH–styrene copolymers

Dimer	Tetrad	Probability formation constant
	—VVVV—	$k_1 = 0.94$
VV	—VVVS—(—SVVV—)	$k_2 = 1.22$
	—SVVS—	$k_3 = 2.30$
	—VVSV—	$k_4 = 0.50$
VS	—VVSS—	$k_5 = 0.26$
	—SVSV—	$k_6 = 0.26$
	—SVSS—	$k_7 = 0.31$
	—VSSV—	$k_8 = 2.44$
SS	—VSSS—(—SSSV—)	$k_9 = 1.30$
	—SSSS—	$k_{10} = 1.00$

(Reproduced by permission of Nauka, Moscow, Russia, from *Vysokomol. Soed. A*, 1993, **35**, 1561)

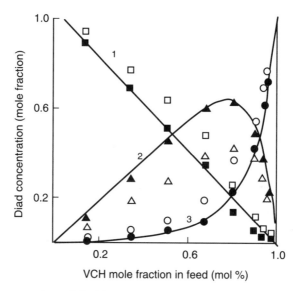

Fig. 8.24 Concentrations of diads in VCH–styrene copolymers: P(SS) (1), P(VS) + P(SV) (2) and P(VV) (3). Closed symbols refer to calculated concentrations of diads; open symbols show the experimental data. (Reproduced by permission of Nauka, Moscow, Russia, from *Vysokomol. Soed A*, 1993, **35**, 1561)

Diad distribution in styrene–glycidyl methacrylate copolymers was examined; the applicability of the boundary effect theory to the copolymers synthesized at both low and high conversions was demonstrated [153]. The boundary effect theory was used to calculate block parameter R, describing the unit distribution, from the data on the yields of monomers in pyrolysis of methyl methacrylate–

styrene copolymers [154]. Quantitative analysis of the sequence of phenolic units (*o*-, *m*- and *p*-cresols) in phenol–formaldehyde polycondensates was also made by pyrolysis GC/MS involving silylation of dimeric and trimeric pyrolysis products [155].

Another approach to sequence distribution analysis is based on a mathematical model describing the decomposition of monosubstituted vinyl copolymers via the free radical degradation mechanism [156]. Unit distributions involving hybrid pyrolysis products and kinetic parameters of relevant decomposition reactions were assessed in these model studies. Using this model, Blazco and co-workers [157, 158] deduced unit distribution and decomposition rate constants for styrene–methyl acrylate and styrene–acrylonitrile copolymers.

In references [159] to [161], when elucidating unit distributions in various copolymers, dimer formation coefficients were assumed to be independent of copolymer composition. A similar approach was applied in the studies of sequence distribution in butadiene–isoprene copolymers [162]. In that study, for a limited set of copolymers with a relatively narrow range of copolymer compositions (the isoprene content varied from 5.5 to 33%), the K_1, K_2 and K_3 coefficients of Eqs. (8.1) to (8.3) were assumed to be almost independent of composition and invariable. The copolymer with the highest isoprene content was used in the calculation. Using the yields of homo- and heterodimers (calculated according to the terminal copolymerization model) adjusting coefficients K were calculated, which were further used to estimate diad concentrations for the entire range of compositions. Figure 8.25 illustrates that the diad concentrations calculated in this manner are in good agreement with those predicted by the theory; this indicates that the copolymers were random.

In reference [163] an attempt was made to evaluate triad concentrations in butadiene–isoprene copolymers. Because the yield of trimers in the pyrolysis of these copolymers was very small and the isomers representing homo- and heterotrimers were poorly resolved in chromatographic separation, a highly sensitive selective ion-detection technique was used. Molecular ions of the following trimers were detected: butadiene–butadiene–butadiene ($M = 162$), isoprene–isoprene–isoprene ($M = 204$), butadiene–butadiene–isoprene + butadiene–isoprene–butadiene + isoprene–butadiene–butadiene ($M = 176$) and butadiene–isoprene–isoprene + isoprene–butadiene–isoprene + isoprene–isoprene–butadiene ($M = 190$). The peak areas of these ions were assumed to be proportional to the concentrations of the corresponding triads. Within the experimental accuracy, the triad concentrations coincided with the calculation based on copolymerization theory (terminal model).

Yet another application of pyrolysis GC/MS is a quantitative evaluation of order imperfections in alternating copolymers. For example, Tsuge *et al.* [164] used this method to assess the disorder in alternating copolymers of styrene with acrylonitrile, methyl acrylate, methyl methacrylate and other comonomers, whereas Mardanov [165] carried out studies with acrylonitrile–VCH copolymers.

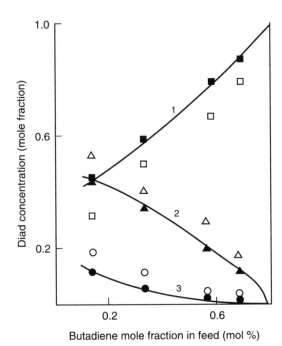

Fig. 8.25 Concentrations of diads in butadiene–isoprene copolymers: *P*(butadiene–butadiene) (1), *P*(butadiene–isoprene) + *P*(isoprene–butadiene) (2) and *P*(isoprene– isoprene) (3). Closed symbols refer to calculated concentrations of diads; open symbols show the experimental data. (Reproduced by permission of Nauka, Moscow, Russia, from *Vysokomol. Soed A*, 1993, **35**, 1561)

Pyrolysis GC/MS may also be used to investigate the tendency to form blocks in block-random copolymers. Such copolymers are obtained, for instance, when copolymerizing ethylene with VCH or other branched α-olefins under non-steady-state conditions. It was found [134, 166] that, during pyrolysis, the VCH block of the macromolecule gives dimers and trimers (e.g. 1,3-dicyclohexyl-propane and its unsaturated analogs), whereas a chain fragment with a random distribution of ethylene and VCH units produces alkylcyclohexanes and their unsaturated analogs. The extent of block character with respect to VCH was deduced from the ratio of the areas of VCH dimers (it was shown that $S_{bl} = \frac{1}{2} K_1 N_{bl}$) to the total area of the zones due to nonylcyclohexane and its unsaturated analogs ($S_r = K_2 N_r$) (where N_{bl} and N_r are the amounts of VCH in the blocks and randomly distributed in the chain). These products show similar retention times and this feature reduces the calculation error. Taking into account the fact that the zones associated with the dimers are due to two VCH units, the extent of block character with respect to VCH was calculated by the following formula:

$$C_{bl} = \frac{2K_2 Q}{2K_2 Q + K_1}$$

Krasnoselskaya *et al.* [167] used pyrolysis GC and mass spectrometric thermal analysis to describe the structure of the graft copolymers synthesized in the reaction of the 'living' acrylonitrile chains with poly(2-vinylpyridine).

8.2.4 SUMMARY

The material reported above demonstrates the considerable utility of pyrolysis GC/MS for investigating the primary structure of copolymers at both qualitative and quantitative levels. This method, however, is not universal; it is mostly applicable to polymers for which the main pathway of thermal degradation involves cleavage of the main chain. It is less suitable for studies of polymers containing side groups which are easily eliminated in heat treatment. In this respect, the investigation of biopolymers by combination of pyrolysis and GC or GC/MS is of interest. Attempts to study the microstructure of oligosaccharides by on-line pyrolysis GC/MS did not lead to quantitative results due to the low volatility of pyrolysis products and undesirable thermal dehydration [168, 169]. More reliable results in investigating cellulose were obtained when on-line pyrolysis was used and the pyrolysis products were characterized by GC/MS after preliminary permethylation followed by reduction with $NaBD_4$ [170]. For the investigation of primary structures of such biopolymers at a quantitative level, the use of pyrolysis MS (field ionization ion source) is more promising [132]. At the same time, amino acid sequence information in proteins may be obtained from pyrolysis GC/MS data [171, 172].

The application of pyrolysis GC/MS is most promising in studies of hydrocarbon polymers for which NMR spectroscopy is not very effective. We direct the reader to a thorough study reported by Kondo *et al.* [173], which compares pyrolysis GC/MS, IR and NMR spectroscopies in elucidating the microstructure of hydrogenated butadiene–acrylonitrile rubbers.

It is worth mentioning that all the literature cited above concerns polymers prepared by usual polymerization processes. However, the other class of polymers obtained through polymer-analogous or intramolecular reactions remained undiscussed, owing to the absence of corresponding material. Nevertheless, it may be supposed that pyrolysis GC/MS can and must be applied to the investigation of microstructures of such polymers, namely the sites of modification and distribution of modified and intact monomer units along the chain. The main problem that may be encountered in such studies is a high thermal lability of groups introduced during modifications which inhibits a cleavage of the main chain. In our opinion, this problem may be solved by preliminary conversion of newly introduced groups into those that are thermally more stable and do not prevent a cleavage of the chain.

8.3   CHEMICAL REACTIONS AS A TOOL FOR ESTIMATION OF UNITS DISTRIBUTION

8.3.1   INTERSEQUENCE REACTIONS BETWEEN FUNCTIONAL GROUPS

When a chemical reaction between functional groups of neighboring units A and B at the boundaries of the A and B sequences is possible, the quantity of the used reagent, as well as the fraction of the converted units and of the units that were not touched by the reaction, are directly connected with the distribution of the units in the chain.

Theoretical analysis of this problem, carried out by Alfrey *et al.* [174], allows one to connect the fraction of the units B that have not reacted with the units A $[f_u(B)]$ using Markovian transitional probabilities:

$$f_u(B) = [\cosh(P_{ab}P_{ba})^{1/2} - (P_{ba}P_{ab})^{1/2}\sinh(P_{ab}P_{ba})^{1/2}]^2 \qquad (8.10)$$

and also the characteristics of the copolymer composition and the Harwood parameter 'run number' R (R is the number of AB + BA diads per 100 units in the chain):

$$f_u(B) = \left\{ \cosh\frac{R^2}{4(A)(B)} \right]^{1/2} - \left[\frac{(A)}{(B)}\right]^{1/2}\sinh\left[\frac{R^2}{4(A)(B)}\right]^{1/2} \right\}^2 \qquad (8.11)$$

A family of curves is presented in Fig. 8.26 of theoretical dependence of the fraction of the unreacted B units on the parameter R for copolymers of various compositions [175]. This dependence proves to be rather strong and can be employed to obtain information on the distribution of units in a copolymer from experiments with reactions between the units A and B. The works of Harwood [175–179] are devoted to the systematic application of this approach and to the elaboration of programs that connect the distribution of units in the chain with the experimentally measured value of $f_u(B)$.

The reaction of intersequence lactonization in copolymers of methyl methacrylate with vinyl halides, which occurs when the copolymer is heated up to 140–200 °C:

has been studied most thoroughly [175, 176].

The fraction of the unreacted units $f(MMA)$ is determined directly from the mass loss measurements. In such a manner Johnston and Harwood [175, 176] have investigated the distribution of units not only in binary copolymers of methyl methacrylate–vinyl halide but also in terpolymer methyl methacrylate–vinyl chloride–styrene. An analogous reaction could be expected to proceed in

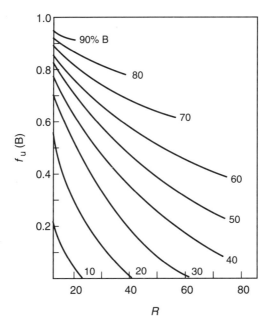

Fig. 8.26 Dependence of the function of unreacted units $f_u(B)$ on the run numer R for copolymers of various composition (%B). (Reproduced by permission John Wiley & Sons Ltd from *J. Polymer Sci. C*, 1969, **22**, 591)

the products of bromination of copolymers of methyl methacrylate with dienes, but it turned out that in this case the formation of γ-lactone takes place completely in the course of bromination. Johnston and Harwood [177] suggested the following mechanism of the reaction:

Determining the proportion of the unreacted units of methyl methacrylate by the PMR method, Johnston and Harwood [177] showed that bromination of copolymers of methyl methacrylate with butadiene is well described by Eqs. (8.10) and (8.11), taking the values of the copolymerization constants 0.73 and 0.18 for the butadiene and methyl methacrylate respectively,

The intersequence reaction with the formation of cyclic carbamide was employed in a very elegant manner [178] in studying the microstructure of copolymers of *tert*-butyl-*N*-vinylcarbamate with phenyl-*N*-vinylcarbamate. After selective hydrolysis of the *tert*-butyl-carbamate with HBr the copolymer was treated with triethylamine. The resulting free amino groups react with neighboring units of the phenylcarbamate, giving cyclic carbamide and free phenol, the quantity of which, determined spectrophotometrically, can be easily correlated with the proportion of the unreacted units. The values of f_u, as calculated from the copolymerization constants, are in good agreement with those measured experimentally. This allowed the authors to come to the conclusion that the hydrolysis of *tert*-butyl-*N*-vinylcarbamate is strictly selective, while the reaction of the amino groups with the units of phenyl-*N*- vinylcarbamate is random in character:

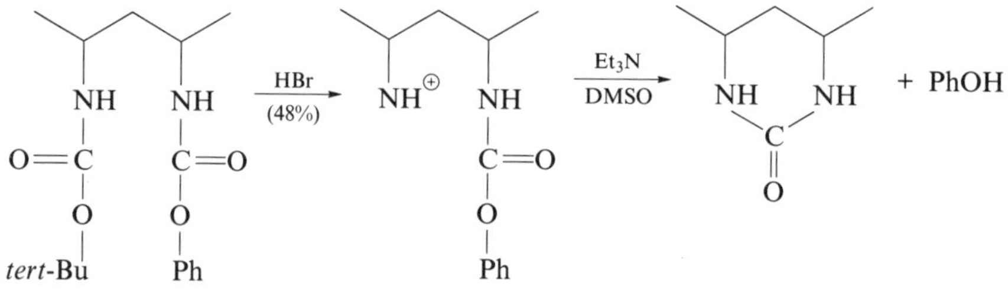

In a similar manner lactonization of the products of hydrolysis of vinyl acetate copolymers with maleic anhydride was employed [180].

The above-stated approach is attractive from the viewpoint of its generality and possibility of using a very broad class of objects. However, this procedure gives a value of R parameter only. Besides, practically uncontrollable errors arise if the process is incomplete or the intersequence reaction mechanism happens to be not fully random, as envisaged by the scheme of Alfrey.

8.3.2 POLYMER-ANALOGOUS REACTIONS

In the case when for a given polymer a reliable procedure of analyzing the units distribution does not exist, it becomes a very important matter to elaborate quantitative methods of converting these macromolecules into ones for which the methods of analysis are developed sufficiently well. In carrying out such reactions one has to take into account two types of possible complications: incomplete course of the reaction, resulting in the formation of copolymers of the initial and converted units, and a change in the stereochemical configuration of the chain as a result of the reaction.

Most extensive use is made of reactions of conversion of poly(methacrylic acid) esters, whose stereoisomerism cannot be established directly by analyzing the PMR spectra since the alkyl signals of the ester group are superimposed on the signals of the chain protons carrying information about the stereo-isomerism. The initial ester is hydrolyzed in the HI medium [181], and poly-(methacrylic acid) which is formed in the reaction is converted into poly(methyl methacrylate) by methylation with diazomethane [182]. Various methods of carrying out the esterification, depending on the structure of the polymer–carboxylic acid and alcohol, are discussed in detail in reference [183]. In particular, the application of such an approach to the products of radical polymerization of a number of n-alkyl methacrylates, from methyl methacrylate to n-hexadecyl methacrylate inclusive, has made it possible to show the stereoisomerism of poly-n-alkyl methacrylates to be independent of the length of the side substituent [184].

In the PMR spectrum of poly(phenyl methacrylate) α-methyl protons are displayed in the form of a single signal. Some authors related this with the idea of a highly syndiotactic chain. However, investigation of the stereoisomerism of poly(methyl methacrylate) obtained from the studied polymer through hydroly-sis and methylation has not confirmed this conclusion [185].

By using the procedure of chemical conversion into poly(methyl methacrylate), Yun *et al.* [186] have investigated the stereoisomerism of poly(phenyl meth-acrylate) obtained with various catalytic systems: benzoyl peroxide, butyllithium, dipiperidylmagnesium, dipyrrylmagnesium. During hydrolysis particular atten-tion should be paid to the reaction to proceed completely. In this case it is preferable to carry out transesterification with sodium methylate, as described in reference [19]. A number of specimens of poly(phenyl methacrylate) with the exactly known distribution of configurations can also be obtained by a reverse procedure: by saponification of a suitable polyalkyl methacrylate and arylation of poly(methacrylic acid). Kempf and Harwood [187] suggested a quantitative procedure of such arylation.

As already known, stereoregularity of polymethacrylonitrile can be estimated directly from the NMR spectrum (^{1}H and ^{13}C) [188, 189], but for a polymer with a high content of isotactic sequences difficulties arise with the selection of the solvent. Hydrolysis in the KOH [190] medium converts the polymer into poly(methacrylic acid) which is then transformed into poly(methyl methacrylate) and further analyzed by the known procedure.

Particular caution should be exercised in carrying out polymer-analogous reactions of poly(acrylic acid) derivatives. Both acid and alkaline hydrolysis of polyacrylonitrile [191–194] is usually accompanied by racemization although esters of poly(acrylic acid) [195] and poly(acrylic anhydride) [196] are racemized only under definite conditions. For instance, poly(isopropyl acrylate) [195] easily racemizes in the presence of bases, but it can be hydrolyzed with the preservation of the stereoisomeric structure in a strong acid medium [193].

The reactions of saponification and esterification with the preservation of the stereoisomeric structure easily proceed in the series:

Ethers of polyvinyl alcohol → polyvinyl alcohol ⇌ esters of polyvinyl alcohol

For analyzing the stereoisomerism, polymers of these series are usually converted into polyvinyl acetate [197–199] for which a PMR analysis procedure has been developed. It should be noted that the advances made in the NMR spectroscopy allow a direct high-accuracy investigation of the isomerism of the triads of polyvinyl alcohol, based on splitting the hydroxyl proton signals at 220 MHz or the signals of methyne carbons in the ^{13}C-$\{^1H\}$ spectra [200].

Reductive dechlorination of polyvinyl chloride (PVC) with lithium aluminum hydride and tri-n-butyltin hydride has been investigated in detail by a group of American scientists [201]. Conditions have been selected that allowed the complete transformation of PVC followed by measuring the content of long branches, methyl branches and unsaturated bonds in the corresponding hydrocarbon chain by the ^{13}C NMR method.

The product of partial reductive dechlorination of PVC is a poly(ethylene-co-vinyl chloride)(E–V copolymer). Jameison *et al.* [202] studied a tri-n-butyltin hydride reduction of various conformers of 2,4-dichloropentane (DCP) and 2,4,6-trichloroheptane (TCH) to elucidate the reaction mechanism. The authors [202] conclude that the attack of tri-n-butyltin radical at the Cl atom is governed by steric and inductive effects. The adjacent methine protons hinder the attack, the effect being stronger for *tttt* conformer of *rr*-TCH. The Cl abstraction rate is enhanced by electron withdrawing γ-Cl substituent. Having used their results related to DCP and TCH the authors succeeded in describing quantitatively the structure of E–V copolymer (obtained by reaction of PVC with tri-n-butyltin hydride) and concluded that compounds tested might be considered to be appropriate low-molecular models for studying the reductive dechlorination of PVC.

Note also the series of polymer analogous transformations recommended by Carlsson *et al.* [203] for investigation of the structure of the products of polyolefins oxidation. The treatment of oxidized polyethylene and polypropylene with various gases enabled conversion of alcohols and hydroperoxides to chloroformates (treatment with phosgene); acids and peracids to methyl esters (diazomethane); acids to acid fluorides (sulfur tetrafluoride); alcohols and hydroperoxides to nitrites and nitrates respectively (nitric oxide). Such transformations facilitate significantly establishment of the structure of oxidized polymers by IR spectroscopy.

Thus, the application of polymer-analogous reactions considerably broadens the 'assortment' of polymers suitable for investigating the distribution of units in the chains.

8.3.3 CHEMICAL DEGRADATION METHODS

With the use of chemical degradation methods, analysis is carried out on the high-molecular and oligomeric products of the reaction, in the course of which breakage of the chain takes place. These methods are oxidative degradation and also selective hydrolysis.

A characteristic example of an application of oxidative degradation is selective oxidation along unsaturated bonds in the ozonolysis of copolymers [204]. Thus, the products of the ozonolysis of copolymers of butadiene with methyl methacrylate are succinic aldehyde (I), succinic acid (II) and dicarboxylic acids containing several methyl methacrylate units (III):

$$-CH_2-CH=CH-CH_2-CH_2-CH=CH-CH_2-(CH_2-\overset{\overset{\displaystyle CH_3}{|}}{\underset{\underset{\displaystyle COOCH_3}{|}}{C}}-)_n-CH_2-CH=CH-CH_2-$$

$$-CH_2-CHO + OHC-CH_2-CH_2-CHO + CH_2-(CH_2-\overset{\overset{\displaystyle CH_3}{|}}{\underset{\underset{\displaystyle COOCH_3}{|}}{C}})_n-CH_2COOH$$

$$\underset{\displaystyle I}{} \quad \underset{\displaystyle COOH}{|}$$

$$HOOC-CH_2-CH_2-COOH \qquad III$$

$$\underset{\displaystyle II}{} \qquad + OHC-CH_2-$$

From the content of succinic acid (and succinic aldehyde) in the ozonolysis products it is possible to estimate the total proportion of butadiene units forming the sequences and the proportion of isolated butadiene units. The proportion of isolated methyl methacrylate units is simply connected with the quantity of the formed 2-methylbutane-1,2,4-tricarboxylic acid (III when $n = 1$). These data provide a possibility for estimating a tendency to the blocks formation or that to the alternation of the units.

The described method has a number of substantial disadvantages: an unsatisfactory material balance, a great number of by-products, etc. Hackathorn and Brock [205, 206] have suggested a different procedure based on quantitative conversion of primary ozonolysis products into ketones under the action of triphenylphosphine. A mixture of ketones is analyzed with the aid of gas chromatography techniques. This method allows working with rather small quantities of the polymer ($\sim 50\,mg$) and gives highly accurate results due to the use of the chromatographic analysis procedure and strictly selective reactions. Using this method alternating copolymers of butadiene with propylene, poly(1,2-co-1,4-butadienes), poly(1,4-co-3,4-isoprenes), and also the content of *cis* and *trans* structures in 1,4-polybutadiene [189] and the content of structures with anomalous addition of the units ('head-to-head', 'tail-to-tail') in a large number of natural and synthetic polyisoprenes have been investigated [193].

The following example illustrates selective oxidation with iodic acid [207] of the products of hydrolysis of copolymers of vinyl acetate with vinylenecarbonate, i.e. copolymers of vinyl alcohol with vinylene glycol:

$$-CH_2-CH-\!\!\!-\!\!\!-(CH-CH)_n-CH-CH-CH_2-CH-CH_2-CH-\!\!\!-\!\!\!-CH-CH-CH_2-$$

with substituents: $OOCCH_3$; O, O ($C{=}O$); O, O ($C{=}O$); $OOCCH_3$; $OOCCH_3$, O, O ($C{=}O$)

$$\downarrow$$

$$-CH_2-CH-(CH-CH)_n-CH-CH-CH_2-CH-CH_2-CH-CH-CH-CH_2-$$
$$\quad\quad OH\quad OH\ OH\quad OH\ OH\quad\quad\quad OH\quad\quad\quad OH\ OH\ OH$$

$$\downarrow$$

$$-CH_2-CHO + (2n+1)\,HCOOH + OHC-CH_2-CH-CH_2-CHO + HCOOH$$
$$\quad\quad\quad\quad\quad\quad\quad\quad\quad\quad\quad\quad\quad\quad\quad\quad OH\quad\quad\quad + OHC-CH_2-$$

Vinylene glycol units which have no neighbors containing vinyl alcohol units are oxidized to formic acid, whereas sequences containing vinyl alcohol units give complex aldehyde alcohols, aldehyde groups originating at the boundaries of the blocks of vinyl alcohol. The ratio of the quantities of the eliminated formic acid and of the consumed iodic acid is unambiguously connected with the sequence distribution of the copolymer. Unfortunately, the products of selective oxidation again react with iodic acid, giving formic acid. The investigators, however, have succeeded in taking into account the amount of formic acid consumed for secondary oxidation.

Enzymatic hydrolysis was employed by Kanakanatt [208] in the investigation of L-glutamic acid and L-tyrosine copolymer. Chymotrypsin catalyzes selective hydrolysis of peptide linkages formed by the acidic end of the tyrosine unit:

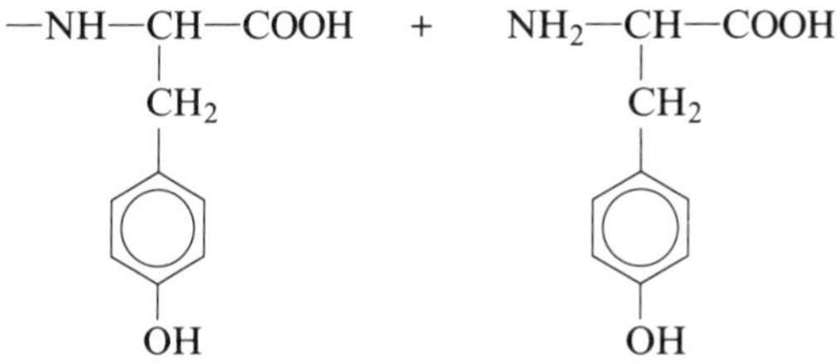

$$-NH-CH-COOH \quad + \quad NH_2-CH-COOH$$

$$| \qquad\qquad\qquad\qquad |$$

$$CH_2 \qquad\qquad\qquad\qquad CH_2$$

(benzene ring with OH)   (benzene ring with OH)

OH   OH

          I

$$O$$
$$\|$$
$$+ \; H(NH-CH-C)_n-NH-CH-COOH \quad + \quad NH_2-CH-$$

$$\qquad\quad (CH_2)_2 \qquad\qquad CH_2 \qquad\qquad\qquad (CH_2)_2$$

$$\qquad\quad COOH \qquad\qquad\qquad\qquad\qquad\qquad COOH$$

(benzene ring with OH)

OH

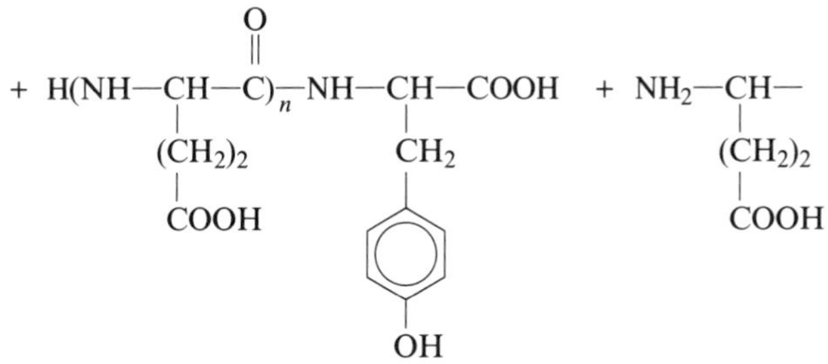

Since a tyrosine block having a length of m units gives after the hydrolysis $(m-1)$ molecules of free tyrosine (I), the difference between the percentage of tyrosine in the copolymer (T) and that of converted into free tyrosine (FT) characterizes the run number R of the copolymer:

$$\frac{R}{2} = T - FT$$

Tada *et al.* [209] used selective alkaline hydrolysis of ester linkages for the analysis of the distribution of units in copolymers of 3,3-bischloromethyl-oxacyclobutane with β-propiolactone:

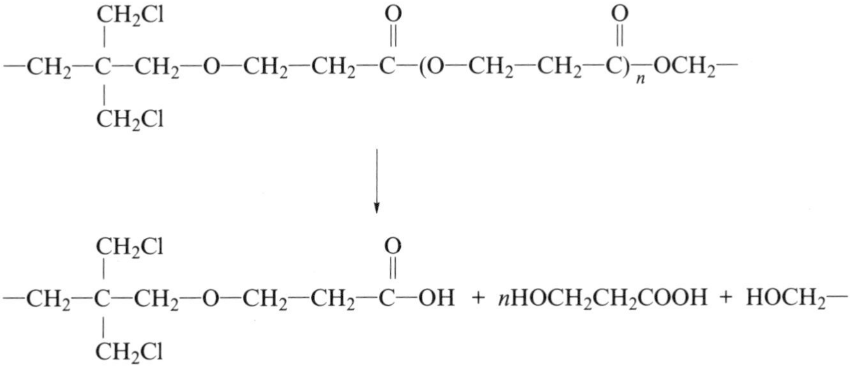

$$CH_2Cl \qquad\qquad\qquad O \qquad\qquad\qquad O$$
$$| \qquad\qquad\qquad\qquad \| \qquad\qquad\qquad \|$$
$$-CH_2-C-CH_2-O-CH_2-CH_2-C-(O-CH_2-CH_2-C)_n-OCH_2-$$
$$|$$
$$CH_2Cl$$

$$\downarrow$$

$$CH_2Cl \qquad\qquad\qquad O$$
$$| \qquad\qquad\qquad\qquad \|$$
$$-CH_2-C-CH_2-O-CH_2-CH_2-C-OH \; + \; nHOCH_2CH_2COOH \; + \; HOCH_2-$$
$$|$$
$$CH_2Cl$$

The amount of forming β-hydroxypropionic acid characterizes the proportion of

β-propiolactone units which are comprised in blocks whose length is two units and over.

It is characteristic for all the methods of chemical degradation that the conducted reaction is directed to a definite type of linkage in the macromolecule. If the selected reaction does not lead to the formation of the initial mixture of the monomers or products of their further transformation, then the composition of the products necessarily carries information about the distribution of units in the initial chain. This information, as a rule, is intermixed with information related to side and secondary processes (the reactions employed being not strictly selective). However, quantitative analysis of degradation products is often more simple and accurate than analysis of the content of corresponding fragments directly in the chain of macromolecules. Thus, the effectiveness of the application of any particular method of chemical degradation is determined, on the one hand, by the accuracy and simplicity of the analysis of the reaction products and, on the other hand, by the possibility of actually taking into account the deviations from strict selectivity.

8.4   STUDYING THE COMPOSITION HETEROGENEITY OF COPOLYMERS BY THE FRACTIONATION TECHNIQUE

Myagchenkov and Frenkel discussed in their book [210] various methods of investigating the composition heterogeneity. The function of compositional distribution of copolymerization products can be constructed from the data obtained by measuring the kinetics of the consumption of each of the monomers in the course of copolymer synthesis. Myagchenkov *et al.* [211] studied in this way heterogeneity of the copolymers of acrylamide with maleic acid, using the polarographic method for separate determination of the content of monomers in the reaction mixture. Analysis of the monomer mixture by means of gas chromatography [212] looks very promising for this purpose. However, since with this approach the polymer itself is not investigated the conclusions on its heterogeneity are valid only if the portion of copolymer being formed at every moment of time is homogeneous within the statistical 'instantaneous' heterogeneity. When this requirement is not met (the presence of several types of active centers, heterophase nature of the system), the application of these methods can yield erroneous results. These methods are, naturally, inapplicable for the products of macromolecular reactions, as it is impossible to obtain any quantitative information on the heterogeneity of the copolymer formed from the kinetics of the low-molecular reagent consumption.

In most cases information on the composition heterogeneity is obtained from the data of investigating the copolymer samples themselves [213–215]. Inasmuch as the composition heterogeneity of copolymers affects their physical and chemical properties, there exist numerous methods for the qualitative estimation of the heterogeneity of samples in composition, based on this dependence

[216, 217]. For a comparative characterization of the composition heterogeneity of a series of samples, turbidimetric titration [218–220] and thin-layer chromatography [221, 222] are often used. The method of thin-layer chromatography can also give quantitative information on the composition heterogeneity of a copolymer [223, 224]. Generally, however, it is not at all easy to carry out a quantitative investigation of composition heterogeneity.

The heterogeneity of block and graft polystyrene–poly(methyl methacrylate) copolymers was studied by Cherkasov and co-workers [225, 226] with the help of the diffusion method. Combining diffusion measurements with the data on sedimentation velocities one can find the dispersion of compositional distribution [227]. This method, however, is only applicable to copolymers whose molecular mass and composition are interrelated.

There is no such limitation in the method of light scattering [228–230] that was used to estimate the dispersion of compositional distribution for numerous statistical, block and graft copolymers [214, 230]. The light-scattering method, however, can only be used to study the heterogeneity of such copolymers whose dissimilar units differ significantly in the increments of refraction index (e.g. copolymers of styrene with alkylmethacrylates). In this case, again, the accuracy of the method is insufficient for quantitative investigations of samples with a relatively small composition heterogeneity [231, 232].

Centrifugation in the density gradient [233, 234] looks very promising for quantitative characterization of composition heterogeneity. This method makes it possible in principle to find the function of compositional distribution. However, centrifugation in the density gradient is as yet applied rather rarely [232, 235–237].

Comparison of the experimentally found composition heterogeneity parameters with the values calculated following the copolymerization theory reveals significant discrepancies. For example, for styrene–methyl methacrylate copolymer whose theoretical value of compositional distribution dispersion is $\sim 10^{-5}$, centrifugation in the density gradient gives the value of dispersion equal to $\sim 10^{-4}$, whereas light scattering—from 10^{-3} to 10^{-1} [232]—depends on the measuring technique.

The most extensive use in investigating the composition heterogeneity has been made of fractionation, based on the dependence between the solubility of copolymers and their composition and molecular mass [215]. This method has the advantages of universality, possibility of isolating relatively homogeneous copolymer samples in a preparative scale, as well as simplicity of experiment. Usually the procedure consists of gradually decreasing the solvent power of the medium (most often by adding a precipitant to the copolymer solution) and isolating the fractions precipitated consecutively in this way. The reverse process, the isolation of consecutively dissolving fractions with a gradual increase in the solvent power of the medium, yields the same results if the same solvent–precipitant system is used in both cases [215, 238].

An important feature of the fractionation of copolymers is the dependence of fractionation results on the nature of the solvent–precipitant system used, which is clearly shown in references [239] and [240]. Thus, according to Fuchs [240], when vinyl acetate–vinyl chloride copolymer (average Cl content 31.3%) is fractionated in the acetone–petroleum ether system, the composition of fractions varied within 30–34% Cl, and in the case of the acetone–methanol pair, within 18–40%. For the methyl acetate–petroleum ether and methyl acetate–methanol systems these limits amounted to 22–33% Cl and 18–40% Cl respectively. However, when methylene chloride was used as the solvent both precipitant, petroleum ether and methanol, produced almost the same composition separation effect. Similar results were obtained for the fractionation of chlorinated polyethylene: in the chloroform–methanol and benzene–methanol systems narrow limits of composition variation were observed in the isolated fractions, and in the chloroform–petroleum ether and benzene–petroleum ether systems these limits were wide.

It is obvious that the different limits of the variation in the average values of the composition of fractions are indicative of considerable differences in the composition heterogeneity of fractions isolated in different solvent–precipitant systems. The curves of compositional distribution plotted from the data of fractionation in different systems naturally differ markedly from one another.

These facts indicate that it is far from easy to obtain quantitative information on the composition heterogeneity of a copolymer. To accomplish this, it is at least necessary to establish a connection between the conditions of fractionation and the heterogeneity of fractions, as well as the nature of the initial copolymer distribution curves plotted from the fractionation data. Appropriate analysis can be performed with the help of a theory of fractionation. The most simple theory of copolymer fractionation as well as the results of an analysis of various procedures of fractionation are given below.

In 1958 Kilb and Bueche [241] derived an equation describing the distribution of graft copolymers between phases during fractionation, but they were unable, however, to develop a calculation procedure necessary for theoretical analysis. Such analysis can be performed with the help of the method proposed by Litmanovich and co-workers [242, 243] for the calculation of molecular and compositional distribution in samples isolated during the fractionation of copolymers.

Let us consider a binary AB copolymer solution in an individual solvent. In thermodynamics of copolymer solutions the following assumptions are made: (a) the entropy variation with copolymer dissolution is equal to that for a homopolymer with the same degree of polymerization [241, 244, 245]; (b) the interaction of solvent with AB copolymer is described with the help of the effective parameter χ_x [228, 244], which is a function of the composition x(x is the mole fraction of A units in copolymer) but not of the macromolecule chain length.

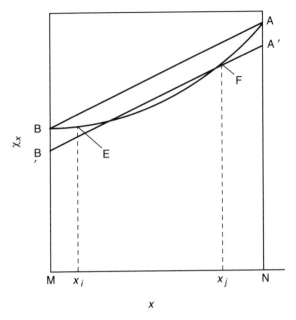

Fig. 8.27 Dependence of parameter χ_x on copolymer composition x. (Reproduced by permission of Khimiya, Moscow, Russia, from Platé, N. A., Litmanovich, A. D. and Noah, O. V., *Makromolekulyarnye Reaktsii* (*Macromolecular Reactions*), Khimiya, Moscow, 1977, p. 156)

Let us make another assumption:

$$\chi_x = x\chi_A + (1 - x)\chi_B \qquad (8.12)$$

where χ_A and χ_B respectively are the parameters of the solvent interaction with units A and B, not dependent on x.

Figure 8.27 shows the dependence of χ_x on x, described by either the BEFA curve or the BA straight line, depending on whether the interaction of the unit in question is or is not influenced by the nature of neighboring units. In the latter case, $\chi_A = NA$ and $\chi_B = MB$ are the parameters of the interaction of homopolymers A and B with the solvent. In the former case (if the composition of macromolecules of the investigated sample varies within $x_i \leqslant x \leqslant x_j$), according to assumption (8.12), the EF portion of the curve is replaced with the B'A' straight line. $\chi_A = NA'$ and $\chi_B = MB'$ will then be the averaged parameters of the interaction with solvent for the A and B units that satisfy Eq. (8.12) in this range of x values. Since the deviations from linearity in the χ_x–x curves are comparatively small [228], the error introduced by the replacement of the curvilinear portion with the straight line will probably remain insignificant for a great number of copolymer–solvent systems.

We assume that with a decrease in temperature the solution will be separated into two phases, the concentrated (precipitate) and diluted (solution), and phase equilibrium will be established. Performing a conventional thermodynamic examination of phase equilibrium in polymer solutions [246] and uncomplicated transformations we get [242]

$$\frac{v'_{r,x}}{v_{r,x}} = \exp\left[r(\sigma + Kx)\right] \tag{8.13}$$

where $v'_{r,x}$ and $v_{r,x}$ are the volume fractions of the component with degree of polymerization r and composition x in the concentrated and the diluted phases respectively (here and further the prime stands for the concentrated phase), and σ and K are the parameters depending on the nature of the copolymer and the conditions of fractionation.

Equation (8.13) describes in a very simple form the distribution of copolymer between the phases depending on its molecular weight and composition. The influence of the composition of macromolecules on their distribution between the phases is characterized by the value of parameter K. At $K = 0$ the separation of macromolecules is controlled only by their dimensions irrespective of their composition. In this case, the fractionation of a copolymer does not differ from the fractionation of homopolymers. The higher the absolute value of K (the solubility of macromolecules of the same length in a given medium, with an increasing number of A units in them, decreases at $K > 0$ and increases at $K < 0$), the stronger the influence of the composition of macromolecules on their distribution between the phases and, therefore, among the fractions.

Equation (8.13) is applicable to describing the fractionation of binary copolymers of different types by various methods, based on the dependence between the solubility of macromolecules and their molecular weight and composition. With the help of Eq. (8.13) one can obtain equations describing the functions of r and x distribution for the nth successive fraction, $w'^{(n)}_{r,x}$, isolated during the fractionation by precipitation and for the sample remaining in solution after the isolation of this fraction, $w^{(n)}_{r,x}$, as well as the function of molecular, $w'^{(n)}_r$, and compositional $w'^{(n)}_x$, distribution for the nth fraction:

$$w'^{(n)}_{r,x} = w^{(n-1)}_{r,x}\left\{1 + \left(\frac{1}{R_n}\right)\exp\left[-r(\sigma_n + K_n x)\right]\right\}^{-1} \tag{8.14}$$

$$w^{(n)}_{r,x} = w^{(n-1)}_{r,x}\left\{1 + R_n \exp\left[-r(\sigma_n + K_n x)\right]\right\}^{-1} \tag{8.15}$$

$$w'^{(n)}_r = \int_0^1 w'^{(n)}_{r,x}\,dx = \int_0^1 w^{(n-1)}_{r,x}\left\{1 + \left(\frac{1}{R_n}\right)\exp\left[-r(\sigma_n + K_n x)\right]\right\}^{-1}dx \tag{8.16}$$

$$w'^{(n)}_x = \int_0^\infty w'^{(n)}_{r,x}\,dr = \int_0^\infty w^{(n-1)}_{r,x}\left\{1 + \left(\frac{1}{R_n}\right)\exp\left[-r(\sigma_n + K_n x)\right]\right\}^{-1}dr \tag{8.17}$$

where $w_{r,x}^{(n-1)}$ is the function of r and x distribution for the copolymer remaining in solution after the isolation of the $(n-1)$th fraction; σ_n, K_n and R_n are parameters depending on the conditions of the formation of the nth fraction, with $R_n = V_n'/V_n$ (V_n' and V_n are the volumes of phases at equilibrium).

The σ_n, K_n and R_n parameters can be found by solving the following set of equations:

$$W'^{(n)} = \int_0^\infty w'^{(n)} \, dr = \int_0^\infty dr \int_0^1 w_{r,x}^{(n-1)} \left\{ 1 + \left(\frac{1}{R_n} \right) \exp\left[-r(\sigma_n + K_n x) \right] \right\}^{-1} dx$$

(8.18)

$$\overline{r_w'^{(n)}} = \left(\frac{1}{W'^{(n)}} \right) \int_0^\infty r \, dr \int_0^1 w_{r,x}^{(n-1)} \left\{ 1 + \left(\frac{1}{R_n} \right) \exp\left[-r(\sigma_n + K_n x) \right] \right\}^{-1} dx$$

(8.19)

$$\overline{x'^{(n)}} = \left(\frac{1}{W'^{(n)}} \right) \int_0^1 x \, dx \int_0^\infty w_{r,x}^{(n-1)} \left\{ 1 + \left(\frac{1}{R_n} \right) \exp\left[-r(\sigma_n + K_n x) \right] \right\}^{-1} dr$$

(8.20)

where $W'^{(n)}$, $\overline{r_w'^{(n)}}$ and $\overline{x'^{(n)}}$ are the weight, weight-average degree of polymerization and average composition of the nth fraction respectively. Equations (8.18) to (8.20) can be solved by numerical methods on a computer. Thus, knowing the distribution function for the initial sample and having determined the weight, average composition and average degree of polymerization for each fraction, one can calculate the distribution functions for samples isolated during the fractionation of copolymers. With the help of the calculation method described above Litmanovich and co-workers examined the regularities governing the separation of copolymers by successive precipitation [243], as well as by cross-fractionation [247] and branched fractionation [248].

The method of cross-fractionation proposed by Rosenthal and White [239] is a combination of several operations of successive precipitation. The initial polymer is first separated into intermediate fractions in such a solvent–precipitant system where, at the same degree of polymerization, the more readily soluble macromolecules are those enriched, for example, with B units. Each intermediate fraction is then separated in another solvent–precipitant system where the more readily soluble macromolecules are those enriched with A units. In branched fractionation [249] the intermediate fractions are further separated in the same solvent–precipitant system that had been used for the separation of the initial copolymer.

For the purpose formulated above, a copolymer with a prescribed $w_{r-x}^{(0)}$ distribution was subjected to 'mathematical fractionation' by different methods, the $w_r'^{(n)}$ and $w_x'^{(n)}$ functions were calculated for the isolated fractions, and with the help of conventional methods the functions of molecular and compositional

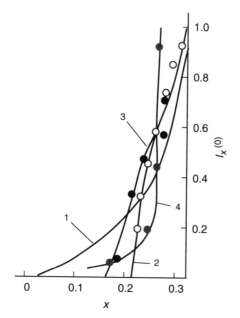

Fig. 8.28 Integral compositional distribution of initial copolymer: prescribed (1) and plotted from the data of fractionation by successive precipitation at K equal to 0.01 (2), 0.03 (3) and -0.01 (4). (Reproduced by permission of Khimiya, Moscow, Russia, from Platé, N. A., Litmanovich, A. D. and Noah, O. V., *Makromolekulyarnye Reaktsii* (*Macromolecular Reactions*), Khimiya, Moscow, 1977, 156)

distribution for the initial sample were plotted based on the fractionation data. In full the results of these investigations are presented in reference [214]; here we only consider the influence of the separation conditions on the nature of the compositional distribution functions of the initial copolymer plotted from the fractionation data.

The integral compositional distribution for the initial copolymer is usually found from the 'corrected cumulative weight of fraction–average composition of fraction' curve. The differential compositional distribution of a sample can be found by graphical differentiation of the integral curve. Figure 8.28 gives the functions of the integral compositional distribution, $I_x^{(0)}$, plotted from the data of 'mathematical' fractionation by successive precipitation at different values of the K_n parameter that characterized the sensitivity of the solvent–precipitant system to the copolymer composition. In all cases the plotted curves differ substantially from the prescribed distribution (curve 1).

Figure 8.29 compares the prescribed differential compositional distribution (curve 1) for the initial copolymer with the curves plotted from the data of cross-fractionation (curve 2) and branched fractionation (curve 3). Curve 3 strongly distorts the heterogeneity of the initial copolymer, whereas curve 2 is

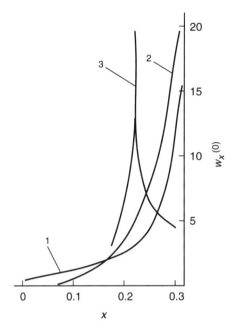

Fig. 8.29 Differential compositional distribution of initial copolymer: prescribed (1) and plotted from the data of cross-fractionation (2) and branched (3) fractionation. (Reproduced by permission of Nauka, Moscow, Russia, from *Vysokomol. Soed A*, 1967, **9**, 1016)

quite close to the prescribed distribution, although it deviates from curve 1. Based on these facts, cross-fractionation is recommended in references [214] and [248] as an effective method of investigating the compositional (as well as the molecular) heterogeneity of copolymers.

The basic principles and conclusions of the theory of copolymer fractionation described above have been confirmed experimentally [250–257]. Teramachi and Nagasawa [256], for example, fractionated an artificial mixture of two styrene–acrylonitrile copolymers containing 31.2 and 20.8 wt% of acrylonitrile in the methylethylketone–cyclohexane system. Figure 8.30 shows the experimental results of fractionation and the curve of the integral compositional distribution for an initial sample, calculated from Eqs. (8.13), (8.18) and (8.20) at the values of parameters $K_n = 0.02$ and $\sigma_n = 0.003$. Taking into account the peculiarities of the initial sample, Teramachi and Nagasawa [256] conclude that the fractionation theory proposed in references [242] and [243] makes it possible to describe satisfactorily the experiment.

Teramachi and Kato [257] fractionated the copolymer of styrene with butadiene by different methods. The curve of the molecular distribution of the initial copolymer, plotted from the data of fractionation by successive precipitation in

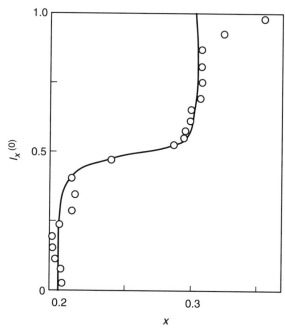

Fig. 8.30 Integral compositional distribution of the mixture of two styrene–acrylonitrile copolymers (x is the weight fraction of acrylonitrile). Points, experimental data; curve, calculation from Eqs. (8.13), (8.18) and (8.20) at $K_n = 0.02$ and $\sigma_n = 0.003$. (Reprinted from *J. Macromol. Sci. Chem.*, 1968, **2**, 1169, by courtesy of Marcel Dekker Inc.)

the benzene–methylethylketone system, proved to be bimodal, which was explained by the regularities of fractionation predicted from theory [243]. Moreover, a comparison of the results of fractionating identical samples by different methods showed that the molecular and compositional heterogeneity of the copolymer could be revealed most fully by cross-fractionation. On the strength of this Teramachi and Kato regard as correct the conclusion made in reference [248] on the advantages of cross-fractionation as a method of investigating the heterogeneity of copolymers.

Bourguignon *et al.* [258] arrived at a similar conclusion having compared the fractionation by different methods of the product of interaction between poly(methyl methacrylate) and $(CH_3)_2NSO_2CH_2Li$ containing 36.6 mol% of substituted units. In this case, too, cross-fractionation proved to be significantly more effective than successive fractionation.

The agreement between the theory and experiment provides grounds to conclude that the above-mentioned calculation method makes it possible to describe correctly, at least to a first approximation, the most important regularities of the fractionation of copolymers and the properties of isolated fractions. The results of comparative estimation of the different fractionation methods can, therefore, be used for a rational choice of a technique to study the heterogeneity of copolymer samples. In particular, the method of cross-fractionation can be recommended for investigating the composition heterogeneity of the products of polymer-analogous transformations.

It should be noted that the deviations of $I_x^{(0)}$ or $w_x^{(0)}$ curves, plotted from fractionation data by the conventional method, from the true compositional distribution functions (see Figs. 8.28 and 8.29) are associated not only with the fractionation conditions but also with the nature of heterogeneity of the initial sample. The difference between the 'corrected cumulative weight of fraction–average composition of fraction' curve and the true $I_x^{(0)}$ curve will be the stronger (all other things being equal) the more asymmetrical is the differential compositional distribution of the initial copolymer. It should be expected that with a symmetrical distribution of the initial copolymer one can obtain by means of cross-fractionation very accurate information on its composition heterogeneity. Teramachi and Kato [259] thus found by cross-fractionation the compositional distribution curve for azeotropic styrene–methyl methacrylate copolymer that is in good agreement with Stockmayer's theory [260].

In polymer-analogous transformations taking place in homogeneous conditions the composition heterogeneity of reaction products, as shown in Chapter 3, is described by symmetrical distribution functions. It is therefore to be hoped that in investigating the composition heterogeneity of such copolymers cross-fractionation will prove to be effective. For a successful investigation of the composition heterogeneity of a copolymer by means of fractionation it is extremely important to choose correctly the solvent–precipitant systems.

The sensitivity of a solvent–precipitant system to the copolymer composition is usually estimated from the data on the values of the precipitation threshold, γ^*, for corresponding samples (γ^* is the volume fraction of precipitant in solution at the moment of the first nondisappearing turbidity observed in the titration of the polymer solution with the precipitant).

Stockmayer *et al.* [228] thus consider it sufficient to determine the γ^* value for the corresponding homopolymers, assuming that a solvent–precipitant system is the more sensitive to copolymer composition the greater is the difference in the γ^* values for the homopolymers. A similar assumption lies at the basis of the method proposed by Elias and Gruber [261, 262] and used in reference [263] for the fractionation of graft styrene–polypropylene copolymers. Such an approach is only justified in the case of a linear dependence of γ^* on copolymer composition x. In experiments, however, a nonlinear dependence of γ^* on x is not infrequently observed [264, 265]. Therefore, proceeding solely from the γ^* values for homopolymers, one can arrive at erroneous conclusions on the sensitivity of the tested solvent–precipitant system to the composition of a copolymer.

Rosenthal and White [239] propose that the sensitivity of a system to the copolymer composition might be estimated from the dependence of γ^* on x, established by titrating the solutions of a number of fractions isolated by means of cross-fractionation of the initial copolymer and having similar values of intrinsic viscosity [η] but different composition x. It is thus necessary to perform extensive preliminary work on the cross-fractionation of the initial copolymer (which, incidentally, also requires selecting solvent–precipitant systems sensitive to copolymer composition). Moreover, due to considerable composition heterogeneity of the samples isolated during the preliminary fractionation, it may happen that the found γ^*-x dependence will not correspond to the true one [214].

When studying the γ^*-x dependence it is necessary to work with samples whose composition is relatively homogeneous. In addition to that, it is important to exclude the influence of the molecular mass of the investigated copolymers upon the determined γ^* values; otherwise the true γ^*-x dependence can be markedly distorted.

Staudinger and Heuer [266] showed the γ^* values to decrease sharply with an increase in the molecular mass of polymers from $\sim 10^3$ to $\sim 10^4$, but to be only very slightly dependent on the molecular weight when it increases further. As exemplified by styrene–methyl methacrylate copolymer, it was shown that in the region of molecular weight $\geq 10^5$ the differences in the molecular weight of samples with the same composition do not affect the value of p^* [250].

To estimate the sensitivity of a solvent–precipitant system the composition of a copolymer it is therefore recommended in references [214] and [250] that a study should be made of the dependence of the precipitation threshold of appropriate copolymer samples in this system on their composition. Only non-fractionated samples or the high-molecular fractions isolated from them, whose molecular masses lie within the range of weak γ^* dependence on the chain length, are to be tested. When working with relatively low-molecular copolymers it is necessary to study the γ^*-x dependence using samples of different composition with similar values of degrees of polymerization. The synthesis of such samples should be conducted under conditions ensuring a relative homogeneity of the composition of copolymers.

The efficiency of this technique of choosing a solvent–precipitant system for the fractionation of copolymers [250] was confirmed experimentally in the works of Litmanovich and co-workers [250–253] and in the work of Danon and Jozefonvicz [254]. The same technique was later applied by Glockner [267] and Galin [258].

For cross-fractionation it is necessary to choose two solvent–precipitant systems, in one of which γ^* increases with increasing x (in the region of x values corresponding to the expected range of the composition of macromolecules in the investigated heterogeneous copolymer) and in the other of which γ^* decreases in the same range. In this way, for example, the systems for the cross-fractionation of N-vinyl-pyrrolidone copolymers with vinyl acetate were chosen [268]. Other examples of investigating the composition heterogeneity of the products of macromolecular reactions by means of cross-fractionation were considered in Chapter 4.

Myagchenkov and Frenkel [210] considered the separation of macromolecules of various compositions by means of complexation with a suitable polymeric matrix as more effective than ordinary fractionation, using an idea formulated in reference [269]. However, real and determined progress in experimental studies of the composition heterogeneity of copolymers is connected with the development of the chromatographic technique. Glockner demonstrated high efficiency of gradient high-performance liquid chromatography (HPLC) in a study of the composition heterogeneity of about 30 copolymers [270]. Very valuable results are obtained by means of cross-fractionation, usually a combination of size exclusion chromatography (separation by size) followed by gradient HPLC. As a result a contour-line map of the distribution in molecular mass and in composition of the copolymer sample may be obtained. Basic and detailed information concerning the methods may be found in the excellent book by Glockner [271].

It is worth noting that, in spite of experimental achievements, efforts have not ceased in an attempt to improve the theory of fractionation. Indeed, theory may help to interpret adequately the experimental results. Recently Ratzsch and co-workers [272, 273] applied principles of continuous thermodynamics to the fractionation of homopolymers that are polydisperse in size. It would be interesting to use their approach in the theory of copolymer fractionation.

REFERENCES

1. Randall, J. C., *Polymer Sequence Determination: Carbon-13 NMR Method*, Academic Press, New York, 1977.
2. Slonim, I. Y., Urman, Y. G. and Kluchnikov, V. N., *Plast. Massy.*, 1988, **1**, 49–52.
3. Koenig, J. L., *Chemical Microstructure of Polymer Chains*, Wiley, New York, 1980.
4. Bovey, F. A. and Jelinski, L. W., *Encyclopedia of Polymer Science and Engineering*, 1987, Vol. 10, pp. 254–327.
5. Bovey, F. A., *High Resolution NMR of Macromolecules*, Academic Press, New York, 1972.
6. Randall, J. C. *NMR and Macromolecules: Sequence, Dynamic and Domain Structure*, ACS Washington, D.C., 1984.
7. Slonim, I. Y. and Urman, Y. G., *Plast. Massy.*, 1993, **2**, 16–19.
8. Slonim, I. Y., Klyuchnikov, V. N., Oreshenkova, T. F., Moryakova, Y. B. and Gruznov, A. G., *Vysokomol. Soed A*, 1992, **34**, 34–7.
9. Gotlib, Y. Y., Darinskii, A. A., Neelov, I. M., Torchinskii, I. A. and Shevelev, V. A., *Vysokomol. Soed A*, 1992, **34**, 45–57.
10. Lundin, A. A. and Khazanovich, T. N., *Vysokomol. Soed A*, 1989, 31, 363–8.
11. Kulagina, T. P., Marchenkov, V. V. and Provotorov, B. N., *Vysokomol. Soed A*, 1989, **31**, 381–6.
12. Spiess, H. W. and Sillescu, H., *J. Magn. Res.*, 1981, **42**, 381–9.
13. Sutjagin, V. M., Lopatinskii, V. P. and Filimonov, V. D., *Vysokomol. Soed A*, 1982, **24**, 1968–73.
14. Platé, N. A. and Stroganov, L. B., *Vysokomol. Soed A*, 1976, **18**, 955–79.
15. Ito, K. and Yamashita, Y., *J. Polymer Sci.*, 1965, **3B**, 625–7.

16. Harwood, J. H. and Ritchey, W. M., *J. Polymer Sci.*, 1965, **3B**, 419–21.
17. Platé, N. A., Stroganov, L. B., Uzhinova, L. D., Golubev, V. B. and Nedorezova, P. M., *Vysokomol. Soed*, 1972, **14**, 440–8.
18. Stroganov, L. B., Platé, N. A., Zubov, B. P., Fedorova, S. U. and Strelenko, U. A., *Vysokomol. Soed A*, 1974, **16**, 2616–27.
19. Bauer, R. G., Harwood, H. J. and Ritchey, W. M., *Polymer Preprints*, 1966, **7**, 973–82.
20. Suzuki, T., Santee, E. R. and Harwood, H. J., *J. Polymer Sci.*, 1974, **12B**, 635–7.
21. Strasilla, D. and Klesper, E. K., *Makromol. Chem.*, 1974, **175**, 535–46.
22. Kapur, G. S. and Brar, A. S., *Makromol. Chem.*, 1992, **193**, 1773–81.
23. Brar, A. S. and Sunita, *J. Polymer Sci.*, 1992, **30A**, 2549–57.
24. Brar, A. S., Arunan, E. and Kapur, G. S., *Polymer J.*, 1989, **21**, 689–95.
25. Aerdts, A. M., de Haan, J. W., German, A. L. and Van der Velden, G. P. M., *Macromolecules*, 1991, **24**, 1473–9.
26. Raghavan, R., Maver, T. L. and Blum, F. D., *Macromolecules*, 1987, **20**, 814–18.
27. Kostka, K. L., Radcliffe, M. D. and Von Meerwall, E., *J. Phys. Chem.*, 1992, **96**, 2289–92.
28. Klesper, E. K., *J. Polymer Sci.*, 1968, **6B**, 313–15.
29. Klesper, E. K. and Gronski, W., *J. Polymer Sci.*, 1969, **7B**, 727–9.
30. Kitayama, T., Ute, K., Yamamoto, M., Fujimoto, N. and Hatada, K., *Polymer Bull.*, 1991, **25**, 683–8.
31. Lochmann, L., Kolavik, J., Doskocilova, D., Vozka, S. and Trekoval, J., *J. Polymer Sci.*, 1979, **17A**, 1727–32.
32. Mitani, K., Ogata, T. and Iwasaki, M., *J. Polymer Sci.*, 1974, **12A**, 1653–69.
33. Frensdorff, H. K. and Ekiner, O., *J. Polymer Sci.*, 1987, **25A**, 1157–75.
34. Ron, E., Mathiowitz, E., Mathiowitz, G., Domb, A. and Langer, R., *Macromolecules*, 1991, **24**, 2278–82.
35. Sugitani, H., Fukasawa, M., Mukoyama, Y. and Isogai, K. *Anal. Sci.*, 1992, **8**, 637–40.
36. Slonim, I. Y. and Urman, Y. G., *YaMR Spektroskopija Geterotsepnych Polimerov* (*NMR Spectroscopy of Heterochain Polymers*), Khimija, Moscow, 1982.
37. Liff, M. I., Lyu, P. C. and Kallenbach, N. R., *J. Am. Chem. Soc.*, 1991, **113**, 1014–19.
38. Davis, D. G. and Bax, A., *J. Am. Chem. Soc.*, 1985, **107**, 2820–1.
39. Paudler, W. W., *Nuclear Magnetic Resonance, General Concepts and Applications*, Wiley, New York, 1987.
40. Kay, L. E., Holak, T. A., Johnson, B. A., Armitage, I. M. and Prestegard, J. H., *J. Am. Chem. Soc.*, 1986, **108**, 4242–4.
41. Kessler, H., Griesinger, C., Kerssebaum, R., Wagner, K. and Ernst, R. R., *J. Am. Chem. Soc.*, 1987, **109**, 607–9.
42. Heffner, S. A., Bovey, F. A., Mirrau, P. A., Tonelli, A. E. and Verge, L. A., *Macromolecules*, 1986, **19**, 1628–34.
43. Barfield, M., Chan, R. J. H., Hall, H. K. and Mou, Y. H., *Macromolecules*, 1986, **19**, 1343–9.
44. Barfield, M., Chan, R. J. H., Hall, H. K. and Mou, Y. H., *Macromolecules*, 1986, **19**, 1350–5.
45. Axelson, D. E. and Russell, K. E., *Prog. Polymer Sci.*, 1985, **11**, 221–82.
46. Abraham, R. J., Hawort, I. S., Bunn, A. and Hearmon, R. A., *Polymer*, 1990, **31**, 728–35.
47. Ivin, K. J., *Pure Appl. Chem.*, 1983, **55**, 1529–31.
48. Hortling, B., Levon, K., Soljamo, K., Pellinen, J. and Lindberg, J. J., *Makromol. Chem.*, 1985, **186**, 131–7.
49. Wang, H. T., Bethea, W. and Harwood, H. J., *Macromolecules*, 1993, **26**, 715–20.

50. Khachaturov, A. S. and Ivanova, V. P., *J. Vsesouznogo Khimicheskogo Obschestva*, 1991, **36**, 230–7.
51. Asakura, T., Demura, M. and Nishiyama, Y., *Macromolecules*, 1991, **24**, 2334–40.
52. Asakura, T., Ando, I., Doi, Y. and Nishiyama, Y., *Polymer J.*, 1987, **19**, 829–37.
53. Schilling, F. C. and Tonelli, A. E., *Macromolecules*, 1980, **13**, 270–5.
54. Segre, A. L., Andruzzi, F., Lupinacci, D. and Magagnini, P. L., *Macromolecules*, 1981, **14**, 1845–7.
55. Aoki, A., *Polymer Commun.*, 1990, **31**, 130–2.
56. Abe, A., Hirano, T. and Tsuruta, T., *Macromolecules*, 1979, **12**, 1092–100.
57. Doddrell, D. M. and Freeman, R., *J. Am. Chem. Soc.*, 1979, **101**, 760–2.
58. Morris, G. A., *J. Am. Chem. Soc.*, 1980, **102**, 428–9.
59. Kalinowski, H. O., Berger, S. and Braun, S., *Carbon-13 NMR Spectroscopy*, Wiley, Chichester, 1988.
60. Bendall, M. R., Doddrell, D. M. and Pegg, D. T., *J. Am. Chem. Soc.*, 1981, **103**, 4603–4.
61. Heatley, F., Ding, J. F., Booth, C., Mobbs, R. H. and Luo, Y. Z., *Macromolecules*, **21**, 2713–21.
62. Heatley, F., Yu, G. E., Sun, W. B., Pywell, E. J., Mobbs, R. H. and Booth, C., *Europ. Polymer J.*, 1990, **26**, 583–92.
63. Gronski, W., Hellmann, G. and Wilsch-Irrgang, A., *Makromol. Chem.*, 1991, **192**, 591–601.
64. Pollock, J. R. and Goff, H. M., *Biochim. Biophys. Acta*, 1992, **1159**, 279–85.
65. Arranz, F., San Roman, J. and Sanchez-Chaves, M., *Macromolecules*, 1987, **20**, 801–6.
66. Urman, Y. G., *Vysokomol. Soed*, 1982, **24**, 1795–807.
67. Brar, A. S. and Sunita, *Makromol. Chem.*, 1993, **194**, 1707–20.
68. San Roman, J. and Levenfeld, B., *Macromolecules*, 1991, **24**, 6083–8.
69. Frey, W. and Klesper, E. J., *Polymer Sci.*, 1987, **25A**, 1409–18.
70. Frey, W., Dernst, C. and Klesper, E. J., *Polymer Sci.*, 1987, **25A**, 3143–57.
71. Abe, A., Tabata, S. and Kimura, N., *Polymer J.*, 1991, **23**, 69–72.
72. Zumbulyadis, N. and O'Reilly, J. M., *J. Am. Chem. Soc.*, 1993, **115**, 4407–8.
73. Rinaldy, P. L., Darlene, R. H., Tokles, M., Hatvany, G. S. and Harwood, H. J., *Macromolecules*, 1992, **25**, 7398–9.
74. Zhou, S. Q., Park, Y. T., Manuel, G. and Weber, W. P., *Polymer Bull.*, 1990, **23**, 491–6.
75. Schilling, F. C., Lovinger, A. J., Davis, D. D., Bovey, F. A. and Zeigler, J. M., *Macromolecules*, 1992, **25**, 2854–9.
76. Mathias, L. J. and Johnson, C. G., *Macromolecules*, 1991, **24**, 6114–22.
77. Hatada, K., Ute, K., Nakano, T., Okamoto, Y., Doyle, T. R. and Vogl, O., *Polymer J.*, 1989, **21**, 171–7.
78. Corley, L. S. and Vogl, O., *Polymer Bull.*, 1980, **3**, 211–16.
79. Kang, S. K. and Jhon, M. S., *Macromolecules*, 1993, **26**, 171–5.
80. Suchoparek, M. and Spevacek, J., *Macromolecules*, 1993, **26**, 102–6.
81. Rinaldi, P. L., Tokles, M., Hatvany, G. S. and Harwood, H. J., *J. Am. Chem. Soc.*, 1992, **114**, 10651–3.
82. Mirau, P. A. *Bull. Magn. Reson.*, 1992, **13**, 109–37.
83. Mirau, P. A. and Bovey, F. A., *Macromolecules*, 1990, **23**, 4548–52.
84. Cheng, H. N. and Lee, G. H., *Polymer Bull.*, 1988, **19**, 89–96.
85. Blumich, B., Hagemeyer, A., Schaefer, D., Schmidt-Rohr, K. and Spiess, H. W., *Adv. Mater.*, 1990, **2**, 72–81.
86. Thompson, R. D., Jarrett, W. L. and Mathias, L. J., *Macromolecules*, 1992, **25**, 6455–9.

87. Bruch, M. D. and Bonesteel, J. A., *Macromolecules*, 1986, **19**, 1622–7.
88. Cheng, H. N., *J. Appl. Polymer Sci.*, 1988, **35**, 1639–50.
89. Heatley, F. and Bovey, F. A., *Macromolecules*, 1669, **2**, 241–2.
90. Afeworki, M., Vega, S. and Schaefer, J., *Macromolecules*, 1992, **25**, 4100–5.
91. Chmelka, B. F., Schmidt-Rohr, K. and Spiess, H. W., *Macromolecules*, 1993, **26**, 2282–96.
92. Stroganov, L. B., Prokhorov, A. N., Galiullin, R. A., Kireev, E. V., Shibaev, V. P. and Platé, N. A., *Vysokomol. Soed A*, 1992, **34**, 146–53.
93. Barmatov, E. B., Stroganov, L. B., Tal'rose, R. V. and Shibaev, V. P., *Vysokomol. Soed A*, 1993, **35**, 1465–72.
94. Kapralova, V. M., Zuev, V. V., Koltsov, A. I., Osetrova, L. V., Skorokhodov, S. S. and Khachaturov, A. S., *Vysokomol. Soed A*, 1992, **34**, 27–34.
95. Stroganov, L. B., Taran, U. A., Platé, N. A. and Seifert, T., *Vysokomol. Soed*, 1974, **16**, 2147–53.
96. Stroganov, L. B., Taran, U. A. and Platé, N. A., *Z. Fiz. Khim.*, 1975, **49**, 2696–9.
97. Platé, N. A., Stroganov, L. B., Seifert, T. and Noah, O. V., *Dokl. Akad. Nauk. SSSR*, 1975, **223**, 396–9.
98. Platé, N. A., Stroganov, L. B. and Noah, O. V., *Vestnik. Akad. Nauk. SSSR*, 1983, **4**, 86–95.
99. Platé, N. A., Noah, O. V. and Stroganov, L. B., *Vysokomol. Soed A*, 1983, **25**, 2243–66.
100. Olonovskii, A. N., Stroganov, L. B., Noah, O. V. and Platé, N. A., *Vysokomol. Soed A*, 1983, **25**, 882–8.
101. Cheng, H. N. and Lee, G. H., *Macromolecules*, 1988, **21**, 3164–70.
102. Cheng, H. N., *J. Appl. Polymer Sci.*, 1988, **36**, 229–41.
103. Zadihanov, R. A., Fedotov, V. D., Spevacek, J. and Straka, J., *Vysokomol. Soed A*, 1992, **34**, 58–66.
104. Mehring, M., *NMR Basic Principles and Progress*, Vol. 11, Springer-Verlag, Berlin, 1976.
105. Spevacek, J., Schneider, B. and Straka, J., *Macromolecules*, 1990, **23**, 3042–51.
106. Voelkel, R., *Angew. Chemie Int. Ed. Engl.*, 1988, **27**, 1468–83.
107. Haeberlen, U., *High Resolution NMR in Solids: Selective Averaging* (*Advances in Magnetic Resonance*), Suppl. 1, Academic Press, New York, 1976.
108. Schaefer, J. and Stejskal, E. O., *Topics in Carbon-13 NMR Spectroscopy*, Wiley, New York, 1979, Vol. 3, pp. 283–324.
109. Blumich, B. and Spiess, H. W., *Angew. Chemie Int. Ed. Engl.*, 1988, **27**, 1655–72.
110. Leisen, J., Boeffel, C., Dong, R. Y. and Spiess, H. M., *Liquid Crystals*, 1993, **14**, 215–26.
111. Blumich, B. and Blumler, P., *Makromol. Chem.*, 1993, **194**, 2133–61.
112. Spiess, H. W., *J. Non-Cryst. Solids*, 1991, **130**, 766–72.
113. Klinowski, B. J., *Nucl. Magn. Reson.*, **16**, 153–90.
114. Rabenstein, D. L. and Guo, W., *Anal. Chem.*, 1988, **60**, 1R–28R.
115. Cholli, A. L., Ritchey, W. M. and Koenig, J. L., *Appl. Spectrosc.*, 1987, **41**, 1418–21.
116. Cholli, A. L. and Thakur, M., *J. Chem. Phys.*, 1989, **91**, 7912–15.
117. Cholli, A. L., Ritchey, W. M. and Koenig, J. L., *Spect. Lett.*, 1988, **21**, 519–31.
118. Winter, H. and Van der Velden, G. P. M., *Macromolecules*, 1992, **25**, 4285–9.
119. Nava, A. D., Thakur, M. and Tonelli, A. E., *Macromolecules*, 1990, **23**, 3055–63.
120. Fedotov, V. D. and Schneider, H., *NMR Basic Principles and Progress*, Vol. 21, Springer-Verlag, Berlin, 1989.
121. Zumbulyadis, N., O'Reilly, J. M. and Teegarden, D. M., *Macromolecules*, 1992, **25**, 3317–19.

122. Frydman, L., Lee, Y. K., Emsley, L., Chingas, G. C. and Pines, A. *J. Am. Chem. Soc.*, 1993, **115**, 4825–9.
123. Zaikin, V. G., Mardanov, R. G. and Platé, N. A. *Polymer Sci.*, 1993, **35**, 1305–20.
124. Kline, G. M. (Ed.), *High Polymers*, Vol. 12, *Analytical Chemistry of Polymers*, Interscience New York, 1962, Part 2.
125. Grassie, N. and Scott, G., *Polymer Degradation and Stabilization*, Cambridge University Press, Cambridge, 1985.
126. Khmelnitskii, R. A., Lukashenko, I. M. and Brodskii, E. S., *Piroliticheskaya Mass-Spectrometriya Vysokomolekulyarnykh Soedinenii* (*Pyrolysis Mass Spectrometry of High Polymers*), Khimiya, Moscow, 1980.
127. Alekseeva, K. V., *Piroliticheskaya Gazovaya Khromatografiya* (*Pyrolysis Gas Chromatography*), Khimiya, Moscow, 1985.
128. May, R. V., Pearson, E. F. and Scothern, D., *Pyrolysis Gas Chromatography*, Chemical Society, London, 1977.
129. Irwin, W. J., *J. Anal. Appl. Pyrol.*, 1979, **1**, 3–122.
130. Mikaya, A. I. and Zaikin, V. G., *Mass Spectrom. Rev.*, 1990, **9**, 115–32.
131. Bamford, C. H. and Tipper, C. F. H. (Eds.), *Comprehensive Chemical Kinetics*, Vol. 14, *Degradation of Polymers*, Elsevier, Amsterdam, 1975.
132. Schulten, H.-R. and Lattimer, R. P., *Mass Spectrom. Rev.*, 1984, **3**, 231–315.
133. Zaikin, V. G. and Mikaya, A. I., *Khimicheskie Metody v Mass-Spectrometrii Organicheskikh Soedinenii* (*Chemical Methods in Mass Spectrometry of Organic Compounds*), Nauka, Moscow, 1987.
134. Mardanov, R. G., Zaikin, V. G., Kleiner, V. I., Krentsel, B. A. and Bobrov, B. N., *Vysokomol. Soed, Ser. B*, 1992, **34**, 73–88.
135. Zaikin, V. G., Mardanov, R. G., Kleiner, V. I. and Krentsel, B. A., *Vysokomol. Soed, Ser. A*, 1990, **32**, 1014–20.
136. Alishoev, V. R., Berezkin, V. G., Tint, L. S. and Mirzabaev, G. A., *Vysokomol. Soed, Ser. A*, 1971, **13**, 2815–20.
137. Voigt, J., *Kunststoffe*, 1964, **54**, 2–11.
138. Van Schooten, J. and Evenhuis, J. K., *Polymer*, 1965, **6**, 561–8.
139. Michajlov, L., Cantow, H.-J. and Zugenmaier, P., *Polymer*, 1971, **12**, 70–81.
140. Zigel, A. N., Ryabikova, V. M., Pirozhnaya, L. N., Popova, G. S. and Madorskaya, L. Ya., *Vysokomol. Soed, Ser. A*, 1991, **33**, 1321–25.
141. Shibasaki, Y., *J. Polymer Sci. Part A1*, 1967, **5**, 521–9.
142. Harwood, J. H. and Ritchey, W. M., *J. Polymer Sci. Part B*, 1964, **2**, 601–12.
143. Minin, V. A., Berkin, A. A., Varshavskaya, A. I., Kovtun, T. S., Karmiloba, L. V. and Enikolopyan, N. S., *Vysokomol. Soed, Ser. A*, 1972, **14**, 9–16.
144. Tsuge, S. and Takeuchi, T., *Analytical Pyrolysis*, Elsevier, Amsterdam, 1977.
145. Okumoto, T., Takeuchi, T. and Tsuge, S., *Macromolecules*, 1973, **6**, 922–30.
146. Ham, G. E. (Ed.), *High Polymers*, Vol. 18, *Copolymerization*, Interscience, New York, 1964.
147. Tsuge, S., Hiramitsu, S., Horribe, T., Yamaoka, M. and Takeuchi, T., *Macromolecules*, 1975, **8**, 721–9.
148. Okumoto, T., Tsuge, S., Yamamoto, Y. and Takeuchi, T., *Macromolecules*, 1974, **7**, 376–83.
149. Platé, N. A., Litmanovich, A. D. and Noa, O. V., *Makromolekulyarnye Reaktsii* (*Macromolecular Reactions*), Khimiya, Moscow, 1977.
150. Mardanov, R. G., Zaikin, V. G., Kleiner, V. I., Krentsel, B. A. and Platé, N. A., *Vysokomol. Soed*, 1990, **32**, 552–6.
151. Zaikin, V. G., Mardanov, R. G., Kleiner, V. I., Krentsel, B. A. and Platé, N. A., *J. Anal. Appl. Pyrol.*, 1990, **17**, 291–304.

152. Nagaya, T., Sugimura, Y. and Tsuge, S., *Macromolecules*, 1980, **13**, 353–7.
153. Kalal, J., Zachoval, J., Kubat, J. and Svec, F., *J. Anal. Appl. Pyrol.*, 1979, **2**, 143–8.
154. Tsuge, S., Kobayashi, T., Nagaya, T. and Takeuchi, T., *J. Anal. Appl. Pyrol.*, 1979, **1**, 133–41.
155. Blazco, M. and Toth, T., *J. Anal. Appl. Pyrol.*, 1991, **19**, 251–63.
156. Varhegyi, G. and Blazco, M., *Europ. Polymer J.*, 1978, **14**, 349–58.
157. Blazco, M. and Varhegyi, G., *Europ. Polymer J.*, 1978, **14**, 625–31.
158. Blazco, M., Varhegyi, G. and Jakob, E., *J. Anal. Appl. Pyrol.*, 1980, **2**, 177–85.
159. Tanaka, M., Nishimura, F. and Shono, T., *Anal. Chim. Acta.*, 1975, **74**, 119–29.
160. Shimono, T., Tanaka, M. and Shono, T., *Anal. Chim. Acta.*, 1978, **96**, 352–8.
161. Shimono, T., Tanaka, M. and Shono, T., *J. Anal. Appl. Pyrol.*, 1979, **1**, 77–83.
162. Mardanov, R. G., Zaikin, V. G., Yakovlev, V. A. and Platé, N. A., *Vysokomol Soed, Ser. B.*, 1991, **33**, 731–4.
163. Zaikin, V. G., Mardanov, R. G., Yakovlev, V. A. and Platé, N. A., *J. Anal. Appl. Pyrol.*, 1992, **23**, 33–8.
164. Tsuge, S., Kobayashi, T., Sugimura, Y., Nagaya, T. and Takeuchi, T., *Macromolecules*, 1979, **12**, 988–92.
165. Mardanov, R. G., Cand Sci (Chem) Dissertation, Institute of Petrochemical Synthesis, Russian Academy of Sciences, Moscow, 1992.
166. Zaikin, V. G., Mardanov, R. G., Kleiner, V. I., Krentsel, B. A. and Bobrov, B. N., *J. Anal. Appl. Pyrol.*, 1993, **26**, 185–90.
167. Krasnoselskaya, I. G., Turkova, L. D., Petrove, S. F., Baranovskaya, I. A., Stepanov, N. G., Klenin, S. I., *et al.*, *Polymer Sci.*, 1992, **34**, 962–70.
168. Helleur, R. J., *J. Anal. Appl. Pyrol.*, 1987, **11**, 297–311.
169. Helleur, R. G., Budgell, D. R. and Hayes, E. R., *J. Anal. Appl. Pyrol.*, 1987, **12**, 367–72.
170. Lomax, Ja., Commandeur, J. M., Arisz, P. W. and Boon, J. J., *J. Anal. Appl. Pyrol.*, 1991, **19**, 65–79.
171. Tsuge, S. and Matsubara, H., *J. Anal. Appl. Pyrol.*, 1985, **16**, 49–64.
172. Boen, J. J. and De Leeuw, J. W., *J. Anal. Appl. Pyrol.*, 1987, **11**, 313–27.
173. Kondo, A., Ohtani, H., Kosugi, Y., Tsuge, S., Kubo, Y., Asada, N., *et al.*, *Macromolecules*, 1988, **21**, 2918–28.
174. Alfrey, T., Lewis, C. and Nagel, B., *J. Am. Chem. Soc.*, 1949, **71**, 3793–5.
175. Johnston, N. W. and Harwood, H. J., *J. Polymer Sci. C*, 1969, **22**, 591–9.
176. Johnston, N. W. and Harwood, H. J., *Macromolecules*, 1969, **2**, 221–4.
177. Johnston, N. W. and Harwood, H. J., *Macromolecules*, 1970, **3**, 20–3.
178. Guilbault, L. J. and Harwood, H. J., *J. Polymer Sci. A-1*, 1974, **12**, 1461–7.
179. Harwood, H. J., *Angew. Makromol. Chem.*, 1968, **4/5**, 279–85.
180. Minsk, L. M., Waugh, G. P. and Kenyon, W. O., *J. Am. Chem. Soc.*, 1950, **72**, 2546–9.
181. Loebl, E. M. and O'Neil, J. J., *J. Polymer Sci.*, 1960, **45**, 538–43.
182. Miller, W. L., Brey, W. S. and Butler, G. B., *J. Polymer Sci.*, 1961, **54**, 329–40.
183. Cohen, H. L., *J. Polymer Sci. A*, 1976, **14**, 7–22.
184. Zeifert, T., Stroganov, L. B., Platé, N. A. and Shibaev, V. P., *Vysokomol. Soed B*, 1974, **16**, 643–4.
185. Ramey, K. C., *J. Macromol. Sci. C*, 1967, **1**, 213–18.
186. Yun, E., Stroganov, L. B., Agasandyan, V. A., Litmanovich, A. D. and Platé, N. A., *Vysokomol. Soed B*, 1972, **14**, 292–9.
187. Kempf, K. G. and Harwood, H. J., *Macromolecules*, 1978, **11**, 1038–41.
188. Kotake, Y., Yoshihara, T., Sato, H., Yamada, N. and Joh, Y., *J. Polymer Sci. B*, 1967, **5**, 163–5.
189. Matsuzaki, K., Kanai, T. and Matsumoto, S., *J. Polymer Sci. A-1*, 1974, **12**, 2377–85.
190. Matsuzaki, K. and Uryu, T., *J. Polymer Sci. B*, 1966, **4**, 255–63.

191. Matsuzaki, K., Uryu, T. and Takeuchi, M., *J. Polymer Sci. B*, 1965, **3**, 835–7.
192. Matsuzaki, K., Uryu, T., Ishida, A. and Takeuchi, M., *J. Polymer Sci. C*, 1967, **16**, 2099–119.
193. McCarty, J. R., *Mod. Plast.*, 1953, **30**, 118–26.
194. Rhodes, J. H., Chiang, R. and Randall, J. C., *J. Polymer Sci. B*, 1966, **4**, 393–5.
195. Ueno, A. and Schuerch, C., *J. Polymer Sci. B*, 1965, **3**, 821–3.
196. Mercier, J. and Smets, G., *J. Polymer Sci. A*, 1963, **1**, 1491–8.
197. Fujii, K., Fujiwara, Y. and Fujiwara, S., *Makromol. Chem.*, 1965, **89**, 278–89.
198. Murachashi, S., Nozakura, S., Sumi, M., Yuki, H. and Hatada, K., *J. Polymer Sci. B*, 1966, **4**, 65–7.
199. Ramey, K. C. and Lini, D. C., *J. Polymer Sci. B*, 1967, **5**, 39–41.
200. Wu, T. K. and Ovenall, D. W., *Macromolecules*, 1973, **6**, 582–4.
201. Starnes Jr, W. H., Schilling, F. S., Abbas, K. B., Plitz, I. M., Hartless, R. L. and Bovey, F. A., *Macromolecules*, 1979, **12**, 13–19.
202. Jameison, F. A., Shilling, F. C. and Tonelli, A. E., in *Reactions on Polymers*, ACS Symposium Series 364, Eds. J. L. Benham and J. F. Kinstle, ACS, Washington, D. C., 1988, Chap. 26.
203. Carlsson, D. J., Brousseau, R., Zhang, C. and Wiles, D. M., in *Reactions on Polymers*, ACS Symposium Series 364, Eds. J. L. Benham and J. F. Kinstle, ACS, Washington, D. C., 1988, Chap. 27.
204. Harwood, H. J., *Angew. Chem.*, 1965, **77**, 1124–33.
205. Hackathorn, M. J. and Brock, H. J., *Polymer Prepr.*, 1973, **14**, 42–58.
206. Hackathorn, M. J. and Brock, H. J., *Rubb. Chem. Technol.*, 1972, **45**, 1295–304.
207. Marder, H. L. and Schuerch, C., *J. Polymer Sci.*, 1963, **44**, 129–37.
208. Kanakanatt, A. T., Thesis, Akron, 1963.
209. Tada, K. Saegusa, T. and Furukawa, J., *Makromol. Chem.*, 1964, **71**, 72–86.
210. Myagchenkov, V. A. and Frenkel, S. Ya., *Kompozitsionnaya Neodnorodnost Sopolymerov* (*Composition Heterogeneity of Copolymers*), Khimiya, Leningrad, 1988.
211. Myagchenkov, V. A., Kurenkov, V. F., Kuznetsov, E. V. and Frenkel, S. Ya., *Vysokomol. Soed B*, 1967, **9**, 251–2.
212. Berezkin, V. G., Alishoev, V. R. and Nemirovskaya, I. B., *Gazovaya Khromatografiya v Khimii Polimerov* (*Gas Chromatography in Polymer Chemistry*), Nauka, Moscow, 1972, Chap. 3.
213. Myagchenkov, V. A. and Frenkel, S. Ya., *Usp. Khim.*, 1968, **27**, 2247–71.
214. Litmanovich, A. D. and Izyumnikov, A. L., in *Novoe v Metodakh Issledovaniya Polimerov* (*Advances in Polymers Characterization*), Eds. Z. A. Rogovin, and V. P. Zubov, Mir, Moscow, 1968, pp. 200–41.
215. Fuchs, O. and Schmieder, W., in *Polymer Fractionation*, Ed. M. J. R. Cantow, Academic Press, New York, London, 1967, Chap. 12.
216. Fuchs, O. and Hellfritz, H., in *Wandel in der Chemischen Technik*, Meister Lucius & Bruning, Frankfurt (Main), 1963, pp. 356–80.
217. Kollynsky, F. and Markert, G., *Makromol. Chem.*, 1968, **121**, 117–28.
218. Melvill, H. W. and Stead, B. D., *J. Polymer Sci.*, 1955, **16**, 505–15.
219. Pyrkov, L. M., Bresler, S. E. and Frenkel, S. Ya., *Zh. Obshch. Khim.*, 1959, **29**, 2750–60.
220. Korotkov, A. A., Shibaev, L. A., Pyrkov, L. M., Aldoshin, V. G. and Frenkel, S. Ya., *Vysokomol. Soed*, 1959, **1**, 443–54.
221. Belenky, B. G. and Gankina, E. S., *Dokl. Akad. Nauk. SSSR*, 1969, **186**, 857–9.
222. Belenky, B. G. and Gankina, E. S., *J. Chromatography*, 1970, **53**, 3–25.
223. Inagaki, H., Matsuda, H. and Kamiyama, F., *Macromolecules*, 1968, **1**, 520–5.
224. Kotaka, T. and White, J., *Macromolecules*, 1974, **7**, 106–16.

225. Klenin, S. I., Tsvetkov, V. N. and Cherkasov, A. N., *Vysokomol. Soed A*, 1967, **9**, 1435–9.
226. Cherkasov, A. N., Klenin, S. I. and Andreeva, G. A., *Vysokomol. Soed A*, 1970, **12**, 1223–32.
227. Cherkasov, A. N., *Vysokomol. Soed B*, 1972, **14**, 115–18.
228. Stockmayer, W. H., Moore Jr, L. D., Fixman, M. and Epstein, B. N., *J. Polymer Sci.*, 1955, **16**, 517–30.
229. Bushuk, W. and Benoit, H., *Canad. J. Chem.*, 1958, **36**, 1616–26.
230. Eskin, V. E., *Rasseyanie Sveta Rastvorami Polimerov (Light-Scattering of Polymer Solutions)*, Nauka, Moscow, 1973, Chap. 5.
231. Baranovskaya, I. A., Litmanovich, A. D., Protasova, M. S. and Eskin, V. E., *Vysokomol. Soed B*, 1967, **9**, 773–6.
232. Lamprecht, J., Strazielle, C., Dayantis, J. and Benoit, H., *Makromol. Chem.*, 1971, **148**, 285–96.
233. Meselson, M., Stahl, F. W. and Vinograd, J., *Proc. Natl Acad. Sci. USA*, 1957, **47**, 581–8.
234. Hermans, J. J. and Ende, H. A., in *Newer Methods of Polymer Characterization*, Ed. B. Ke, Interscience, New York, London, Sydney, 1964, Chap. 13.
235. Dayantis, J., *J. Chim. Phys.*, 1968, **65**, 648–54.
236. Nakazawa, A. and Hermans, J. J., *J. Polymer Sci. A-2*, 1971, **9**, 1871–85.
237. Krivobokov, V. V. and Frenkel, S. Ya., *Vysokomol. Soed B*, 1980, **22**, 128–30.
238. Matsumoto, M. and Takayama, G., *Chem. High Polymers (Tokyo)*, 1961, **18**, 169–74.
239. Rosenthal, A. J. and White, B. B., *Ind. Eng. Chem.*, 1952, **44**, 2693–6.
240. Fuchs, O., *Verhandlungsber Kolloid-Ges.*, 1958, **18**, 75–81.
241. Kilb, R. W. and Bueche, A. M., *J. Polymer Sci.*, 1958, **28**, 285–94.
242. Topchiev, A. V., Litmanovich, A. D. and Shtern, V. Ya., *Dokl. Akad. Nauk. SSSR*, 1962, **147**, 1389–91.
243. Litmanovich, A. D. and Topchiev, A. V., *Neftekhimiya*, 1963, **3**, 336–42.
244. Lautout-Magat, M. and Magat, M., *Z. Phys. Chem. (Frankfurt)*, 1958, **16**, 292–301.
245. Lautout-Magat, M., *J. Polymer Sci.*, 1962, **57**, 421–8.
246. Flory, P. J., *Principles of Polymer Chemistry*, 2nd edn, Cornell University Press, Ithaca, N.Y., 1957, Chaps. 12 and 13.
247. Litmanovich, A. D. and Shtern, V. Ya., *J. Polymer Sci. C*, 1967, **16**, 1375–82.
248. Kudryavtseva, L. G. and Litmanovich, A. D., *Vysokomol. Soed A*, 1967, **9**, 1016–21.
249. Howard, G. J., *J. Polymer Sci.*, 1959, **37**, 310–13.
250. Kudryavtseva, L. G., Litmanovich, A. D., Topchiev, A. V. and Shtern, V. Ya., *Neftekhimiya*, 1963, **3**, 343–7.
251. Litmanovich, A. D. and Shtern, V. Ya., *Dokl. Akad. Nauk. SSSR*, 1964, **154**, 1429–31.
252. Eskin, V. E., Baranovskaya, I. A., Litmanovich, A. D. and Topchiev, A. V., *Vysokomol. Soed*, 1964, **6**, 896–900.
253. Litmanovich, A. D. and Shtern, V. Ya., *Vysokomol. Soed*, 1965, **7**, 1332–4.
254. Danon, J. and Jozefonvicz, J. *Europ. Polymer J.*, 1969, **5**, 405–17.
255. Benoit, H., *Ber. Bunsenges Phys. Chem.*, 1966, **70**, 286–96.
256. Teramachi, S. and Nagasawa, M., *J. Macromol. Sci. Chem. A*, 1968, **2**, 1169–79.
257. Teramachi, S. and Kato, Y., *J. Macromol. Sci. Chem. A*, 1970, **4**, 1785–96.
258. Bourguignon, J. J., Bellissent, H. and Galin, J. C., *Polymer*, 1977, **18**, 937–44.
259. Teramachi, S. and Kato, Y., *Macromolecules*, 1971, **4**, 54–6.
260. Stockmayer, W. H., *J. Chem. Phys.*, 1945, **13**, 199–207.
261. Elias, H. G., *Makromol. Chem.*, 1961, **50**, 1–19.
262. Elias, H. G. and Gruber, U., *Makromol. Chem.*, 1964, **78**, 72–99.
263. Juranicova, V., Florian, S. and Berek, D., *Europ. Polymer J.*, 1970, **6**, 57–62.

264. Litmanovich, A. D., Shtern, V. Ya. and Topchiev, A. V., *Neftekhimiya*, 1963, **3**, 217–21.
265. Glockner, G., Francuskiewicz, F. and Muller, S., *Faserforsch und Textiltech*, 1975, **26**, 287–93.
266. Staudinger, H. and Heuer, W., *Z. Phys. Chem.*, 1943, **171**, 129–80.
267. Glockner, G., Francuskiewicz, F. and Muller, K. D., *Plaste und Kautschuk*, 1971, **18**, 654–6.
268. Agasandyan, V. A., Kudryavtseva, L. G., Litmanovich, A. D. and Shtern, V. Ya., *Vysokomol. Soed A*, 1967, **26**, 2634–6.
269. Papisov, I. M. and Litmanovich, A. A., *Vysokomol. Soed A*, 1977, **19**, 716–22.
270. Glockner, G., *J. Appl. Polymer Sci. Appl. Polymer Symp.*, 1992, **51**, 45–54.
271. Glockner, G., *Gradient HPLC of Copolymers and Chromatographic Cross- Fractionation*, Springer-Verlag, Berlin, 1991.
272. Ratzsch, M. T., Kehlen, H. and Tschersich, L., *J. Macrom. Sci. Chem. A*, 1989, **26**, 921–35.
273. Ratzsch, M. T., Enders, S., Tschersich, L. and Kehlen, H., *J. Macrom. Sci. Chem. A*, 1991, **28**, 31–46.

Conclusions

In this book we have tried to show that the area of macromolecular reactions today is not only an 'organic side' of polymer chemical transformations but is really a whole new branch of thermodynamics, kinetics and statistics of intra- and intermacromolecular and polymer-analogous reactions.

In many cases the quantitative approach to the description of macromolecular reactions has been developed. It became clear that some qualitative particularities of these reactions exist that differentiate them from chemical transformations of small organic molecules. In addition to that an extremely important field— intermacromolecular reactions—appeared which is open for further studies and there is no doubt that new discoveries will come there. The first success in the chemistry of interpolymer complexes, as described in Chapter 7, gives much ground and hope for researchers who are involved in the development of various aspects of this field.

Of special importance and interest are peculiarities of intermacromolecular reactions when one of the components is a biopolymer—DNA, RNA and their analogs as well as proteins—because this is exactly the area important for understanding many chemical processes taking place in living cells and organisms.

Intermacromolecular reactions also embrace problems of network formation from branched and oligomeric systems including interpenetrating networks and crosslinked macromolecules—thermosetting plastics and vulcanized rubbers. A clear understanding of the physical picture of how the reaction with macromolecular reagent proceeds is of importance for governing such processes when a material with a desirable set of properties is created. Synthesis of so-called 'regular' networks with equal distances between crosslinks and preparation of membranes having uniform and regularly arranged pores are examples of some tasks in the field of intermacromolecular reactions between linear chains. Due to the lack of space we were not able to touch these problems here.

As concerns the quantitative approach to the description of macromolecular reactions, it should be noted that the neighboring groups effect has been rather well elaborated. Theory now exists which permits the kinetics of polymer analogous reaction, units distribution and composition heterogeneity to be calculated and predicted. The experimental methods for evaluation of the reaction rate constants has been developed. For that type of reaction it is possible to estimate any parameter of a units distribution or composition heterogeneity including such ones as can not be determined experimentally.

Of practical value could be calculations related to chlorination and sulfo-chlorination of polyolefins, partially hydrolyzed polyacrylic and polymethacrylic esters, polyacrylamide, poly(vinyl acetate), partially acetalizated poly(vinyl alcohol), partially quaternized poly(vinyl pyridines) as biocatalyst models, etc.

It should be pointed out that the neighboring group effect, although being a general feature of the macromolecular reaction, is manifested in the 'pure' state only rarely and to a greater or less extent is accompanied by conformational, stereochemical and other effects described above. The authors here have tried to show exactly what has already been done in the theory and experimental approach to study macromolecular reactions and what are further problems to be solved. As to reactions of quasi-isolated macromolecules, further goals are consideration of the role of remote units and stereochemistry of the chain; here there is nothing very complicated that can not be solved using modern computer techniques. Some examples of such theoretical approaches are described in Chapters 3 and 5. The principal difficulty, however, is to determine experimentally all individual rate constants in the general case when there are six or more of them.

A new important advance in the field discussed in the book is the calculation of the principles of the theory of the interchain effect. It is difficult to overestimate the significance of this branch of the theory of macromolecular reactions for chemical modifications of polymers in their bulk state, in particular in blends and composites. It should also be noted that a description of interdiffusion in reacting polymer blends leads to a further development of statistical physics of polymer systems.

One can surely say that to solve at best various problems of theory and experimental approaches to study and to carry out chemical modifications of polymers, the domain of chemical transformations of macromolecules should be developed and should reach not only the existing level in the field of polymerization and polycondensation reactions but should go further than that because of its great scientific and practical value in modern chemistry. The understanding of the authors of the current situation is that rich 'organic' and technological experience in the area of chemical modification of polymers is now meeting the expanding wave of theoretical considerations of how they behave. They should behave like giant molecules when they enter into collisions and interact with small and large chemical species. This results in mutual fertilization to the benefit of fundamental and applied polymer science.

Index

Note: Figures are indicated by *italic page numbers*, Tables by **bold numbers**.

Index compiled by P. Nash